Future of Utilities – Utilities of the Future

How Technological Innovations in Distributed Energy Resources will Reshape the Electric Power Sector

Future of Utilities – Utilities of the Future

How Technological Innovations in Distributed Energy Resources will Reshape the Electric Power Sector

Edited by

Fereidoon P. Sioshansi
Menlo Energy Economics
Walnut Creek, CA
United States of America

AMSTERDAM • BOSTON • HEIDELBERG • LONDON
NEW YORK • OXFORD • PARIS • SAN DIEGO
SAN FRANCISCO • SINGAPORE • SYDNEY • TOKYO
Academic Press is an imprint of Elsevier

Academic Press is an imprint of Elsevier
125 London Wall, London EC2Y 5AS, UK
525 B Street, Suite 1800, San Diego, CA 92101-4495, USA
50 Hampshire Street, 5th Floor, Cambridge, MA 02139, USA
The Boulevard, Langford Lane, Kidlington, Oxford OX5 1GB, UK

British Library Cataloguing-in-Publication Data
A catalogue record for this book is available from the British Library

Library of Congress Cataloging-in-Publication Data
A catalog record for this book is available from the Library of Congress

ISBN: 978-0-12-804249-6

This book is dedicated to Clark Gellings
for his valuable contributions
and leadership

Contents

Author Biographies   xi
Foreword   xxv
Preface   xxvii
Introduction   xxxi

Part I
What is Changing, What are the Implications   1

1. What is the Future of the Electric Power Sector?   3

    Fereidoon P. Sioshansi
    Menlo Energy Economics, Walnut Creek, CA,
    United States of America

2. Value of an Integrated Grid   25

    Clark W. Gellings
    Electric Power Research Institute, Palo Alto, CA,
    United States of America

3. Microgrids: Finally Finding their Place   51

    Chris Marnay
    Microgrid Design of Mendocino, Comptche, CA,
    United States of America

4. Customer-Centric View of Electricity Service   75

    Eric G. Gimon
    Energy Innovation LLC, San Francisco, CA, United States of America

5. The Innovation Platform Enables the Internet of Things   91

    John Cooper
    Siemens PTI, Business Transformation & Solution Engineering, Austin, TX,
    United States of America

6.   Role of Utility and Pricing in the Transition   109
      Tim Nelson*, Judith McNeill†
      *AGL Energy, North Sydney, NSW, Australia; †Behavioral, Cognitive and Social
      Sciences, Institute for Rural Futures, University of New England, Armidale,
      NSW, Australia

7.   Intermittency: It's the Short-Term That Matters   129
      Daniel Rowe, Saad Sayeef, Glenn Platt
      CSIRO Energy, Mayfield West, NSW, Australia

Part II
Competition, Innovation, Regulation, and Pricing   151

8.   Retail Competition, Advanced Metering Investments,
      and Product Differentiation: Evidence From Texas   153
      Varun Rai*, Jay Zarnikau†
      *The University of Texas at Austin, Austin, TX, United States of America;
      †The University of Texas at Austin and Frontier Associates, Austin, TX,
      United States of America

9.   Rehabilitating Retail Electricity Markets: Pitfalls
      and Opportunities   175
      Ralph Cavanagh*, Amanda Levin†
      *NRDC, San Francisco, CA, United States of America; †NRDC Washington,
      DC, United States of America

10.   Residential Rate Design and Death Spiral for Electric
      Utilities: Efficiency and Equity Considerations   193
      Rasika Athawale, Frank Felder
      Center for Energy, Economic, & Environmental Policy,
      Edward J. Bloustein School of Planning and Public Policy,
      Rutgers University, New Brunswick, NJ, United States of America

11.   Modeling the Impacts of Disruptive Technologies
      and Pricing on Electricity Consumption   211
      George Grozev*, Stephen Garner†, Zhengen Ren*, Michelle Taylor†,
      Andrew Higgins‡, Glenn Walden†
      *CSIRO, Clayton, VIC, Australia; †Ergon Energy, Fortitude Valley, QLD, Australia;
      ‡Dutton Park, QLD, Australia

12.   Decentralized Reliability Options: Market Based
      Capacity Arrangements   231
      Stephen Woodhouse
      Pöyry Management Consulting, Oxford, UK

13. Network Pricing for the Prosumer Future:
     Demand-Based Tariffs or Locational Marginal Pricing?   247
     Darryl Biggar*, Andrew Reeves[†]
     *Australian Competition and Consumer Commission,
     Melbourne, VIC, Australia; [†]Former Chairman, Australian Energy Regulator,
     Hobart, Tasmania, Australia

14. The Evolution of Smart Grids Begs Disaggregated
     Nodal Pricing   267
     Günter Knieps
     University of Freiburg, Freiburg, Germany

Part III
Utilities of the Future: Future of Utilities   281

15. New Business Models for Utilities to Meet the
     Challenge of the Energy Transition   283
     Paul Nillesen*, Michael Pollitt[†]
     *PwC, Amsterdam, The Netherlands; [†]Cambridge University,
     United Kingdom

16. European Utilities: Strategic Choices and Cultural
     Prerequisites for the Future   303
     Christoph Burger, Jens Weinmann
     European School of Management and Technology (ESMT), Berlin, Germany

17. Thriving Despite Disruptive Technologies:
     A German Utilities' Case Study   323
     Sabine Löbbe*, Gerhard Jochum[†]
     *Reutlingen University, Reutlingen, Germany; [†]Büro Jochum, Berlin, Germany

18. The Future of Utility Customers and the Utility
     Customer of the Future   343
     Robert Smith*, Iain MacGill[†]
     *East Economics, Sydney, Australia; [†]Centre for Energy and Environmental
     Markets (CEEM) and School of Electrical Engineering and
     Telecommunications, UNSW, Sydney, Australia

19. Business Models for Power System Flexibility:
     New Actors, New Roles, New Rules   363
     Luis Boscán*, Rahmatallah Poudineh[†]
     *Department of Economics, Copenhagen Business School, Frederiksberg,
     Denmark; [†]Oxford Institute for Energy Studies,
     Oxford, United Kingdom

20. **The Repurposed Distribution Utility: Roadmaps to Getting There**   383

    Philip Q. Hanser, Kai E. Van Horn
    The Brattle Group, Cambridge, MA, United States of America

21. **Distributed Utility: Conflicts and Opportunities Between Incumbent Utilities, Suppliers, and Emerging New Entrants**   399

    Kevin B. Jones, Taylor L. Curtis, Marc de Konkoly Thege,
    Daniel Sauer, Matthew Roche
    *Institute for Energy and the Environment, Vermont Law School,
    South Royalton, VT, United States of America*

22. **The Fully Integrated Grid: Wholesale and Retail, Transmission and Distribution**   417

    Susan Covino, Andrew Levitt, Paul Sotkiewicz
    PJM Interconnection, LLC, Audubon, PA, United States of America

Subject Index   435

Author Biographies

Rasika Athawale is a Research Analyst at CEEEP, Rutgers University where she manages evaluation and research activities for the New Jersey Board of Public Utilities. Her research interests are regulation and economics of the power sector, business models of utilities, and financial modeling.

Her prior work experience includes KPMG India and PricewaterhouseCoopers India as a Senior Consultant where she advised clients on business strategy, and prior to which she worked in the utility tariff design team at Reliance Energy, Mumbai.

She holds PGDBM with specialization in Finance from Sydenham, Mumbai University and BTech in Chemical Engineering from Nagpur University.

Darryl Biggar is the Special Economic Advisor for regulatory matters at the Australian Competition and Consumer Commission and the Australian Energy Regulator where he provides advice and carries out research in the economics of public utility regulation and electricity markets. He has multiple publications on regulation, market power, and investment in electricity networks including a textbook, with M. Hesamzadeh, on the *Economics of Electricity Markets*, published in 2014.

Prior to the ACCC, he worked at the OECD in Paris and for the New Zealand Treasury.

He holds PhD from Stanford University in Economics and MA in Mathematics from the University of Cambridge.

Luis Boscán is PhD fellow at the Department of Economics, Copenhagen Business School (CBS). His research focuses on the design of contracts and markets to enable power system flexibility, supported by *ForskEL*, a Danish fund in environmentally friendly technologies.

Prior to joining CBS, he worked as Economic Analyst at the Central Bank of Venezuela working on global oil and gas markets and macroeconomic policy.

He holds an undergraduate degree in Economics from Universidad del Zulia (Venezuela), MSc in Systems Modeling and Simulation from Universidad de Los Andes (Venezuela), and MSc in Energy Economics and Policy from the University of Surrey (United Kingdom).

Christoph Burger is a faculty member and senior associate dean of executive education at ESMT European School of Management and Technology (ESMT), Berlin. His research focus is on energy/innovation and decision making/negotiation

He is the author of the *ESMT Innovation Index—Electricity Supply Industry* and *The Decentralized Energy Revolution—Business Strategies for a New Paradigm*. Before joining ESMT, he worked in the industry and as a consultant.

Christoph studied business administration at the University of Saarbrücken, Germany, the Hochschule St. Gallen, Switzerland, and economics at the University of Michigan, Ann Arbor, USA.

Ralph Cavanagh is a senior attorney and codirector of NRDC's energy program, a Visiting Professor of Law at Stanford and UC Berkeley, a lecturer at the Harvard Law School, and served on US DOE's Advisory Board.

His focus has been mobilizing the utility industry to invest in clean energy solutions, led by cost-effective energy efficiency, and ensuring that progress is not blocked unintentionally by regulatory policies. He has received numerous awards in recognition of his years of service including from the National Association of Regulatory Utility Commissioners, the Yale Law School, and the Bonneville Power Administration.

He is a graduate of Yale College and the Yale Law School.

John Cooper launched North American consulting services for Siemens Business Transformation. Since joining Siemens in 2013, John has been responsible for engaging pioneer utilities in a managed transition to a new more sustainable business model.

Being coauthor of *The Advanced Smart Grid: Edge Power Driving Sustainability, 2nd edition*, he has led numerous innovative projects as a thought leader in the Smart Grid industry, also authoring many white papers, articles, and blog posts since his early days helping to launch the pioneer Smart Grid at Austin Energy in 2004–05.

He has BA in Government and MBA with honors from the University of Texas at Austin.

Susan Covino is a Senior Consultant, Emerging Markets at PJM. She previously served as Manager for Demand Side Response.

She has been actively involved in the development of demand response in the PJM region and nationally. She has served as PJM's liaison to the Mid-Atlantic Distributed Initiative (MADRI) and has led the development of multiple iterations of the Demand Response Road Map for the PJM Region. She has led PJM stakeholder outreach in demand response, organizing three demand response and smart grid symposia. She has served as board chair of the Association for Demand Response and Smart Grid. She is currently the lead for PJM's role as technical advisor to MIT's "Utility of the Future" initiative.

She earned BA with a double major in Economics and History from the University of Connecticut and Juris Doctor degree from Dickinson School of Law.

Taylor Curtis is an associate researcher and community solar clinician at Vermont Law School's Institute for Energy and the Environment.

She consults with Vermont communities regarding net-metered solar projects. She also works to educate policymakers and regulators about smart grid and distributed generation deployment.

She graduated in 2012 with Masters in Environmental Law and Policy, while helping to compose *The Watt? An Energy 101 Primer*. She is a 2015 JD/Energy Law candidate at VLS and plans to use her degree to advocate for sustainable and ethical energy initiatives.

Marc de Konkoly Thege is a research associate at Vermont Law School's Institute for Energy and the Environment, researching distributed generation deployment in Brazil and the evolution of smart grid regulation in the United States.

Before attending VLS, he began his career in the energy sector working for the New York branch of Green Mountain Energy Company, a retail electricity supplier.

He graduated in 2012 with a Bachelor of Arts in International Relations from George Washington University in Washington, DC. He is a 2015 candidate for an Energy Regulation and Law master's degree from Vermont Law School.

Frank Felder is Director of the Center for Energy, Economic and Environmental Policy and Associate Research Professor at the Edward J. Bloustein School of Planning and Public Policy at Rutgers University.

He directs applied energy and environmental research including energy efficiency, renewable portfolio standards, modeling of state energy policies, and restructured electricity markets. He has published widely on market power and mitigation, wholesale market design, reliability, transmission planning, and rate design. He was a submarine officer in the US Navy.

He holds undergraduate degrees from Columbia College and the School of Engineering and Applied Sciences and a masters and doctorate from M.I.T in Technology, Management, and Policy.

Stephen Garner joined Ergon Energy in 2011 as a research analyst. His interests involve developing data processing and analytics tools to help the business understand the distribution network and implement appropriate new technologies.

He is a data scientist, quant strategist, MATLAB developer, and choice architect. Prior to Ergon, he worked in finance and funds management. The majority of his career has involved highly quantitative work in engineering, geophysics, and software development.

He has BSc with Japanese Language, BSc Hons 1 and PhD in Microelectronic Engineering (Electrical Geophysics), all from Griffith University in Brisbane, Australia.

Clark Gellings is a fellow at the Electric Power Research Institute, responsible for technology strategy in energy efficiency, demand response, renewable energy, and clean technologies.

He joined EPRI in 1982 progressing through management and executive positions including seven vice president positions. Prior to EPRI, he was at Public Service Electric and Gas Company in New Jersey. He is the recipient of numerous awards, has served on numerous boards and advisory committees.

He has Bachelor of Science in Electrical Engineering from Newark College of Engineering in New Jersey, Master of Science in Mechanical Engineering from New Jersey Institute of Technology, and Master of Management Science from the Stevens Institute of Technology.

Eric Gimon consults as a technical expert, research scholar, and policy adviser with Energy Innovation. He is a main contributor to America's Power Plan, working especially on questions of renewable energy integration.

Prior to working with Energy Innovation, he was a technical adviser for Vote Solar, a nonprofit solar advocacy organization. He was a AAAS fellow with two offices in the Department of Energy, transitioning out of a 15-year career in high energy physics.

He holds BS/MS from Stanford University in Mathematics and Physics, as well as PhD in Physics from the University of California in Santa Barbara.

George Grozev is a principal research scientist and a team leader at CSIRO Land and Water, Australia. His background is in operations research with experience in energy market modeling and network theory.

His current projects include analysis of customer load profiles for distribution networks, data analysis, and simulation of distributed energy resources, including solar PV, batteries, and electric vehicles under different scenarios and tariffs. He led the development of a simulation tool for Australia's national electricity and gas markets.

He has PhD from the Institute of Engineering Cybernetics and Robotics, Bulgarian Academy of Sciences and Bachelor degree in Engineering Physics from St. Kliment Ohridski University, Sofia, Bulgaria.

Philip Hanser is a principal of The Brattle Group with over 30 years of consulting experience in the energy industry. He has appeared as an expert witness before the Federal Energy Regulatory Commission and numerous state public utility commissions, environmental agencies, utility boards, arbitration panels, and federal and state courts.

Prior to Brattle, he was manager of Demand-Side Management Program at the Electric Power Research Institute. He held teaching positions at several universities including Columbia University and is a Senior Associate in the Mossavar-Rahmani Center at the Harvard Kennedy School.

He had undergone undergraduate training in economics and mathematics at the Florida State University and doctoral studies in economics and statistics at Columbia University.

Andrew Higgins is a principal research scientist at CSIRO Land and Water where he manages the development of decision support tools to forecast technology uptake across building stock under different policy interventions. He also leads a large research portfolio in freight transport logistics, particularly in northern Australia agriculture.

Before joining CSIRO in 1996, he completed PhD and Post Doctoral Research Fellowship at the Queensland University of Technology through the Mathematics and Civil Engineering departments.

Gerhard Jochum is a senior consultant at BÜRO JOCHUM based in Berlin and a board member in the energy industry. Currently, he acts as member of the Boards of GASAG Berliner Gaswerke Aktiengesellschaft; GDF SUEZ Energie Deutschland AG; Repower, Poschiavo, Switzerland; and STEAG GmbH.

Prior to his current position, he was Member of the Management Board of Energie Baden-Württemberg AG, Karlsruhe; CEO of swb AG, Bremen; and Director of VSE AG, Saarbrücken. Before starting his management career in the energy industry, he was managing partner of a consulting company focused on the energy business.

He holds a degree in economics.

Kevin Jones is the Deputy Director and Professor of Energy Technology and Policy at the Institute for Energy and the Environment at Vermont Law School where he leads the Smart Grid Project and the Energy Clinic.

Previously, he worked at the Long Island Power Authority, Navigant Consulting, and the City of New York. He is the coauthor of *A Smarter, Greener Grid: Forging Environmental Progress Through Smart Energy Technologies and Policies.*

He has a Doctorate from Rensselaer Polytechnic Institute; a Master of Public Affairs from the LBJ School of Public Affairs, University of Texas at Austin; and BS from the University of Vermont.

Günter Knieps is Professor of Economics at the University of Freiburg, Germany. Prior to joining Freiburg, he held Chair of Microeconomics at the University of Groningen, the Netherlands. He is member of the Scientific Council of the Federal Ministry of Economics and Energy and the Ministry of Transport and Digital Infrastructure.

His main research interests include study of network economics; deregulation; competition policy; industrial economics; and sector studies on energy, telecommunications, and transportation. He has published widely in academic and professional journals.

He has diplomas in Economics and Mathematics and PhD in Mathematical Economics from the University of Bonn, Germany.

Amanda Levin is an Energy and Climate Advocate with the Natural Resources Defense Council, focusing on alternative rate designs and the integration of distributed energy resources on the grid.

Prior to joining the NRDC, she worked as a research associate at the Congressional Research Service where she was responsible for handling congressional requests related to ecosystem restoration projects, trans-boundary water disputes, and water infrastructure.

She has both BA and MA in Public Policy, with a concentration in Energy and Natural Resources, from Stanford University.

Andrew Levitt is a Senior Strategist, Emerging Markets at PJM. He identifies, monitors, and assesses emerging technologies including but not limited to solar and batteries. He also analyzes the impact of emerging technologies on PJM markets, operations, and planning.

He previously managed the vehicle-to-grid research and development projects for NRG Energy. This work included overcoming technical and regulatory barriers to unlock the grid value of electric vehicles. This project featured a partnership between NRG and Honda.

He earned BS in Physics from the University of Toronto and MS degree in Marine Policy from the University of Delaware's Center for Carbon-Free Power Integration.

Sabine Löbbe is a professor and researcher at Reutlingen University, School of Engineering/Distributed Energy Systems and Energy Efficiency, Reutlingen Research Institute, Germany and lectures in the master program at the University of Applied Sciences HTW Chur, Switzerland. Her consulting company advised utilities in strategy and business development and in organizational issues.

Prior to her current position, she was Director for Strategy and Business Development at swb AG Bremen, Project Manager at Arthur D. Little Inc., and Project Manager at VSE AG, Saarbrücken.

She holds a Doctorate in Business Administration from the university in Saarbrücken and studied business administration in Trier, Saarbrücken, and EM Lyon/France.

Iain MacGill is an Associate Professor in the School of Electrical Engineering and Telecommunications at the University of New South Wales, Australia, and Joint Director of the University's Centre for Energy and Environmental Markets (CEEM). His teaching and research interests include electricity industry restructuring and the Australian National Electricity Market, sustainable energy technologies, and energy and climate policy.

Iain leads CEEM's research in Sustainable Energy Transformation including energy technology assessment and renewable energy integration, and in Distributed Energy Systems including smart grids, distributed generation, and demand-side participation. He has published and consulted widely in these and related areas.

He has a Bachelor of Engineering and Masters of Engineering Science from the University of Melbourne, and PhD on electricity market modeling from UNSW.

Chris Marnay is a retired Staff Scientist from LBNL, where he worked for 29 years. He currently leads an independent consultancy, Microgrid Design of Mendocino, but retains affiliate status with LBNL's China Energy Group.

He has lectured widely on microgrid principles, economics, and demonstrations; chaired the first 10 annual International Microgrid Symposiums; and is Convenor of CIGRÉ Working Group 6.22: Microgrid Evolution Roadmap. He was a Japan Society for the Promotion of Science Fellow in 2006 and a member of the team that proposed the CERTS Microgrid concept.

He obtained his BA, MS, and PhD from UC Berkeley. His dissertation proposed mitigating intermittent air quality events by biasing generator dispatch against dirtier sources.

Judith McNeill is an economist and Senior Research fellow in the Institute for Rural Futures at the University of New England in Armidale, Australia. She is engaged in investigating the impact of carbon policy on energy supply in rural Australia, climate change adaptation, infrastructure finance, and ecological economics.

Prior to academic life, she conducted research for the Australian Parliament and the Office of National Assessments in Canberra. She has also held economic policy positions in the Australian and Northern Territory Treasuries.

She has Masters Degree and PhD in Economics from the University of New England, Armidale.

Tim Nelson is the Head of Economics, Policy, and Sustainability at AGL Energy—one of Australia's largest energy utilities—where he is responsible for sustainability strategy, greenhouse reporting, economic research, corporate citizenship program, and greenhouse gas policy. He is an Adjunct Associate Professor with Griffith University. Before joining AGL Energy, he was an economic adviser to the NSW Department of Premier and Cabinet and the Reserve Bank of Australia.

He has advised governments and utilities on energy and climate change policy and published in Australian and international journals. He holds a degree in economics, is a Chartered Secretary, and is currently completing a PhD in Energy Economics.

Paul Nillesen is a partner with PwC based in Amsterdam where he works in the firm's Global Energy Utilities & Mining Practice and the coleader of PwC's

Global Renewables Practice. He is also a member of the International Energy Agency's Renewable Energy Working Group.

He is specialized in regulatory economics, strategic analyses, demand analyses, business simulation, and modeling primarily focused on energy and utilities. He has advised companies, governments, regulators, and national and international institutions and published numerous articles in energy and regulatory economics.

He holds First Class Master's degree in Economics from Edinburgh and Oxford and PhD in Economics from Tilburg University.

Michael Peevey was President of the California Public Utilities Commission for 12 years, until his retirement in Dec. 2014. At the CPUC, he spearheaded a number of initiatives including the 33% Renewables Portfolio Standard, banning the import of coal generated power, promotion of electric vehicles, continued emphasis on energy efficiency, the California Solar Initiative, and the rollout of smart meters.

Prior to the CPUC, he was CEO of New Energy, Inc., now the nation's largest independent electricity supplier, and prior to that, he was President of Southern California Edison Company.

He holds an undergraduate degree and MA in Economics from the University of California, Berkeley.

Glenn Platt leads the Grids and Energy Efficiency Systems Program within CSIRO Energy, developing technologies for reducing the carbon emissions and increasing the uptake of renewable energy. His interests include solar cooling, electric vehicles, smart grids, the integration of large-scale solar systems, and understanding consumers' uptake of low carbon energy options.

Prior to CSIRO, he worked in Denmark with Nokia Mobile Phones on the standardization and application of cutting-edge mobile communications technology and prior to that, for various Australian engineering consultancies.

He holds PhD, MBA, and electrical engineering degrees from the University of Newcastle Australia and is an Adjunct Professor at the University of Technology, Sydney.

Michael Pollitt is Professor of Business Economics at Cambridge Judge Business School, University of Cambridge; Assistant Director of the Energy Policy Research Group (EPRG) at the University of Cambridge; and a Fellow and Director of Studies in Economics and Management at Sidney Sussex College, Cambridge. He was coleader of the Cambridge-MIT Electricity Project.

His interests include the regulation of network utilities. He has published nine books and numerous articles on energy policy and business ethics. He is coeditor of the Economics of Energy and Environmental Policy.

He holds MA in Economics from the University of Cambridge and MPhil and DPhil in Economics from the University of Oxford.

Rahmatallah Poudineh is Lead Research fellow for the Electricity Research Program at the Oxford Institute for Energy Studies where he is focused on industrial organization and regulation and economics of the power sector.

His research has appeared in *Energy Policy, Energy Economics*, and *the Energy Journal*. He was the recipient of the best PhD thesis award for his contribution to the field of energy economics.

He holds PhD in Energy Economics from Durham University, MSc in Energy Economics and Policy from University of Surrey, a graduate Diploma in Economics from Queen Mary University of London, and BSc in Aerospace Engineering from Tehran Polytechnic.

Varun Rai is an Assistant Professor at the LBJ School of Public Affairs at the University of Texas at Austin where he studies the interactions between the social, behavioral, economic, technological, and institutional components of the energy system.

Before joining UT Austin, he was a research fellow at the Program on Energy and Sustainable Development at Stanford University. He serves as a Commissioner at Austin Energy and is on the Editorial Board of *The Electricity Journal* and *Energy Research & Social Science*.

He received his PhD and MS in Mechanical Engineering from Stanford University and bachelor's degree in Mechanical Engineering from the Indian Institute of Technology, Kharagpur.

Andrew Reeves is the former Chair of the Australian Energy Regulator. During his 6 years with the AER, he had oversight of the economic regulation of the electricity and gas networks in the Australian National Energy Market and consumer protection where the AER implemented major changes in the economic regulation of the networks including determination of network pricing.

Prior to AER, he was the Tasmanian Energy Regulator and Commissioner of the Government Prices Oversight Commission, following a career in minerals and energy policy. He was an Associate Commissioner of the Australian Competition and Consumer Commission.

His qualifications include engineering [BE (Hons)] with postgraduate qualifications in Business Administration and Economics.

Zhengen Ren is a senior research scientist at CSIRO Land & Water where he manages the development of decision support tools to simulate energy consumption for residential building stock and in development of AccuRate—a benchmark house energy rating tool widely used in Australia.

He has extensive experience in building energy performance simulation and has published in numerous journals, including *Energy Policy, Energy and Buildings, Buildings and Environment*.

He has PhD and was a Research fellow at the Queen's University of Belfast. He has BA and MA from School of Energy and Power Engineering, Xi'an Jiaotong University, China.

Matthew Roche is a class of 2016 JD candidate at Vermont Law School; a research associate for the Institute for Energy and the Environment at VLS; an intern at BCM Environmental and Land Law in Concord, New Durham; and a Staff Editor for the *Vermont Journal of Environmental Law*.

Prior to attending VLS, he worked for the DOE in the Clean Transportation program at the Northeast Ohio Clean Cities Coalition and for the National Park Service as a green team intern.

He has BA in History and Political Science from Notre Dame College.

Daniel Rowe is an Engineering Analyst and Project Leader within the CSIRO's Grids and Energy Efficiency Systems Program. His work involves energy for buildings, energy efficiency, renewable energy, and grid integration. He currently leads CSIRO's Plug and Play Solar off grid power system optimization and Residential Solar Cooling deployment projects.

His research interests include energy management technologies, solar cooling technology, solar forecasting, and grid integration techniques. He led the publication of CSIRO's landmark report *Solar Intermittency: Australia's Clean Energy Challenge*.

He has B Eng in Electrical Engineering from The University of Newcastle. He is an International Representative on the board of Young Engineers Australia.

Daniel Sauer is a research associate at Vermont Law School's Institute for Energy and the Environment where he conducts research and analysis on the evolution of smart grid regulation in the United States and distributed generation deployment in Chile.

Prior to attending VLS, he worked as a policy research intern for RCW Advocacy, a Governmental Affairs Law firm in Lansing, Michigan followed by a year of service with AmeriCorps in Detroit, Michigan.

He has BS in Environmental Policy and Public Administration from Central Michigan University and is pursuing Master's of Environmental Law and Policy at Vermont Law School.

Saad Sayeef is a Research Scientist in the Grids and Energy Efficiency Research Program within CSIRO Energy, working in renewable energy integration and energy efficiency. He was the lead author of a 2012 report on Solar Intermittency: Australia's Clean Energy Challenge.

Prior to joining CSIRO, he was a Research Fellow at the University of Wollongong where he worked on the control of wind turbines and energy storage for remote area power supply systems.

He holds PhD in Electrical Engineering from the University of New South Wales, Australia and Bachelor of Engineering in Electrical and Electronics Engineering from the University of Auckland, New Zealand.

Fereidoon Sioshansi is founder and president of Menlo Energy Economics, a consulting firm advising clients on energy-related issues. For 25 years, he has been the editor and publisher of ***EEnergy Informer***, a monthly newsletter with international circulation.

His prior work experience includes working at Southern California Edison Co., EPRI, NERA and Global Energy Decisions, acquired by ABB. Since 2006, he has edited eight volumes on different subjects including evolution of global electricity markets, energy efficiency, smart grid, and distributed generation.

He has BS and MS in Civil and Structural Engineering and MS and PhD in Economics from Purdue University.

Robert Smith is a consulting economist with East Economics and has over 25 years experience working in industry economics, electricity market design, regulation, economic evaluation, energy efficiency, and demand management.

His interests include applied economic analysis and understanding how economics, technology, incentives, regulation, and customers' behavior interact to create change.

He has graduate degree in econometrics, masters in economics from the University of NSW, and postgraduate qualifications in finance from the Securities Institute of Australia.

Paul Sotkiewicz is Senior Economic Policy Advisor in the Market Services Division at PJM Interconnection where he provides analysis on PJM's market design and performance including the implications of various federal and state policies. He has led initiatives to reform scarcity pricing as mandated by FERC Order 719, examine transmission cost allocation and the potential effects of climate change policy on PJM's energy market, and develop proposals for compensating demand response resources under FERC Order 745.

Prior to his present post, he served as the Director of Energy Studies at the Public Utility Research Center, University of Florida and an economist at the Federal Energy Regulatory Commission.

He received BA degree in History and Economics from the University of Florida and MA and PhD degrees in Economics from the University of Minnesota.

Michelle Taylor manages the Technology Development group at Ergon Energy. She has more than 25 years' experience in energy storage, photovoltaics, and power electronics. She leads a team looking to how new technologies will impact the network business and customers. She has worked on the Australian and New Zealand Standards committee for energy storage, inverter energy systems, and renewable energy.

Previous roles include managing the Energy Solutions group, which provided stand-alone power systems in regional Australia incorporating photovoltaics, batteries and diesel generators, and principal engineer in Technology Innovation.

She holds Bachelor Degree in Electrical Engineering from University of New South Wales, Australia.

Peter Terium was born in 1963 in Nederweert, the Netherlands. He studied to become chartered accountant at the Nederlands Institut voor Registeraccountants in Amsterdam and worked at the same time as an independent auditor for the Dutch Ministry of Finance. In 1985 he became an audit supervisor at KPMG, Eindhoven, the Netherlands.

From 1990 to 2002, he worked in various senior international finance and controlling positions in the packaging industry for Schmalbach-Lubeca AG, Ratingen, Germany.

He joined the RWE Group on Jan. 1, 2003: he started his time there as Head of Group Controlling at RWE AG and in 2004 also became a member of the Executive Board of RWE Umwelt AG. He was significantly involved in restructuring and selling RWE Umwelt AG.

In Jul. 2005 he was appointed Chief Executive Officer of RWE Trading. His responsibilities included the merger of RWE Trading GmbH and RWE Gas Midstream GmbH into RWE Supply & Trading GmbH.

In this position he made a significant contribution to the realization of RWE's growth strategy. As of 2009 he took responsibility for the integration process of Essent. Up to Dec. 31, 2011 he held the position of Chief Executive Officer of RWE's Dutch subsidiary. He was appointed Member of the Executive Board and Deputy Chairman of the Board of RWE AG on Sep. 1, 2011, and since Jul. 1, 2012 he has held the post of Chief Executive Officer of RWE AG.

Kai Van Horn is currently an Associate with The Brattle Group in Cambridge, MA, where he advises Brattle's clients on a range of energy-sector issues including the potential reliability and economic impacts of renewable integration, and effective congestion hedging in wholesale electricity markets.

He holds a BS in Multidisciplinary Engineering from Purdue University, and an MS and PhD in Electrical Engineering, both from the University of Illinois at Urbana-Champaign. His research interests include electricity market design, PMU-based operational reliability tool development, and technical and policy issues related to the integration of renewables into the electricity system.

Glenn Walden is Manager, Emerging Markets at Ergon Energy. He is an electrical engineer with more than 32 years' experience including the areas of distribution, generation, renewables, remote area power supply, project management, business management, and business development.

The Emerging Markets team monitors the industry and global environment for emergent trends and business opportunities; evaluates products and guides their development, industry engagement, and collaboration to position Ergon Energy to meet the service, sustainability, and technical challenges in future electricity supply.

He holds bachelor degree (with honors) in Electrical Engineering from James Cook University, Australia and Master of Business Administration degree.

Jens Weinmann is Program Director at the European School of Management and Technology (ESMT) in Berlin. Before joining ESMT, he worked as a manager at economic consultancy E.CA Economics.

His research focuses on the analysis of *decision making* in regulation, competition policy, and innovation, with a special interest in energy and transport.

He graduated in Energy Engineering at the Technical University of Berlin, received his PhD in Decision Sciences from London Business School, and held fellowships at the Kennedy School of Government, Harvard University and the Florence School of Regulation, European University Institute.

Stephen Woodhouse is Director at Pöyry Management Consulting. He heads Pöyry's Market Design group, which deals with all aspects of energy market policy regulation and design, for private and public-sector clients. He specializes in the economics of transmission and interconnection; market regulatory policy across Europe and the United Kingdom and Irish electricity and gas markets. He leads Pöyry's business development in the interrelated areas of intermittency, smart grids, and market design.

Prior to joining Pöyry, he was an Economic Modeler for Ofgem.

He has MA and BA in Economics from the University of Cambridge.

Jay Zarnikau is president of Frontier Associates, a consulting firm specializing in the design and evaluation of energy efficiency, pricing, solar photovoltaic, and "smart grid" programs. He is also an Adjunct Professor of Public Policy and Statistics at The University of Texas, where he teaches statistics and research methods.

He formerly served as a Program Manager at the Center for Energy Studies, University of Texas at Austin and was the Director of Electric Utility Regulation at the Public Utility Commission of Texas. He has published on energy pricing, resource planning, renewable energy, and energy efficiency.

He has PhD degree in Economics from University of Texas at Austin.

Foreword

For over a century the electric utility business was mostly stable, predictable, and, at times, dull. Not surprisingly, the business of regulating utilities was also stable, predictable and, for the most part, equally dull.

Utility executives had to make sure that the lights stayed on, most of the time, and that investors were reasonably rewarded. Regulators had to make sure prices were just and reasonable while keeping utilities reasonably profitable.

The utility business model—nobody called it that in the old days—was predicated on two time-tested principles: predictable sales growth and flat or falling per unit prices. Regulators set and periodically adjusted the cents per kilowatt-hour multiplier, which was applied to kilowatt-hour sales. It worked like a charm for too long. Ratepayers, as they were called, were billed on flat tariffs times the number of kilowatt-hours consumed. With flat or falling cents per kilowatt-hours—adjusted for inflation—sales continued to grow, which meant that costs could be spread among more kilowatt-hours, which meant lower per unit costs, which encouraged increased consumption.

This happy state of affairs—simplified to its core—lasted roughly to the turn of the millennium, when the fundamentals began to change. Sales growth slowed and, in some cases, has come to a virtual halt, while retail tariffs began to rise.

Significant technological change, mostly on the customer end of the supply chain, began to become noticeable. Not only could consumers manage and control their usage better, they could increasingly meet some of their own consumption through distributed self-generation. In a relatively short period of time, distributed energy resources became a buzz word, transforming *consumers* to *prosumers*—active participants in the market, rather than passive consumers of bulk kilowatt-hours from the grid.

To everyone's surprise and within a stunningly short span of time, a growing number of prosumers in high retail regions of the world could generate some or virtually all their needs, at prices on par or cheaper than buying it from the grid. Add speculation about the falling costs of storage, microgrids and other technologies, and you can see the current interest in utility of the future, the topic of this book.

What do these changes portend and how can or should the industry and its regulators respond? As an ex-regulator, my views no longer much matter. Yet, I would hope that my experience as a utility executive and a regulator might be relevant, which is why I have agreed to share a few remarks at the request of this book's editor:

- First, as observed by Andrew Reeves, himself an ex-regulator and a contributor to this book, regulators cannot dictate innovation, but they can create a hospitable environment for it to thrive, and can certainly avoid stifling it. This, in my view, should serve as a guiding principle for regulators in deciding how to regulate and, equally important, when not to.

- Second, the current debate about the relative merits of centralized versus distributed business paradigm may miss the point. As the editor and Clark Gellings point out elsewhere in this book, the integrated grid of the future should enable these two paradigms not only to coexist, but actually to complement each other—at least until the cost of storage falls to a level where the grid is no longer critical. In my view, this will take a while, if not a lot longer. For some customers, it may never materialize, and we need to think carefully about a potentially bifurcated future.

- Third, perhaps like many other regulators, I believe that the incumbents, who have invested a fortune, in good faith, in an infrastructure to serve ratepayers, deserve due process before we pull the rug from under their feet. Perhaps my view is tainted because I was a utility executive before I became a regulator, in which case you can cross this one off. Incumbents, as a minimum, should be given regulatory clarity to transition to where they should be heading.

- Finally, regulatory proceedings in New York, Hawaii, California, as well as in Australia, Europe, and Japan, where distributed energy resources have made significant inroads, provide useful pathways. Many of the challenges faced by regulators around the world are universal—which suggests that we can learn a lot through collaboration and exchange of views. Not invented here syndrome should be avoided, if at all possible, which makes a collected contribution such as this useful, given the breadth of topics and the diversity of views.

Michael Peevey
President, California Public Utilities Commission, 2003–14

Preface

After the computer mainframe industries in the 1980s and then the conventional camera industries in the late 1990s, to name but two, it is now the turn of the electric utilities to take their place on the anvil of technological and societal change, and to reshape themselves for the future.

As a CEO, leading such a change of course for a company with over 100 years of history and ranking #7 by market value at the European level with 23 million customers, I greatly welcome the publication of this comprehensive and forward-looking book, since all leaders of utilities need the widest possible repository of concepts to identify and implement the ones most relevant to their own situation. It is no longer sufficient to look only within the borders of a utility's own market boundaries in order to obtain fruitful ideas for the future, so the breadth of geographical coverage afforded by this publication is of particular value to its readers.

The challenge of transforming a competent utility which grew up during the monopoly era into the high-performing, nimble utility of the future, is truly Herculean but is one to which the whole company is rising. As compared to the computer and camera industries, which were victims of technological change alone, utilities have to weather very much more: a perfect storm which entails additionally major regulatory changes leading to the introduction of subsidized capacity into an already well-supplied market, static to declining demand and, on top of all this, further reduction in the thermal power plant earnings due to abrupt downward changes in the global commodity markets.

But, despite challenged balance sheets, utilities are holding their own. Necessity is the mother of invention and in the case of RWE, and no doubt other affected utilities around the world, the most remarkable changes in the attitudes and approach of staff are to be found. Creativity, flexibility, and new working practices are flourishing as never before, all of which are driving the transformation of the incumbents to position themselves, as suggested by the title of this book, into "Utilities of the Future."

As I look at the different chapters covered, I see how well they correspond to the challenges utilities are facing and the corresponding changes they are in process of making. The first and overriding question, which Fereidoon Sioshansi picks up in his opening chapter, concerns the size and utility turnover of the future market. The Global Recession of 2008 has left much deeper tracks behind than expected and, at a European level, the gross demand recorded in 2013 was still 4% below the 2008 peak, with demand net of nonutility renewables being considerably lower. Electricity is a premium product and has the potential to play an essential role in decarbonizing the wider energy economy. It is therefore important that this product—an important ecological driver—should not be inappropriately loaded with taxes and other surcharges, which together, would hinder the electrification process in the heating and transportation sectors. Moreover, such moves into these sectors would at least partly compensate for the future loss of demand through further efficiency initiatives, and thus help maintain good utilization of the existing infrastructure.

Next, as owners of substantial distribution networks, mainly in Germany but also in Central East European countries, we can only underline Clark Gellings' attention given to the value of an integrated grid. As a result of the phenomenally rapid build-up of renewables within RWE's home territory, the company has been at the forefront of transforming the grids from a former one-way conduit, for which they were originally designed, to bidirectional carriers, with the capacity of the new renewable sources, at certain hours, totally overshadowing the peak demand. The distribution grids will be the host to eventually more than 50% of total power production and it is therefore vital that they be made fit for the purpose. Equally, it is important that the tariff structures should be developed in such a way that the economic and societal benefits of the grids are appropriately supported, and that it does not become artificially attractive for customers to go off-grid, topics extensively covered in the book.

The new power on the consumer side of the meter, also covered in the chapters, is something we are experiencing at first hand and see as an opportunity rather than a threat. With our customer base of 23 million, approaching 10% of all European customers, we have an enormous insight into their changing requirements, and it is one of our priorities to take full advantage of this, and respond appropriately. Rather than simply selling and billing kilowatt-hours, we are working toward much more comprehensive, high-value relationships which will include helping our customers achieve their decentralized generation objectives and achieve the maximum value from each unit of energy.

Issues of market design and consumer rate design are two major preoccupations. Our company supports market-based capacity arrangements, and is therefore pleased that the subject has been covered through Stephen Woodhouse's contribution as one of a number of possible approaches. Retail design is an important but socially-sensitive issue: in order for the market to function effectively, it is desirable that the different pricing signals should be communicated to the end-consumers to allow them to react, but the low-income classes must not be unduly burdened through such a transition.

Lastly, I would like to identify with the chapters covering new value pools and innovation, both of which lie right at the heart of our transformation to a Utility of the Future. As the owners of thermal plants in Europe have seen one of their largest value-pools largely evaporate, the replacement through new sources of revenue is absolutely critical. The contribution of Nillesen and Pollitt and Cooper on business transformation and the Internet of Things (IoT) runs very much in line with our own initiatives. At RWE, we are proud to have been ranked in first place by the European School of Management and Technology (ESTM) innovation index in 2012 and, not surprisingly, we identify strongly with the proposals made in **Chapter 16: European Utilities: Strategic Choices and Cultural Prerequisites for the Future** by Burger and Weinmann, while embracing disruptive technologies as espoused by Löbbe and Jochum in **Chapter 17: Thriving Despite Disruptive Technologies: A German Utilities' Case Study**.

Overall, I believe that assembling such a wide range of ideas, as attempted in this book, is helpful in enabling utilities to make their journeys away from the monopoly world, which, for over at least a decade, has no longer offered them any kind of protection, forward to the exciting new world of the future. At least for a considerable period, many of the existing assets (such as thermal power plants) will still be required, but are being adapted to be flexible partners to the renewable capacity. The networks will be required for as far ahead as one is currently able to see, but will need to be substantially adapted. And the customers will always be there, looking to extract the maximum value from every unit of power they buy or generate. The successful utility of the future will be the one which manages to put all these pieces together—something RWE is aspiring to do. As noted by Mike Peevey, California's former top utility regulator, there is a lot we can learn from each other's experiences.

Peter Terium
Chief Executive Officer
RWE AG
Essen, Germany

Introduction

There are many who believe that the electric power sector is entering, or has already entered, a new phase in its evolution, requiring significant changes in its operations, business model, culture, and how it is regulated—issues further examined in this book.

The debate about the future of power industry is heated, with varied opinions on the ultimate outcome. One side says the dreaded utility *death spiral* is here, or near, and the industry is heading into a fatal stall from which there is no escape. According to a recent study by **Accenture**, for example, by 2025, the annual revenues of US utilities may potentially be as much as $48 billion *lower* than they would have been otherwise due to the rapid growth of **distributed generation** and gains in **energy efficiency** (Fig. 1A). The same study says demand on the grid may fall by as much as 15% relative to the "status quo" scenario.

Likewise, European utilities may experience equally sizable revenue erosion over the same time horizon, for similar reasons (Fig. 1B). The Accenture study is one among many who reach similar conclusions, basically projecting that, over time, consumers in mature markets around the world will buy less power from the grid, as appliances and electric devices become more efficient, as do homes and buildings, which reduce demand for electricity. At the same time, consumers in many of the same markets will generate more of what they need from distributed self-generation, particularly rooftop solar PVs, whose costs continue to decline, relative to grid-supplied tariffs.

Other studies[1] conclude that with rapidly falling solar PV prices, an increasing number of consumers will be able to meet some or virtually all their electric service needs through distributed self-generation. In his book, *Power Shift*, Robert Stayton (2015)[2] predicts a future where solar energy will power the global economy, displacing fossil fuels by 2060. While there is disagreement on how soon and how fast, the prognosis is similar for many parts of the United States, Europe, Japan, and Australia, to name a few.

US electric power consumption appears to have peaked, with little sign of a rebound, reflecting declining demand growth rates over time. Future economic

---

1. For example, refer to chapter: New Business Models for Utilities to Meet the Challenge of the Energy Transition summarizing the results of a study by PwC, a consulting firm. Another 2015 survey on attitudes of consumers and businesses by Deloitte, a consulting firm, found that 79% of businesses said reducing electricity costs is essential to staying competitive, while 55% said they generate some portion of their electricity needs from onsite generation.
2. Stayton, R., 2015. Power Shift. Sandstone Publishing, Santa Cruz, CA.

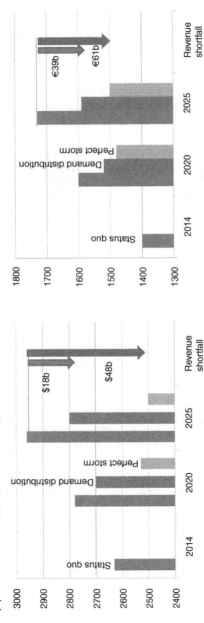

FIGURE 1   Projections of utility sector revenue erosion in the US (A) and Europe (B) due to distributed energy resources under alternative scenarios. *Refers to electricity purchase from the grid, excluding self-generation. **Includes EU's 10 largest economies. Analysis by Accenture suggests significant revenue shortfall potential based on reduced purchase from the grid under two alternative scenarios; the greatest risk to demand and revenues in Europe are in the period to 2020 with potential loss of 235 TWh in the perfect storm scenario. (*Data Source: How can utilities survive energy demand disruption, Accenture, 2014.*)

growth no longer requires a commensurate growth in electricity consumption—as was historically the case. As more economies mature, they face a similar predicament as the United States and other OECD economies. Accordingly, US electricity consumption, by many estimates, has already peaked.

The other side says this too will pass, or that the demise of incumbent utilities is highly exaggerated. **Moody's**, the credit rating agency, for example, acknowledges that utilities will look different, and need to make a number of adjustments to survive, but says don't write them off, yet.

A more sensible debate, however, is on the future evolution of **centralized** versus **decentralized** utility paradigm. Nearly everyone agrees that the centralized generation, transmission, and distribution model is seriously challenged by an emerging decentralized version, with *prosumers*[3] exercising increased control over how much power is locally produced, consumed, stored or exchanged with other *prosumers* using the network that connects them.

This leads to transactive energy, where future prosumers may increasingly interact with *each other* using an open platform while relying on the distribution and transmission grid to facilitate the transactions.[4] Tesla has unveiled a battery storage system that allows solar PV customers to virtually cut the copper cord altogether.[5]

But, even here, there are fundamental disagreements about how the future grid will enable consumers and prosumers to remain connected, and—more important—who will pay for the critical services provided by the so-called integrated grid (see Chapter 2: Value of an Integrated Grid). While some consumers may be able to coast along more or less on their own, most are likely to remain dependent on the grid for essential services, especially reliability and backup.

One side argues that the grid is a **social good**, and should be paid for as such. It is like having and funding a public library, even if some citizens never use it. These people would typically propose that since the costs of maintaining and upgrading the grid are mostly fixed, everyone who is connected should pay a fixed monthly fee for its upkeep.

Such arguments quickly boil down—or disintegrate—into debates about what portion of a typical customer's monthly electricity bill should be fixed versus variable. David Owen, former Ex VP of Edison Electric Institute, the lobbying arm of the US investor-owned utilities, for example, was quoted in an article in the *Wall Street Journal* (Feb. 3, 2015) pointing out that, "it costs most utilities $40 to $60 a month to serve a home." He added, "So customers who use (he meant to say buy) little power shift costs to others." As explained later, there is an important distinction between how much electricity consumers "use" versus how much they "buy."

---

3. The term prosumer, now in wide use, refers to consumers who are also producers.
4. For example, see Transactive Energy by Ed Cazalet and Steve Barrager, 2014.
5. (http://reneweconomy.com.au/2015/tesla-launches-home-business-and-utility-battery-storage-range-74034).

Customers, of course, might "buy" little power either because they have invested in efficiency and/or in distributed self-generation, most commonly rooftop solar PVs. The combination of the two is referred to distributed energy resources (DERs).

According to this line of reasoning, consumers who *buy* little power from the grid are somehow evil and should be penalized for their antisocial behavior. Not surprisingly, many utilities are in favor of fixed fees or minimum monthly fees—specifically targeting prosumers, that is, those who consume little and produce a lot.

The *Wall Street Journal* article describes a radical proposal submitted to regulators in state of Wisconsin in 2014:

> *"Last May (2014), Madison Gas & Electric Co. asked state utility regulators to let it charge residential customers $68 a month by 2017 as a fixed monthly fee for electricity service, covering 77% of the utility's fixed costs, versus the existing $10.50 fee, which covered 12%. In return, the utility agreed to cut the price of electricity in half, to 7 cents a kilowatt-hour. The proposal created such uproar that the utility withdrew the request. Instead, it got approval to begin charging about $20 a month, enough to cover 23% of its fixed costs, and to slightly reduce its electricity price."*

Why the uproar? Because, broadly speaking, customers dislike large fixed or minimum fees. Many believe that electricity is like gasoline or groceries. If you don't buy much, you don't pay much. If you don't buy any, you pay none.

Low income consumer advocates, energy efficiency, and solar PV fans don't like large fixed charges or minimum fees either because they are deemed to be unfair to low income, energy frugal, and consumers who have invested in solar PVs.

Despite the uproar and the objections, fixed fees and minimum charges appear to be on the rise in a number of states, generally favored by incumbent utilities who are concerned about gradual erosion of revenues due to the rise of DERs. In other parts of the world, fixed monthly fees or capacity charges are more common, so the issue is not whether but how much of a fixed fee.

In early Feb. 2015, the *Houston Chronicle* reported that some utility customers in Texas have been *surprised* to find they are charged a fee for using *less* than a specified amount of electricity—the minimum charge level. This is even more ironic, since many of the same utilities are actively promoting energy efficiency, usually prompted by the regulators.[6]

The *Houston Chronicle* examined rate plans of 49 retailers in the Houston area—a total of more than 300 plans—and found that more than 200 included the minimum-use fee. In some cases, the plan offered a discount to customers who used *more* than a specified amount of power. Retailing is competitive in Texas, which means retailers can offer any plan they think will appeal to customers.

---

6. Refer to Rai & Zarnikau and Cavanagh & Levin in this volume.

The *Houston Chronicle* reported that, "While the fees varied, they averaged $10.67 and most were triggered if customers used less than 1,000 kilowatt-hours of power." Average residential consumption in Texas is about 1200 kWh/month.

Most retailers defend the practice, saying the minimum fees are needed to pay for fixed costs that occur, regardless of whether customers use a lot of electricity or very little.

Jake Dyer, a consumer advocate with the Texas Coalition for Affordable Power, disagrees. He was quoted in the *Houston Chronicle* saying, "You don't pay a minimum-use fee when you step into a grocery store. You don't pay a minimum-use fee when you shop for any other product. Most businesses price their product in a way that the people who actually buy it will pay for their fixed-cost infrastructure."

As the preceding discussion illustrates, the debate about fixed versus variable fees is controversial, as are the arguments about solar versus nonsolar customers and net energy metering (NEM). If solar customers are assessed a penalty or fixed fee because they are buying less from the grid, then consumers with super-efficient appliances, zero-net energy (ZNE) homes, or consumers without central air conditioners, or big appliances, or EVs should also pay a penalty.

The mindset of some industry executives is to challenge the frugal prosumer: "How dare you *not* use lots of kWhrs when we have invested in this entire expensive infrastructure to serve you?"

Others, certainly in the United States, are beginning to not only accept, but have already internalized the transformation of the power sector. Speaking at the Edison Electric Institute's 2015 Annual Convention,[7] Ted Craver, the CEO of Edison International, said:

> *"Southern California Edison is focused on creating a modern, bidirectional grid that can accommodate California's growing solar generation capacity any time of day,"*

adding,

> *"If a customer spends $15,000 to $20,000 to put solar panels on his roof, then the utility better be prepared to take on the electricity when the customer has excess and, in turn, provide electricity when he needs more. If SCE can't accommodate that customer, the customer will invest a little more money on energy storage and disconnect from the grid completely. Then we have lost a customer for good."*

Utterances such as this would have been unthinkable a few years earlier. But as further explained in the book's Preface by Peter Terium, RWE's CEO, and described in Chapters 6 and 16 by Nelson & McNeil and Burger & Weinmann, utilities in Europe, as well as Australia, are responding to the underlying

---

7. Quote attributed to Mr Craver appeared as editorial by Teresa Hansen, editor in chief of Power Grid International, Jul. 2015, available at http://www.elp.com/articles/powergrid_international/print/volume-20/issue-7/departments/from-the-editor/grid-modernization-discussions-take-center-stage-at-eei-convention.html

changes in the business environment in ways that would have been considered radical.

This book, which is focused on the topic of the utility of the future, examines the ramifications of the rapid transformation of the industry and its traditional business model, which was predicated on continued demand growth coupled with flat or falling retail tariffs—neither of these apply.

Among the main issues covered in the book is a discussion of a rational way to address these changes and find a constructive path forward. Clearly, very few prosumers would wish to go completely off-grid since the costs of enjoying the same level of reliability and service provided by the existing grid is rather high—don't believe anyone who says otherwise. Until the cost of storage and microgeneration declines to make such options truly affordable, we must find a compromise that allows the centralized and decentralized models not only to coexist, but rather to complement each other.

Mauricio Gutierrez, COO of NRG Energy,[8] the largest US independent power producer (IPP) and a major player in distributed generation was recently quoted in Power Engineering (Jan. 2015) saying, "I think there is a way competitive energy companies and the utility space can work collaboratively," adding, "I don't think it's an either or."

It is rather self-serving for NRG—who has relatively little to lose and a lot to gain from the growth of distributed generation—to say this. Yet the basic premise that the future is likely to be a synergistic combination of the two may turn out to be spot on—at least until the cost of storage declines to the point where physical connectivity to the grid becomes a moot point.

Having set the context, the book is organized into three parts, with a brief summary of each chapter below as an easy guide for the reader.

1 PART I. WHAT IS CHANGING, WHAT ARE THE IMPLICATIONS?

As the title suggests, this part includes a discussion of the significant developments taking place within the power industry and on its periphery and how these changes are affecting the utility business and the business model.

In **Chapter 1: What is the Future of the Electric Power Sector?, Fereidoon Sioshansi** explores how rapid changes taking place within the electric power sector, including the rapid uptake of distributed generation and renewables, microgrids, storage, zero-net energy buildings, and developments on the customer side of the meter are impacting the traditional utility business.

The chapter's main focus is to ask if, in fact, there will be a future for electric utilities as we currently know them.

The chapter's main conclusion is that for the foreseeable future, the majority of customers will most likely be better off remaining connected to the grid

---

8. He became CEO of NRG in Dec 2015.

and benefiting from its myriad of services. When it comes to centralized versus decentralized, the future is most likely not an either or—this suggests fair and equitable solutions must be found to make the two paradigms to complement, rather than compete.

In **Chapter 2: Value of an Integrated Grid, Clark W. Gellings** points out that the *integrated* grid leverages the optimal combination of local generation, energy storage, energy efficiency, and new uses of electricity, with central generation and utility-scale storage to provide society with reliable, affordable, and sustainable electricity.

The author explains that such an integrated grid is characterized by connectivity, rules enabling interconnection, and innovative rate structures that enhance the value of the power system to all consumers while blending the most valuable generation, storage, power delivery, and end-use technology services, including DERs, tailored to meet local circumstances.

The chapter emphasizes that simply being connected to the grid—a *fait accompli* in developed economies—does *not* constitute the formation of an integrated grid, but rather argues that individual consumers and society at large benefit from integration beyond services they currently receive from simply being connected.

In **Chapter 3: Microgrids: Finally Finding Their Place, Chris Marnay** notes that a few existing microgrids performed exceptionally well during the 2011 Japanese earthquake, Hurricane Sandy, and other recent natural disasters. Together with microgrids' other desirable features, these experiences have accelerated interest in their adoption into the electricity supply chain.

The author points out that microgrids can improve the effectiveness of power delivery by providing targeted high quality power to sensitive loads and low quality power to loads that can function on it. Moreover, microgrids can manage small-scale renewables, controllable loads, and other small-scale DERs locally, while participating in wholesale energy and ancillary services markets at scale, thereby relieving the megagrid. The implications are fundamental to the industry because the evolution of microgrids effectively changes the characteristics of the megagrid, redefining the requirements it must meet.

In this context, the chapter argues that the principle of universal service must be rethought, and microgrids need regulatory recognition so their capabilities can provide the most benefit.

In **Chapter 4: Customer-Centric View of Electricity Service, Eric Gimon** explores the nature of the services coming through the customer/grid interface, from the point of view of the customer. While the electricity grid is being redefined by the rise of new mass-scale distributed energy technologies, these are typically examined for how they impact the grid from a perspective tied to utility-side needs.

The author asks what customers have traditionally gained from being connected to the grid, and how the value proposition offered by the grid has changed. He reconsiders the compromises customers make and seeks definitions for new grid services more directly aligned with the needs of empowered *prosumers*.

This chapter's main conclusion is that the transactions happening at the customer/grid interface can most helpfully be understood via the language from the commerce and finance world. These hinge on the currency-like features of electricity and lead to a future "fractal" grid.

In **Chapter 5: The Innovation Platform Enables the Internet of Things, John Cooper** envisions a new role for the electric utility in the 21st century, well beyond a slow evolution from affordable, highly reliable electricity distribution, to supporting the integration of DER and renewables. By becoming a platform for innovation, utilities support the emerging Internet of Things (IoT).

To explain the visionary challenge for utilities and to think beyond their current paradigm, the author describes the infrastructure to services transformation experienced in the telecom and information industries, the role that platforms play in today's economy, and the essential nature of innovation in the emerging era of the IoT.

The chapter's main conclusion is that, since infrastructure inevitably matures into the foundation for services that support greater personalization and value, electric utilities should seize the chance to redefine themselves for another century-long run of success as the providers of the essential platform to enable that transformation.

In **Chapter 6: Role of Utility and Pricing in the Transition, Tim Nelson** and **Judith McNeill** point out that new technology has increased the economic viability of embedded generation and storage for many consumers. However, coexistence rather than complete grid-substitution is the reality facing utilities and new entrants.

The authors argue that pricing grid and nongrid services will be a key determinant in the success of utilities in navigating the transition. Incumbent utilities in Australia are repositioning themselves, given the very high uptake rates of distributed generation, by adjusting prices of grid and nongrid services to reflect consumer preferences.

The chapter's main conclusion is that reliable supply will be of significant value to consumers, and incumbents are likely to offer a diverse range of products, such as solar leasing, demand tariffs, and storage with prices that reflects the "insurance premium" value consumers place on reliable supply.

In **Chapter 7: Intermittency: It's the Short Term That Matters, Daniel Rowe, Saad Sayeef**, and **Glenn Platt** point out that while solar photovoltaic energy systems hold much promise, there is great concern as to how *intermittency* may affect the rollout and benefits. Already today, such concerns are causing restrictions on system design and performance.

This chapter explores the significant disagreement regarding what the actual impact of intermittency is, examining the areas where it has been demonstrated to cause significant problems, and the potential solutions to these issues that are being investigated around the world.

The chapter's main conclusion is that there is significant concern, but equally significant disagreement about the effect of intermittency on electricity network operation. Additional work is needed to find the genuine intermittency limits, but, in the meantime the efficiency and effectiveness of combined gridside and loadside mitigation options must be better understood to understand opportunities for utilities of the future to respond to solar intermittency as part of new business models in a changing grid.

2 PART II. COMPETITION, INNOVATION, REGULATION, AND PRICING

Part II of the book examines the rapid pace of technological advancements taking place on the customer-end of the business, and the need for new policies, regulations and pricing schemes not only *not* to stifle innovation, but rather to accelerate them.

In **Chapter 8: Retail Competition, Advanced Metering Investments, and Product Differentiation: Evidence From Texas, Varun Rai** and **Jay Zarnikau** examine how retail competition, coupled with smart grid investments, is leading to expanded customer choice in retail electricity rates and services in Texas.

A comparison of service offerings and pricing plans between areas with and without retail competition in Texas suggest that slim margins in a competitive retail market lead to a need for product differentiation to achieve higher margins.

The authors observe that the first wave of product differentiation through green pricing has now run its course, with the retail market now entering a second wave, with peak rebate programs, free nights/free weekend programs, prepay electricity programs, smart thermostats, and other promotions to differentiate a retailer's products from the rest of the pack. Future differentiation is likely to promote energy storage and Green Button technologies.

In **Chapter 9: Rehabilitating Retail Electricity Markets: Pitfalls and Opportunities, Ralph Cavanagh** and **Amanda Levin** examine customer power in retail markets. "Free electricity" offerings and rewards have proliferated in active markets. At the same time, customers are increasingly investing in DERs. The authors analyze how retail providers are incorporating these technologies, and consider how retail competition can obstruct or unleash their full value.

The authors first delve into current offerings and the nascent incorporation of DERs in the retail electricity market, expanding into discussion on the potential DERs hold in all retail markets to transform the provider–customer relationship and catalyze novel services.

The authors find that market focus on low per kilowatt-hour costs is hampering DER adoption, along with system reliability. The chapter details strategies to make these markets more consistent with society's reliability and environmental objectives. It then considers whether "fixing" retail competition is worth the effort, in light of alternatives.

In **Chapter 10: Residential Rate Design and Death Spiral for Electric Utilities: Efficiency and Equity Considerations, Rasika Athawale** and **Frank Felder** explore the ramifications of federal and state policies promoting distributed generation and energy efficiency on different types of residential consumers in the US context of a US state.

The authors quantify the amount of revenue imbalance created as a result of rate design and regulatory policies that favor the owners of distributed resources, and create disincentive for nonparticipants.

The chapter examines the net effect on winners and losers under different scenarios of rate adjustments, along with the impacts on sales tax for New Jersey.

In **Chapter 11: Modeling the Impacts of Disruptive Technologies and Pricing on Electricity Consumption, George Grozev, Stephen Garner, Zhengen Ren, Michelle Taylor, Andrew Higgins,** and **Glenn Walden** examine scenarios for rapid uptake of new DERs and energy storage, and explore a hypothesis that adoption of residential "network tariffs" will reduce or eliminate the distortionary effects of volume-based tariffs on residential energy consumption patterns.

The authors apply an integrated modeling approach with high spatial and temporal resolution to Townsville in Northern Queensland, Australia. The methodology examines annual consumption patterns for specific dwelling types, using network-based tariffs, feed-in-tariffs, and TOU rates for solar PV generation and battery storage.

The analysis demonstrates that electricity consumption[9] could drop by more than 10% in the next decade, potentially more in combination with new tariffs. The scenario results demonstrate that cost reflective tariffs can improve network utilization, and potentially put downward pressure on retail prices.

In **Chapter 12: Decentralized Reliability Options: Market Based Capacity Arrangements, Stephen Woodhouse** notes that in the European power markets with fast-growing wind and solar generation, the EU's market coupling arrangements are incompatible with the growing trend for national capacity remuneration schemes (CRMs).

The author outlines the use of reliability options (ROs) as a form of CRM, which is more compatible with efficient spot price formation and effective market coupling. However, in existing RO designs, there is a single "spot" market, whereas European electricity markets have physical day-ahead, continuous intraday and balancing markets, with market coupling applying from day-ahead. The author describes a decentralized RO model for use in European markets.

The chapter concludes that decentralized ROs may be a more appropriate form of CRM for European markets, allowing capacity with different degrees of flexibility to find their own place in the market and lowering regulatory risk, compared with centralized arrangements.

In **Chapter 13: Network Pricing for the Prosumer Future: Demand-Based Tariffs or Locational Marginal Pricing?, Darryl Biggar** and **Andrew**

---

9. This refers to purchase from the network, not necessarily total consumption.

Reeves look at the pricing and hedging arrangements that will be necessary to sustain the utility of the future, as elaborated in accompanying chapters.

The authors point out that there is no single retail contract that is best for all retail customers. Instead, an efficient outcome will involve retail customers choosing from a range of retail contracts, with different degrees of price smoothing and control over local devices and appliances.

The chapter's main conclusion is that where there is retail competition, policymakers should focus on ensuring that retailers can hedge the risks they face and getting the wholesale prices right. As long as retailers can hedge, retailers can convert those price signals into a range of contracts which provide the end-customers with the combination of price smoothing and direct control those customers desire.

In **Chapter 14: The Evolution of Smart Grids Begs Disaggregated Nodal Pricing, Günter Knieps** provides a critical appraisal of the evolution of smart grids, from the perspective of their integration and compatibility with the complementary electricity networks.

The disaggregated nodal pricing framework further described in chapter 14 is applied in the context of integrating smart grids into the electricity environment, taking into account the day ahead and intraday markets for injection into, and extraction from, electricity networks. Make or buy decisions to build and operate a smart grid platform, as well as decisions to extract and inject electricity from an electricity network, should take into account the node-specific opportunity costs imposed on electricity networks by injection or extraction.

The implementation of disaggregated nodal pricing leads to an optimal allocation of scarce transmission capacities. This holds not only within networks, but also regarding incentive compatible integration of smart grids into electricity networks.

3 PART III. UTILITIES OF THE FUTURE: FUTURE OF UTILITIES

Part III of the book, as the title suggests, explores how the fundamental role and function of the utility of the future may differ from those of the past. In fact, the term "utility" may no longer be appropriate.

In **Chapter 15: New Business Models for Utilities to Meet the Challenge of the Energy Transition, Paul Nillesen** and **Michael Pollitt** examine how the traditional value chain is changing and is likely to evolve over time, and what this means for traditional utilities.

The authors describe the underlying drivers causing the change in the value chain, and where new value pools are likely to arise, and how these pools can be exploited using innovative business models. Their analysis draws on existing examples of new entrants and traditional utilities, but also on the experiences from adjacent and different industries.

The chapter concludes that utilities need to determine their inherent capabilities, and match them to the changing landscape. This inside–outside approach, known as Capability Driven Strategy, is likely to limit, rather than expand, the number of strategic options available to traditional players.

In **Chapter 16: European Utilities: Strategic Choices and Cultural Prerequisites for the Future, Christoph Burger** and **Jens Weinmann** point out that European energy utilities are choosing diverging paths in redefining their business models. Some opt for an internationalization strategy, with emphasis on renewable or conventional generation, whereas others envision themselves as providers of complex service solutions in a "smart" energy world.

The authors explain that companies that have chosen the former path provide a convincing narrative, whereas companies that focus on the latter face the challenge of bridging the revenue gap until the mass market is ready, and to realign their workforce from "think big" to "think small," from selling to managing energy. Exogenous factors such as the degree of state ownership or regulation also influence utilities' strategies.

The chapter's conclusion is that for utilities in a smart energy world, embracing the empowered customers is the key to success.

In **Chapter 17: Thriving Despite Disruptive Technologies: A German Utilities' Case Study, Sabine Löbbe** and **Gerhard Jochum** examine alternative future strategies for utilities in the context of disruptions taking place within the German power sector.

The authors point out that, currently, utilities' strategy development and implementation is based on bottom-up approaches focusing on today's dynamic and uncertain regulatory environment. In Germany, at least, these disruptive and unpredictable developments are part of the energy transition process, offering different strategic challenges and opportunities over time. In this environment, utilities need to identify future markets, opportunities, and niches, forecasting customers' needs while considering competitor's likely moves.

The chapter explains that such strategies will enable companies to unlock these opportunities in a world moving toward distributed generation and energy efficiency. Success factors consist of timing of strategic initiatives, development of core competences, acquisition and integration of resources and capabilities.

In **Chapter 18: The Future of Utility Customers and the Utility Customer of the Future, Robert Smith** and **Iain MacGill** consider what electricity customers will want, the choices they will face, and how these choices will shape the utility of the future.

Focusing on residential customers, the chapter examines the electricity industries' evolution from its early tailored customer focus to becoming a monopoly provider of "universal" essential public goods, then moving toward a commodity market business model, and now to facing changing customer options, including self-generation, energy efficiency, "enabled" appliances, local energy storage, and electric vehicles.

Providing a counterweight to the utility industry's introspective focus on new technologies, new entrants focus on business models, and policy makers focus on more economically efficient pricing, the chapter envisions a future utility which is repositioned, rather than replaced by noncommodity electricity services, in an environment where customers' interactions, not technologies of themselves, determine what the future holds.

In **Chapter 19: Business Models for Power System Flexibility: New Actors, New Roles, New Rules, Luis Boscán** and **Rahmatallah Poudineh** note that increasing the share of renewables to significant levels poses a number of planning and operational challenges, increasing the need for additional flexibility on the demand side. At the same time, technological advances in the Smart Grid space offer new opportunities, which offer additional customer flexibility to the existing supply chain.

The authors explain that new market participants, such as software developers working in innovative automation and control solutions, and aggregators who serve as intermediaries between small-scale suppliers of flexibility and markets, are starting to play an active—and promising—role in the evolving electricity market. Similarly, existing players are changing their traditional roles by becoming involved in new activities, in order to address many of these challenges.

This chapter reviews the possible ways that flexibility can be incorporated into the business models of future utilities.

In **Chapter 20: The Repurposed Distribution Utility: Roadmaps to Getting There, Philip Q. Hanser** and **Kai E. Van Horn** discuss the decision-making processes utilities face as they consider a transformation into the Repurposed Distribution Utility (RDU). While the RDU has been widely heralded as the future of the distribution utility, little is known about the transformation.

The authors explore the challenges utility planners face when integrating DERs, and the planning and operational changes required to make the RDU transition. Furthermore, they present recent examples of planning and operational changes from utilities in the midst of the RDU transition.

The chapter's main contribution is a framework that provides a systematic means of conceptualizing and addressing the complex questions associated with the RDU transformation.

In **Chapter 21: Distributed Utility: Conflicts and Opportunities Between Incumbent Utilities, Suppliers, and Emerging New Entrants, Kevin B. Jones, Taylor L. Curtis, Marc de Konkoly Thege, Daniel Sauer,** and **Matthew Roche** analyze the transition utilities are facing across the United States from emerging distributed resources. State commissions in New York, California, Massachusetts, Minnesota, and Hawaii are leading this evolution, redesigning policy models to reshape the industry.

The authors explore the challenges presented by New York's Reforming the Energy Vision proceeding, while comparing this approach to developing policy in the other states. The chapter evaluates the future of the utility, supplier, and distributed resource sectors, as incumbent players and new entrants battle for a competitive advantage.

Encouraged by growth in clean, efficient, distributed generation and expanding demand side alternatives, the "Utility of the Future" will increasingly incentivize vibrant distributed resource participation. The chapter concludes with advice for both domestic and international regulators.

In **Chapter 22: The Fully Integrated Grid: Wholesale and Retail, Transmission and Distribution, Susan Covino, Paul Sotkiewicz,** and **Andrew Levitt** explain that DERs affect markets and operations from the largest generator down to the smallest load. Currently, there is no relationship between price formation in wholesale and retail markets, nor do transmission operators have visibility to the presence of DER.

The authors envision a future in which the distinction between wholesale and retail markets becomes blurred; in which dynamic nodal wholesale prices joined to variable distribution rates inform customer decisions. Transmission system operators (ISOs) and evolving distribution system operators (DSOs) will share information and coordinate closely to enhance the reliability and resiliency of the grid.

The chapter explores the impacts of growing DER penetration on wholesale market operations and planning, while examining the role of wholesale markets in optimizing the attributes of all grid-connected resources including DERs.

Part I

What is Changing, What are the Implications

Chapter 1

What is the Future of the Electric Power Sector?

Fereidoon P. Sioshansi

Menlo Energy Economics, Walnut Creek, CA, United States of America

1 INTRODUCTION

In 2014, the New York Public Service Commission (NYPSC) initiated an ambitious regulatory proceeding called Reforming the Energy Vision (REV) to redefine the state's future electric utility business model, further described in chapter by Jones et al. The effort, spearheaded by Audrey Zibelman, the chair of the Commission with the support of New York Governor, Andrew Cuomo, involves a thorough examination of the evolving role of utilities, particularly the distribution network, at a time of rapid change and technological innovation.

New York, of course, is not alone in reexamining the industry's future. Similar proceedings are underway in California, Hawaii and a number of others states—and regulators in Europe, Australia, Japan, and a number of other countries are also addressing similar issues, but with a local flavor and with different degrees of urgency.

To those not familiar with the issues, this may seem perplexing, especially since the general public views utilities as stable, stodgy, slow moving enterprises that have been around forever—not particularly liked, but ultimately necessary. Why all the hype about new utility business models and the need for new forms of regulations?

As the following chapters of this book explain, the reasons are rather obvious to those within the industry, as well as those who regulate it. For a start, the utility business is no longer a growth industry—something that was taken for granted for a very long time.

As the experience of the Empire State illustrates (Table 1.1), electricity demand growth has fallen precipitously since WWII, and not just in New York. On the other hand, retail electricity prices, historically flat or falling, adjusted

Future of Utilities - Utilities of the Future. http://dx.doi.org/10.1016/B978-0-12-804249-6.00001-4

TABLE 1.1 Average Electricity Sales Growth in New York, 1996 to Present Plus Projections to 2024, in Percentage

Period	Average sales growth (%)
1966–76	3.8
1976–86	1.5
1986–96	1.4
1996–2006	0.9
2003–13	0.3
2014–24	0.16

Get used to it: no growth in utility business—New York or elsewhere.
Source: NY Public Service Commission, Feb. 26, 2015.

for inflation, are now flat or rising,[1] even in the United States (Fig. 1.5), where shale gas is plentiful and cheap, certainly by international standards.

Rising retail rates is not limited to the United States. As noted in chapter by Nelson and McNeil, rising retail prices have been especially pronounced in Australia, mostly due to increased network costs. In many parts of Europe, including Germany, retail tariffs have also risen, even when wholesale prices have been falling, as explained in chapter by Burger and Weinmann.

In New York, the electricity demand[2] is projected to grow at a tepid average annual rate of 0.16% through 2024, which is among the underlying reasons for redefining the utility business model. The same low or flat demand projections are evident in other states (Fig. 1.1) and in other mature economies, as far away as in Australia.

While sales growth has stagnated, retail tariffs are expected to rise, partly due to the need for upgrading and modernizing the distribution network. The US power sector expects *more*, not less, investment going into the network, as illustrated by recent trends in Fig. 1.2. Historically the industry thrived by investing lavish amounts, routinely expanding and upgrading its infrastructure. Since the volume of sales was growing, the costs could be spread among more kilowatt-hours sold, which meant that the average consumer did not feel the pinch.

As described by Averch and Johnson (1962), under traditional rate of return regulation, all else being equal, regulated utilities historically had an incentive to *overinvest* in capital if/when they could, and there are indications that some in the industry still operate on this basis, if they can get away with it (EEnergy Informer, 2015b).

---

1. According to the US Labor Department, electricity rates in the United States rose 97.5% over the past 30 years, while the price of consumer goods and services increased 127.8%.
2. Since prosumers can meet some or most of their internal consumption from self-generation and/ or through energy efficiency, a distinction is necessary between total consumption and the volume purchased from the grid or net load.

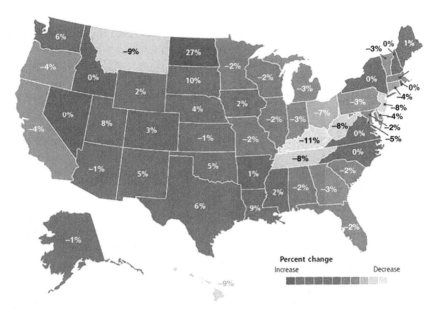

FIGURE 1.1   Electricity demand growth rates in different states. *(Source: QER Report: Energy Transmission, Storage, and Distribution Infrastructure, Apr. 2015.)*

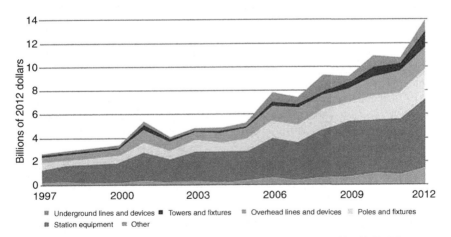

FIGURE 1.2   US investment in transmission and distribution, 1997–2012. More, not less, grid investments by US utilities. *(Source: Quadrennial Energy Review, US Department of Energy, Apr. 2015.)*

To frame the discussion, between 1960 and 2000, electricity sales in New York grew 422%, while retail tariffs, adjusted for inflation, declined from 20.8 to 11.32 cents/kWh.

These were the industry's *golden years*, where demand grew, while prices fell through the miracle of *economies of scale*. The increased investments could be spread among the rising volume of sales, which meant lower prices per unit for the average consumer that, in turn, encouraged more consumption—straight from the *Economics 101* chapter.

As covered further in this chapter and in following chapters of this book, these fundamentals are now reversed. While sales are stagnant or falling, retail tariffs are expected to be rising, encouraging conservation and distributed self-generation. Adding to the industry's woes is the fact that a growing number of consumers in high retail tariff jurisdictions can now produce some, or virtually of all, their electric needs through distributed self-generation, primarily by investing in solar panels on their roofs.

What makes rooftop photovoltaics (PVs) compelling is supportive policies that either pay a generous feed-in-tariff (FiT), or offers a credit for all generation in excess of domestic consumption—the so-called net energy metering (NEM) schemes currently prevalent in much of the United States. Among the thorny issues facing regulators in many parts of world is deciding how much to pay for such intermittent consumer self-generation because it erodes utility revenues without a commensurate reduction in utility fixed costs, which are significant.

Other technological advancements, notably in storage, microgrids, and home energy management allows more consumers to become virtually self-sufficient, in the sense that they no longer need to rely on the incumbent utility for kilowatt-hours, while still depending on the grid for reliability, load balancing, and a host of other valuable services—topics discussed in a number of chapters in this book.

This chapter is organized as follows: Section 2 describes how the traditional utility business model is eroding due to falling demand and rising tariffs, a phenomenon pronounced in developed economies. Section 3 describes how the rise of distributed energy resources (DERs) means new roles and new rules for "utilities." Section 4 speculates what these fundamental changes may mean for the future of utilities, or utilities of the future—the book's title—followed by the chapter's conclusions.

2 DOUBLE WHAMMY: FALLING DEMAND, RISING TARIFFS

The utility business was never glamorous or exciting but, historically, incumbents, many of whom were vertically integrated companies, could count on predictable sales and revenue growth. As electricity consuming devices gradually get more efficient and as mature economies slowly move away from energy-intensive industries, however, electricity demand growth is slowing, or in some cases, actually falling, which—assuming all else remains constant—leads to flat or declining revenues. This phenomenon is not limited to the United States, but applies to virtually all mature OECD economies, to some more than others.

This section briefly explains why net load[3] on many networks is flat or falling. More important, however, is that, as the sales volume declines, the industry's significant fixed costs must get spread across a declining base, resulting in rising tariffs—all else being equal (see chapter by Athawale and Felder.).

Electricity demand growth rates are rapidly approaching zero, in a number of advanced economies, as illustrated for the United States (Fig. 1.3 on the left) and New Zealand (Fig. 1.3 on the right).

The reasons vary from country to country, but broadly speaking, advanced economies are gradually *deindustrializing*, as they move away from electricity-intensive segments toward services and high value-added manufacturing. Such economies can generate enormous wealth while using relatively little energy.

Other reasons include the fact that buildings are getting progressively more energy efficient, as do lights, motors, and electric appliances, while distributed generation, notably in the form of solar rooftop PVs, is becoming cost competitive relative to rising utility tariffs, and this is eating into the industry's sales and revenues.

The evidence, while piecemeal, is nevertheless sobering. For example, Germany's primary energy (not electricity) demand declined 4.8% in 2014, compared to 2013. Moreover, this was *not* a single year anomaly. German economy appears to be humming along while using less energy since 2006. The same applies to electricity in nearly all mature economies.

The most extreme case of dropping energy demand—increasingly met with rising share of renewables—may be observed in Denmark, a country set on phasing out its reliance on fossil fuels *entirely* by 2050.

Another country experiencing falling electricity demand is Australia. Successive forecasts of future demand by the Australian Energy Market Operator (AEMO) portray a persistent drop in demand over time—every year's projections are lower than the prior one (Fig. 6.2 in the chapter by Nelson and McNeil). The latest projection shows a virtually flat demand through 2024 (last line on the bottom of the graph).

A combination of factors—the usual suspects—are responsible for the demand decline, including rising retail tariffs, depressed industrial demand, energy efficiency gains, plus the recent uptake of solar PVs—a pronounced phenomenon in the last few years.[4]

---

3. Net load refers to how much a customer may be buying form the network net of what is self-generated, and/or is not needed, due to investments in energy efficiency. This becomes important for, say, a consumer in a zero net energy building where self-generation may meet total energy consumption on an annual basis, but not necessarily at all times. Under a volumetric tariff, such a customer's net load may be virtually zero and if there are no fixed charges, the total annual bill may be virtually zero. In this context, net load on the network may be flat or falling, while fixed costs of maintaining the network are virtually unaffected.

4. For further details of why electricity demand is flat or falling refer to EEnergy Informer (2015a).

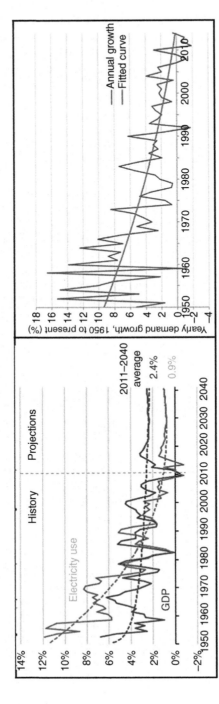

FIGURE 1.3 Electricity demand growth and GDP for United States (left) and New Zealand (right). New Zealand not dissimilar to the United States, or the United Kingdom, or other OECD economies. *(Source: US Energy Information Administration, Annual Energy Outlook 2013. Early Release for the United States.)*

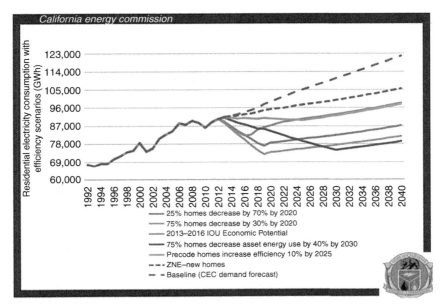

FIGURE 1.4   California projects flat or falling electricity demand through 2040. *(From "California Residential and Commercial Building Energy Use Characteristics," by California Energy Commission.)*

The story is much the same in California, the most populous state in the United States.

According to the California Energy Commission (CEC), stringent building standards and zero net energy requirements (Box 1.1) for new residential homes by 2020 and new commercial buildings by 2030, plus continued energy efficiency efforts, are expected to flatten electricity consumption through 2040, compared to the baseline forecast as shown in Fig. 1.4.

Likewise, in 2013, ISO New England, the grid operator for the six-state region in Northeast United States, projected practically flat demand for electricity for at least a decade,[5] with much of the decline attributed to improvements in energy efficiency.[6] In describing its forecast, ISONE said, "The more consumers generate power for themselves, the less they need to buy from the grid." (EEnergy Informer, 2015a).

The story in England is much the same as in New England, where the latest government statistics show that per capita electricity consumption is down 10%,

---

5. For details, refer to EEnergy Informer, Dec. 2013.
6. The flat projected demand, which includes the effect of energy efficiency gains and self-generation is shown as the line on bottom of graph. The line on top is demand growth without the impact of energy efficiency as noted in EEnergy Informer, Dec 2013.

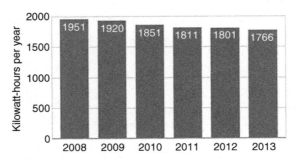

FIGURE 1.5   **Per capita electricity consumption in the United Kingdom, 2008–13 in kilowatt-hour per year.** *(Source: Energy Saving Trust analysis of DECC figures.)*

compared to 5 years ago, despite the fact that people are buying bigger TVs, refrigerators, and more appliances (Fig. 1.5).

According to Michael Pollitt, " … electricity demand (in UK) peaked in 2005. Since then electricity demand has fallen by 9% to 2013. While industrial demand has fallen by 16%, there has been a large drop in domestic consumption, around 10%." (EEnergy Informer, 2015a).

The explanation? Basically, everything is getting more energy efficient over time. The latest model fridge-freezer sold in the United Kingdom, for example, uses 73% less electricity, compared to units sold 20 years ago, saving consumers £100 (roughly $150) a year. Similarly, today's more efficient light bulbs use 30% less electricity than in 2008. As light-emitting diodes become more popular and affordable, further reductions in electricity consumption can be expected over time.

National Grid's future scenarios for the United Kingdom, illustrated in Fig. 1.6, show flat or barely growing projections, except for a scenario where heating and the transportation sector are virtually electrified in order to decarbonize the economy.

England and New Zealand are literally on opposite sides of the globe, with vastly different climates and customer base. Yet both are reporting *falling* electricity demand. The picture for New Zealand (Fig. 1.3 on the right) is nearly identical to that of the United States—or, in fact, any other mature OECD economy—showing a persistent drop in growth rate since the 1950s, and asymptotically heading to zero.

Similar trends are evident in Japan, a country that has managed to keep the lights on while losing, or shutting down, virtually all of its 55 nuclear reactors, following the Fukushima disaster in 2011.

As illustrated in Fig. 1.7, Japan's electricity demand shows similar patterns, much of it due to Japan's prolonged economic stagnation. With a dormant economy and a gradually aging and declining population, there is little chance for any increases in electricity consumption in the future.

For the United States the evidence is mixed and the experts disagree about the longer-term prognosis. The Energy Information Administration (EIA)

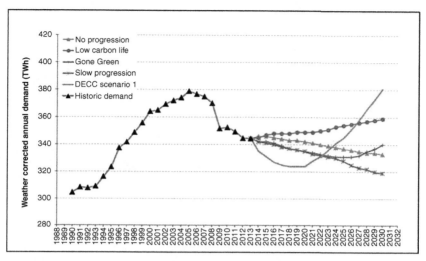

FIGURE 1.6   **Projections of demand in the United Kingdom are flat or falling.** UK demand projections, little or moderate growth, except for Gone Green scenario. *(Source: Pöyry Management Consulting, based on data from Department of Energy and Climate Change (DECC), National Grid.)*

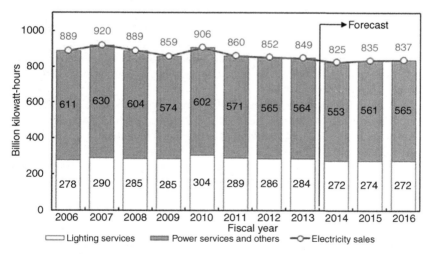

FIGURE 1.7   **Japan's electricity consumption 2006–2013 with projections.** Japanese life goes on without nuclear reactors since Fukushima. *(Source: EEnergy Informer, 2015a.)*

officially predicts 0.9% demand growth through 2025. But many experts believe that it could be far less, especially if states continue to push utility-sponsored energy efficiency programs and/or states and the federal government continue to push for higher appliance energy efficiency standards and more stringent building codes.

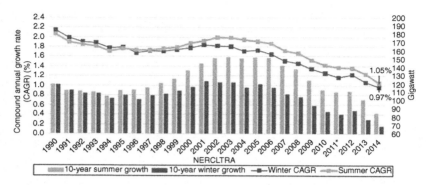

FIGURE 1.8   NERC assessment shows declining US demand growth. NERC-wide demand: 10-year growth rates (summer and winter) at lowest levels on record. *(Source: 2014 Long-term reliability assessment, NERC, Nov. 2014.)*

A study by the Lawrence Berkeley National Laboratory (LBL), for example, concluded that, through increased energy efficiency measures, the IEA's projected 0.9% rate of growth in electricity demand could be significantly reduced, resulting in a virtually flat, if not gradually declining, demand (EEnergy Informer, 2015a).

A recent report by the Institute for Electric Innovation (IEI) reported that US customer-funded energy efficiency programs have seen a 30% boost compared to 2010 and could double by 2025, reaching $14 billion. Existing programs clipped an estimated 126 TWh from the US electricity consumption in 2012 and there is plenty of room for more.

Looking at the track record since 2000, some analysts believe that US electricity consumption has already peaked at 3.77 trillion kWh in 2008, before the global financial crisis, and is unlikely to ever exceed that level.

According to the latest long-term reliability assessment by North American Electric Reliability Corporation (NERC), the rate of growth in demand has been on a slow decline since 2002. NERC notes that the latest growth rates are "the lowest on record" (Fig. 1.8, emphasis added).

These and other anecdotal evidence, while nonconclusive, suggest that the power sector's demand growth days are mostly behind us, at least in the mature economies of the world.

The same phenomenon is reflected in the International Energy Agency's latest World Energy Outlook, where the latest projections of global *energy* (not electricity) demand to 2040 shows a slightly *declining* trajectory for the OECD countries as a whole with China's demand flattening out in the 2030s (Fig. 1.9).

The bottom line is that advanced OECD economies are on a trajectory of more efficient energy consumption, while many are also experiencing a rapid uptake of distributed self-generation. As illustrated for Australia, solar PVs are

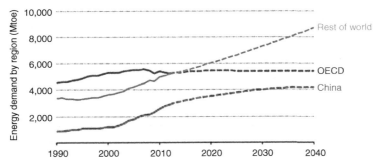

FIGURE 1.9   Total demand in OECD is projected to fall. Total energy demand declining in OECD, flat for China. *(Source: World Energy Outlook 2014, International Energy Agency, Nov. 2014.)*

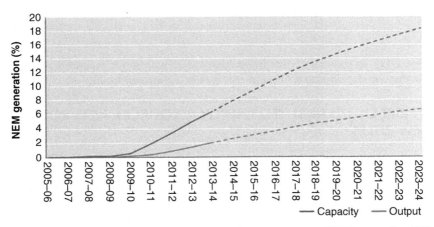

FIGURE 1.10   Australian PV capacity and generation as percentage of NEM generation. Call it customer power down under; solar PV generation capacity and output in Australia. *(Source: State of the energy market 2014, Australian Energy Regulator.)*

expected to reach 18% of capacity and 6% of output, by 2024—among the highest rates in the world (Fig. 1.10).

Put energy efficiency gains *and* distributed generation together, and it is easy to see why demand growth is disappearing in many parts of the world, with more countries to follow.[7]

It is, of course, a different story in the rapidly growing global economies, but they too will eventually reach a plateau, followed by less energy guzzling economic growth. Recent evidence from China, for example, shows a significant drop in electricity demand growth, in the past few years

7. Further discussion of energy efficiency's potential may be found in Sioshansi (2013).

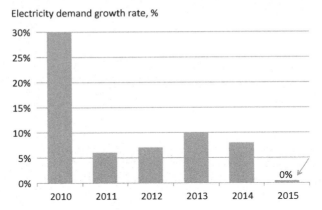

Electricity demand growth rate, %

FIGURE 1.11   **China's electricity demand growth rates have fallen from their historical levels.** *(Data Source: Based on data from CEIC.)*

(Fig. 1.11). In summary, what is already evident within rich countries is likely to extend to developing economies—it is only a matter of time.

3   RISE OF DISTRIBUTED ENERGY RESOURCES: NEW ROLES, NEW RULES

Needless to say, utility executives in parts of the world where demand growth is tepid, retail tariffs are rising, and a growing number of *prosumers*—who are consuming *less* due to energy efficiency while producing *more* through small-scale distributed generation—are facing difficult times. Among other things, they must decide whether to fight off the rapid uptake of solar PVs, or, failing that, to join the competition in assisting more consumers to become *prosumers*—as a number of utilities have decided to do (see chapter by Nelson and McNeil).

For regulators, the challenge is how to regulate best, if they will regulate at all, as the stodgy, predictable, and boring industry is going through a literal renaissance, as further described by developments in New York that are closely watched by other regulators facing similar issues across the United States and elsewhere.

For prosumers, however, the future is full of new and exciting opportunities that, for the first time, *empower* them to do things that were barely imaginable a mere decade ago; most importantly, the possibility to generate clean, sustainable, nonpolluting solar power on the rooftop, at prices that meet, or in some cases beat, grid-supplied juice. And they have more options to control their consumption and to manage their usage.

If the cost of energy storage falls as rapidly as the cost of solar PVs did, connection to the grid may become a *nice to have*—for backup, reliability, and load balancing. It will no longer be necessary to buy a large number of kilowatt-hours from generators, or pay distributors for delivery of electrons.

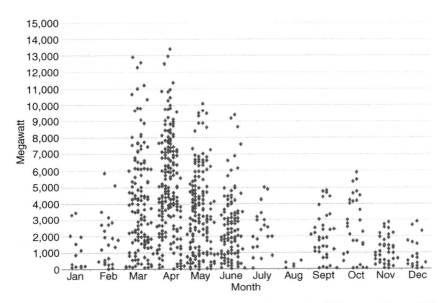

FIGURE 1.12   Projected overgeneration in California assuming 40% renewable portfolio standard by 2024. *(Source: CAISO.)*

In fact, some people predict that, with the rapid proliferation of renewables—which have zero marginal cost—electrons will become cheaper over time, perhaps costing *zero* in a not too distant future. In many markets, wholesale electricity prices frequently drop to extremely low levels, or even negative, when there is too much generation relative to load—a condition called *overgeneration.*

California's grid operator, for example, expects to have an increasing number of such episodes by 2024, under a hypothetical 40% renewable portfolio standard[8] (Fig. 1.12). Similar episodes are becoming commonplace in other renewable rich markets, including Germany, Denmark, Texas, and others.

As several chapters in this book further explain, the utility business is becoming anything but *business as usual* (Box 1.1). How the incumbent stakeholders and regulators respond to these changes will matter a lot. *Prosumers,* however, will increasingly do what is best for them, circumventing both utilities and the regulations, if they get in the way of what is technically possible and economically advantageous.[9]

---

8. Refer to chapters by Gimon and Smith & MacGill.
9. Refer to chapters: Customer-Centric View of Electricity Service; The Future of Utility Customers and the Utility Customer of the Future by Gimon, and Smith and MacGill, for perspective on customers of the future.

BOX 1.1 ZNE Homes: Here Today, Affordable Tomorrow[p]

In 2008 the state of California set a goal for *new* residential buildings to reach a zero net energy (ZNE) target starting in 2020, with the same extended to *new* commercial buildings by 2030—with ZNE defined as "a general term applied to a building with a net energy consumption of zero, over a typical year." To cope with fluctuations in demand, zero energy buildings export electricity to the grid when there is a surplus, and draw electricity when not enough electricity is being produced. The amount of energy provided by onsite renewable energy sources is equal to the amount of energy used by the building.

When initially proposed, skeptics said the ZNE goal could not be met, certainly not by 2020, and not even remotely at a cost anyone could afford. But indications are that reaching the target may not be as onerous as some had initially thought. Like any energy efficient car or appliance, you pay a little more when you buy it, but it saves you a bundle over time in lower energy costs (Fig. 1.13).

On Earth Day, Apr. 22, 2015, Meritage Homes, a major US homebuilder based in Scottsdale, Arizona, said it was developing Sierra Crest, a small ZNE community consisting of 20 homes in California.

The homes in Sierra Crest will have high-efficiency solar photovoltaic (PV), HVAC systems, water heating equipment, heat pumps, and integrated fresh air ventilation. In addition, each will have spray foam insulation, highly insulated windows, energy-efficient lighting, smart chargers, and smart appliances. Together, these features are expected to reduce a typical home's energy use by as much as 60%, compared to a house built to the latest California building code.

Because the homes are highly efficient, they need a relatively small solar PV capacity to reach the ZNE target. "In a conventional home, the size of PV ... sufficient to make a home ZNE can be between 7–10 kW. Because of the included advanced energy efficiency, our ZNE homes can achieve (this target) with 3.5–4.5 kW," according to CR Herro, Vice President of Environmental Affairs at Meritage Homes.

Skeptics dismiss such talk as pure speculation and hyperbole. They argue that utilities have been around for over a century and will manage to overcome the challenges facing them today, either by beating or joining their competitors.

While some may, in fact, survive, and possibly thrive, many will be too slow to change or handicapped by the same regulations that have shielded them from competition for so long.

For one thing, the speed and sheer magnitude of change is unprecedented. As noted in Introduction, a recent report by Accenture says US utility revenues can be expected to *fall* some $48 billion by 2025, due to the double whammy of increased solar PV self-generation coupled with energy efficiency gains (Fig. 1 in Introduction). Given that the US power industry collects roughly $360 billion per annum, this would amount to a 13% loss in the sector's revenues. Not devastating, perhaps, but neither pleasant. European incumbents are projected to suffer roughly similar loss of revenues.

FIGURE 1.13   High-performance home built by Garbett Homes near Salt Lake City, Utah. This is a certified U.S. DOE Zero Energy Ready Home. *(Source: Courtesy of the Department of Energy).*

4   WHAT IS THE FUTURE?

As the preceding discussion suggests, there is a lot of speculation about the longer-term implications of these trends, and what it means for the future of utilities—or the utilities of the future.

The search for the "utility" of the future, like the search for Shangri La, however, is elusive, mostly because no one truly knows what it is, or where to find it. But, given how much is at stake, there is no shortage of people trying.[10]

Among those trying is a report by PwC Global Power & Utilities titled, The Road Ahead: Gaining Momentum From Energy Transformation. It describes the main drivers of industry transformation and how they may interact with one another over time, further described in chapter by Nillesen and Pollitt.

Opinions among others looking into the future of the power sector vary. A sample, not necessarily in any order is given as follows:

- Future is decentralized. Many experts believe that the future will increasingly be distributed and decentralized with many consumers/prosumers operating in a *grid-light, grid-assisted* or, potentially, *off-grid* mode—where individuals or groups of them will live in a zero net energy environment with distributed generation, storage, and/or relying on semiautonomous microgrids—for some of the reasons described in chapter by Marnay. The University of California,

---

10. Among the noteworthy efforts, there is an ongoing study at MIT on this very topic.

BOX 1.2 UCSD's Microgrid: Win, Win, Win

The popularity of microgrids is on the rise, not only because they provide an extra layer of reliability—for example, when the super grid fails—but also because they save money.

An example appeared in Lucas (2015), an excerpt of which is reproduced here.

"Some organizations, such as military bases, may have specific reasons to want to be independent of outside suppliers, but for most of them the main motive is to save money. The University of California, San Diego (UCSD), for example, which until 2001 had a gas plant mainly used for heating, changed to a combined-heat-and-power (CHP) plant which heats and cools 450 buildings and provides hot water for the 45,000 people who use them. The system generates 92% of the campus's electricity and saves $8m a year. As well as 30MW from the CHP plant, the university has also installed more than 3MW in solar power and a further 3MW from a gas-powered fuel cell. When demand is low, the spare electricity cools 4m gallons (15m liters) of water for use in the air-conditioning—the biggest load on the system—or heats it to 40° to boost the hot-water system. Universities are ideal for such experiments. As autonomous public institutions they are exempt from fiddly local rules and from oversight by the utilities regulator. And they are interested in new ideas.

Places like UCSD not only save money with their microgrids but advance research as well. A server analyses 84,000 data streams every second. A company called ZBB Energy has installed innovative zinc-bromide batteries; another company is trying out a 28kW supercapacitor—a storage device far faster and more powerful than any chemical battery. NRG has installed a rapid charger for electric vehicles, whose past-their-prime batteries are used to provide cheap extra storage. And the university has just bought 2.5MW-worth of recyclable lithium-ion iron-phosphate battery storage from BYD, the world's largest battery manufacturer, to flatten peaks in demand and supply further.

In one sense, UCSD is not a good customer for the local utility, San Diego Gas & Electric. The microgrid imports only 8% of its power from the utility. But it can help out when demand elsewhere is tight, cutting its own consumption by turning down air-conditioners and other power-thirsty devices and sending the spare electricity to the grid.

UCSD is one of scores of such microgrids pioneering new ways of using electricity efficiently and cheaply through better design, data-processing technology and changes in behavior. The IEA reckons that this approach could cut peak demand for power in industrialized countries by 20%. That would be good for both consumers and the planet."

As described in more detail in chapter by Marnay, microgrids, initially considered appropriate for remote locations, are decidedly moving mainstream—and in the middle of the super grid. Not only do they not impose on the super grid, they in fact enhance its resilience. As the example of UCSD illustrates, they are good for the customer, local utility, and the grid. A true example of win, win, win.

San Diego (Box 1.2) and the West Village of University of California Davis already operate in such fashion. There is speculation that more will follow.

Would a distributed/self-sufficient future work for everyone? Probably not. But a subgroup of customers, including university campuses, major hospitals, shopping malls, office complexes, military installations, especially in remote areas, plus communities that are too far from an existing grid, may very well find such a model not only suitable, but cost effective and reliable. How much would such microgrids depend or *liaise* with the super grid is a topic of intense interest.

- Future is integrated. Others believe that while the future of *utilities* may be unpredictable, the future of the integrated grid is indeed bright, as further described in chapter by Gellings.

    According to these people, the growth of distributed generation and intermittent renewables will significantly *increase* the value of the integrated grid that, as Gellings explains, is rather different than being merely connected to the grid, as most customers currently are.

- Future is not *necessarily* distributed. Others are convinced that, while there will be more distributed generation, the future of the power sector is *not* necessarily distributed, or even decentralized, certainly not for all. This line of argument may be found in a blog by Severin Borenstein,[11] who finds many flaws with the distributed and decentralized utility model.

    Among the reasons many consumers are becoming prosumers is not the inherent efficiency or low cost of distributed generation, but the way it is assisted or subsidized through incentives, such as net energy metering, and tariffs that are not cost reflective. In California, where roughly half of all US rooftop solar PVs may currently be found, high tiered tariffs, more than anything else, explain the popularity of solar PVs, and why solar leasing companies, such as SolarCity, are headquartered there. As explained in a number of chapters in the book, redesign of tariffs to make them more cost reflective and redesign of FiT and NEM schemes will go a long way to rationalize the true costs and benefits of distributed generation.

- The future is energy storage. The future, according to some, depends on the cost and practicality of energy storage. The future can certainly be more distributed if prosumers could store their excess generation for use at other times. Tesla's Powerwall batteries,[12] for example, may play a critical role in such a future.

    Currently, most prosumers get a free ride, courtesy of nonsolar customers, by using the grid as an infinite battery, at virtually zero cost. Batteries— or, more broadly speaking, energy storage—will enable prosumers to operate in a grid-assisted or grid-parallel mode, hence reducing the free ridership. While this may be attractive to consumers who wish to operate independently of the grid or feel empowered or whatever, it is not for everyone, even if the cost of storage becomes trivial.

- The future is renewable. The future, according to many, is definitely increasingly renewable, which tends to be intermittent, and perhaps mostly solar, as described in a recent MIT report on future of solar energy.[13]

---

11. Severin Borenstein Blog dated 4 May 2014, at https://energyathaas.wordpress.com/2015/05/04/is-the-future-of-electricity-generation-really-distributed/.
12. Tesla's Powerwall storage device is mentioned in several chapters in the book.
13. The Future of Solar Energy, 2015, MIT.

There are very few experts who would disagree with such a future; it is only a matter of time and scale, as costs of renewables continues to drop, while there are more restrictions on use of fossil fuels, partly to limit carbon emissions. As already noted, countries like Denmark are on track to phase out the use of fossil fuels from their economy by 2050, essentially—and they may as well succeed. Hawaii is also pressing ahead with a 100% electric sector, as California's governor has suggested a 50% renewable target by 2030. To meet ambitious renewable targets requires an integrated and highly intelligent grid, as previously described.

- The future is regulation. The future, according to many, is what regulators and policy makes want it to be, given the political nature of the electricity service that is viewed as a public good, like clean air or water.

  Looking at examples from many parts of the world, it is clear that politicians tend to meddle in not only the price of the electricity service, but its composition, how it is procured, delivered, and who subsidizes whom in the process. While technology and economics will continue to play a role, so will regulations and policy—whether one likes it or not.

- The future is bifurcated. The future of power business may become increasingly bifurcated among the haves and have nots, with growing disparity of service needs and grid dependency.

Much of the current debate about the solar versus nonsolar customers is centered around equity and fairness, and who is subsidizing whom. As some customers move away from total dependence on the grid, those who remain totally dependent have to bear higher costs, as their numbers shrink. This has the prospect of more inequality, not just in terms of income, literacy, or opportunity, but also in terms of grid dependency.

Depending on one's point of view and assumptions, one or more of these futures or a combination of them may emerge—and it is unlikely to be the same future for all regions.

Utility executives are already beginning to make assumptions about what they believe the future entails, and are devising strategies accordingly, as described in Box 1.3.

One of the critical drivers of change, many experts agree, is the emergence of platforms and technologies that support their rapid spread—enhanced by the growing number of smart meters, smart devices, home energy management systems, wireless sensing, automation, and artificial intelligence, as described in chapter by Cooper.

Consumers can already get better information on who offers what services in a given area, just as they do when they are looking for cheap airline seats. The electric equivalent of Kayak is nearly here, even if not widely used.

Aggregation of individual customer loads, traditionally an arduous chore, is getting easier by the day—as are the ways to encourage the aggregated load to respond to prices or incentives.

BOX 1.3 NRG Versus E.ON

For those looking for what future "utilities" may look like, NRG may be a good place to start.[a]

What makes NRG an interesting company is that it owns some 53 GW of merchant generation capacity—more than any other IPP in the United States, and more than many countries' total installed capacity. In the past few years, NRG has been busy acquiring strategic assets in related areas, investing in new ventures, and expanding into new business lines—putting its eggs in as many baskets as it can manage, while keeping its thermal generation business intact.

Since the company does not have a franchise service area or a captive customer base, it has little to lose and, potentially, a lot to gain from the mayhem that is afflicting traditional US utilities. NRG is well-positioned to cherry-pick bits and pieces of profitable businesses selectively, wherever it can find them. And it has been relentlessly expanding in areas where it believes it has a compelling value proposition.

Six years ago, power generation accounted for 98% of NRG's revenue. By 2013, the number was down to roughly 50%. How did NRG achieve such a feat? Not through divesting its fossil fuel generation assets—as E.ON did in Germany—but through aggressive acquisitions of nongeneration assets, investing in several retail businesses, and creating an alternative energy unit.

While E.ON is treating its thermal plants as toxic assets, NRG sees nothing wrong with the generation business. In fact, NRG sees itself as a *marketing company that happens to be selling power*, with a focus on distributed solar, storage, and smart energy systems. Visiting its website, one gets the misleading impression that it is engaged in everything except power generation.

NRG's Home Solar division, for example, is aiming to boost its capacity from 75 to 2400 MW, by 2022. It says the cost of installed solar will drop by another 20% in coming years. NRG's Distributed Generation is in distributed energy, while its ESCO business is pursuing energy services.

NRG plans to transform its existing commercial and industrial business into a full service, high margin energy provider, using demand-side management, fuel cells, onsite generation, and microgrids. Crane wants the profit from these businesses to grow fourfold to nearly $500 million, by 2022. It plans to get into utility-scale renewables business while developing 2500 MW of commercial and industrial solar by 2022.

Crane acknowledges that, sooner or later, all thermal generators will be operating in a carbon-constrained world, either because of regulatory constraints, or from consumer pressure—most likely a mixture of the two. He does not wish to wait for the regulations to impose such constraints on NRG, but rather to stay ahead of the game.

He reminds investors that NRG will transition away from centralized, fossil fueled and polluting business to a decentralized, clean, and resilient future. Likewise, he believes that transportation should move away from fossil fuels to an electrified future. He drives a Tesla.

eVgo, one of NRG's subsidiaries is investing in electric vehicle charging stations with the aim to boost its market share in the nascent business from 10% of the available EVs to 50%, within next 3 years.

According to Crane, "Electricity is on the cusp of a massive transformation," adding, "If you want to win today and tomorrow, you have to have a strategy that is at least bifurcated" – that is to say, between conventional power and alternative energy.

SolarCity Corporation, the biggest US residential solar installer, currently has a third of the US market, offering a zero-down installation plan that most homeowners prefer, even if buying the panels outright is a better way to go. NRG, currently fifth, is planning to be No. 2 by the end of 2015.

Crane recently stated that NRG already has "an embedded SolarCity within it," adding, "Everyone is beginning to believe that residential solar is this trillion-dollar market that currently has about 1% market penetration." Crane is focused on capturing a big share of the remaining 99%. He said, "We can offer home solar customers so much more than panels on the roof."

Why is Crane excited about distributed generation, electric vehicles, and solar energy? Because he knows that conventional power generators will face declining demand growth, as consumers increasingly produce more of their own electricity, while investing in energy efficiency.

Speaking at a briefing for investors and analysts, in mid-Jan. 2015, Crane said, "Winning in the future is more complex than ploughing full steam into the past like our peer group," referring to the Neanderthal utilities. "We are trying to protect our investors from future shock."

At the briefing, Crane said dramatic change was inevitable and that the same rules that created massive wealth to innovators in the Internet and telecom booms—and which destroyed the wealth of those that didn't adapt—will apply to the electricity industry.

Crane said the history of the telecom industry tells us that the incumbent who fully embraces the future technology, while continuing to compete aggressively to win in the present industry paradigm, will be best positioned.

NRG is doing its best to squeeze as much out of its existing legacy assets, gas and coal-fired generation, while putting lots of eggs in multiple baskets, not knowing which will be a good bet. That may be the best strategy, perhaps a blueprint for the "utility of the future."

a. As an example of how fast things change, when this box was initially drafted, NRG was considered a company to emulate. By the time the proofs arrived, its CEO had departed due to investor pressure as a result of poor performance.

Through aggregation, not only can customers respond better to prices, but they can pool together and bid in wholesale markets—something that only very large industrial customers can manage today.

Beyond these, there are those who predict that passive consumers will not only become active prosumers, but will be able to interact with *each other* using the distribution and possibly high voltage transmission network to transact—an idea espoused by Cazalet and Barrager in *Transactive Energy*.

5   CONCLUSIONS

As the preceding discussion suggests, it is imminently clear that the electric power sector, long shielded from competition, and the last in adopting the latest communication and information technologies, is undergoing rapid change.

However, what is not clear, is how, when—or in fact *will*— the incumbents react and reposition themselves in time and with decisiveness. The short answer to "where they are," "what future," or perhaps, "if there is a future" for electric utilities, depends on how the incumbents react and how soon.

REFERENCES

Averch, H., Johnson, L., 1962. Behavior of firm under regulatory constraint. Am. Eco. Rev. 52, 1052–1069.

Lucas E., 2015. Let There be Light. Special Report: Energy and Technology. The Economist, Jan 17, 2015.

EEnergy Informer, 2015a. The New Normal: Flat Electricity Demand. EEnergy Informer, Feb 2015.

EEnergy Informer, 2015b. When All Else Fails Gold Plate the Network. EEnergy Informer, Jun. 2015.

Sioshansi, F. (Ed.), 2013. Energy Efficiency: Towards the End of Demand Growth. Academic Press, Waltham, MA.

Chapter 2

Value of an Integrated Grid

Clark W. Gellings

Electric Power Research Institute, Palo Alto, CA, United States of America

1 INTRODUCTION

The Integrated Grid leverages the optimal combination of local generation, energy storage, energy efficiency, and new uses of electricity integrated with central generation and storage, in order to provide society with reliable, affordable and sustainable electricity. It requires a modernized grid, characterized by connectivity, rules enabling interconnection, and innovative rate structures that enhance the value of the power system to all consumers. The Integrated Grid blends the most valuable generation, storage, power delivery, and end-use technology tailored to meet local circumstances.

The important distinction here is that virtually all consumers in the developed world are connected to the grid. As such, the grid is "interconnected" with all consumers, and their devices and appliances, including distributed energy resources (DER) of all types. The distinction is that being connected to the grid does not constitute the formation of an "integrated grid." Individual consumers and society at large benefit from integration, beyond what they receive from being simply connected.

The chapter first highlights the value of grid connectivity to consumers in Section 2, followed by Section 3, which describes the added value which integration brings to the Grid. Section 4 summarizes the expanded connectivity enabled by the Integrated Grid. Section 5 highlights the advantages of an integrated grid in enabling additional penetration of DER. Section 6 speculates how distribution utilities might react to these changes, followed by the chapter's conclusions.

2 THE BENEFITS OF GRID CONNECTIVITY TO CONSUMERS

The primary motivation for electricity consumers in connecting to the grid is to receive electric service. Simply stated: the grid provides electrical support to consumer installations. Connection to the grid, called interconnection or grid connectivity, provides electricity with high levels of power quality and reliability, as well as voltage and frequency regulation, and mitigation of harmonic

Future of Utilities - Utilities of the Future. http://dx.doi.org/10.1016/B978-0-12-804249-6.00002-6

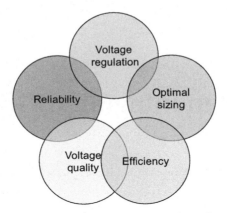

FIGURE 2.1 **Primary benefits of grid connectivity to consumers.** *(Source: The Integrated Grid Phase II, 2014)*

distortion. It also provides these services at a relatively reasonable cost, while imposing only modest environmental harm. Fig. 2.1 illustrates the primary benefits of grid connectivity.

These benefits accrue to consumers who choose to generate their own electricity as well, with the additional benefit of leveraging the grid to provide stability to the local consumer power system, and to enable the added resulting efficiency. In the case of electricity, energy flows through the same electrical grid, whether locally or centrally connected and, without that grid, a consumer generating locally will not receive nearly the same level of service. For consumers who self-generate, without connectivity to the grid, the reliability and capability of the consumer's power system would be diminished, unless substantial investments were made in the local power system to achieve equivalence of service quality to what the grid provides.

3 THE ADDED VALUE OF INTEGRATION

The Integrated Grid enables high penetration of local energy resources, and benefits broadly all consumers of electricity who are connected to the grid, beyond that which an interconnected grid might provide. Fig. 2.2 illustrates the potential benefits of local generation to all consumers who are part of an integrated grid.

The Integrated Grid enhances reliability and affordability by reducing the risk of grid instability due to sudden drops in a large amount of central and/or local resources, enables the integration of both local generation, and local or distributed storage, minimizes outages, and provides ancillary services [frequency response, nonspinning reserve, and participation in Demand Response (DR) programs]. It can reduce cost by optimizing from a combination of local and central resources. It also enables a higher penetration of local resources. It can

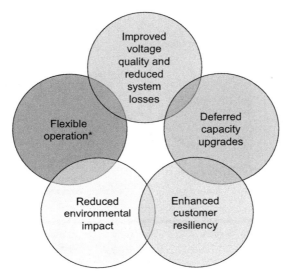

FIGURE 2.2   Primary benefits of local generation to all consumers.
*Enhances reliability and impacts distribution operations, DER integration, minimizes reconductoring, provides ancillary services (frequency response, nonspinning reserve and DR).

be configured to provide incentives for local resources to provide services to the grid. Further examples of this value are described later. The Integrated Grid[1] enables true synergy between consumers and the grid. It enables a number of potential benefits to accrue to all consumers connected to the grid, provided the grid has been modernized to accommodate local resources.

3.1   Grid Support

Photovoltaic (PV) systems and other distributed generation and storage systems offer natural support for distribution systems. For example, Fig. 2.3 illustrates the coincident demand from PV, as compared to a typical distribution load profile. The system can support utility needs, and mitigate distribution system demand. However, the extent to which this is possible is very dependent on local circumstances. The overall need for generation capacity on the power system may not be impacted substantially until there is large penetration of local generation, and depending on the modernization of the grid. Local generation, however, will still reduce the need for the operation of central generation, even if it does not reduce capacity requirements.

---

1. The Integrated Grid Approach has been referred to as part of the concept ElectriNet[SM] in previous literature.

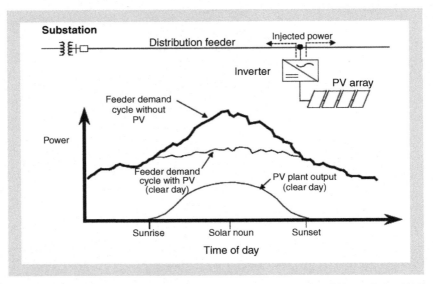

FIGURE 2.3 Grid support value of distributed PV. *(Source: Distributed Photovoltaics, 2008)*

3.2 Optimizing Distribution Operations

The effectiveness of central voltage control can be enhanced by connectivity of interconnected local generation systems. Relatively high voltage and fluctuations can be caused by local generation installed with no voltage control—these subsequently interfere with the effective operation of customer and distribution equipment, leading to premature failure of components and appliances. In an Integrated Grid, the voltage can be controlled to mitigate over-voltage problems and, hence, reduce the adverse resulting costs.

The Integrated Grid would enable coordinated control of high-penetration PV systems, as illustrated in Fig. 2.4. Here advanced inverters are used to enhance voltage control, and balance the ratio of real and reactive power needed to reduce losses and improve system stability.

3.2.1 Microgrids

One technology configuration related to the integrated grid which will enable enhanced distribution operations is the Microgrid (see chapter by Marnay). There is increasing interest in the potential value which Microgrids could provide, as part of the grid. It is the Integrated Grid which would enable the connectivity of microgrids to the conventional grid, and enable that value. Microgrids are defined as: "…a group of interconnected loads and DER within clearly defined electrical boundaries that act as a single controllable entity with respect to the grid. A microgrid can connect and disconnect from the grid to enable it to operate in both grid-connected and island-mode" (Ton, 2014). As part of an integrated grid, a

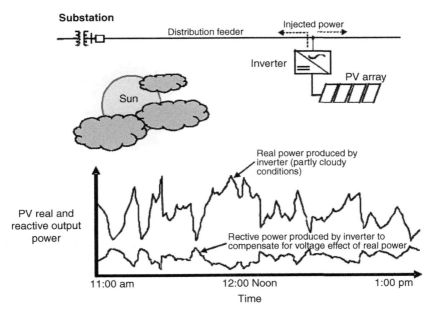

FIGURE 2.4   Advanced inverter for reactive compensation. *(Source: Distributed Photovoltaics, 2008)*

microgrid enables the effective optimization of multiple interacting components spread across a defined geographic space. To be integrated, the components are connected and monitored with advanced sensing, control, and communications technologies configured to meet the needs of a variety of dynamic load types, and operate under a range of grid conditions (Siemens, 2015).

The benefits of Microgrids are listed in Table 2.1. The business case for their development is becoming stronger, as further described in the following chapter.

The Integrated Grid offers a means by which Distributed Generation and Storage technologies can be leveraged. Simply stated, DER owned or leased by consumers which allow some utility control will have a higher value proposition from three value streams:

1. Incorporating DER into utility planning and operations can increase reliability and flexibility.
2. Dispatchable or curtailable DER can be used for grid control, reliability, and resiliency, potentially creating additional value for the consumer.
3. Sophisticated controls can optimize benefits to consumers and the grid.

3.2.2 Distributed Storage

Often referred to as "The Holy Grail" of power system technologies, electric energy storage can make a substantial difference in the functionality of the power

TABLE 2.1 Benefits of Microgrids

Energy cost savings	Resiliency and lower O&M costs	Potentially reduces peak load
Greenhouse gas reductions	Meeting environmental regulations	Back-up power supply
Economic development	Potentially lower cost than upgrading existing distribution infrastructure	Assure power quality
Reducing T&D capital costs and grid congestion	Integration of smart grid technologies	Enhance reliability
Future-proofing energy infrastructure	Facilitate the integration of DER and CHP	Demand management and load leveling
Reducing the economic impact of extended outages	Reduced T&D losses	Voltage reduction and overload protection

Source: Olearczyk, M.G., 2015. Literature Review – Microgrid Implementations in North America, EPRI, Jan. 2015 (unpublished).

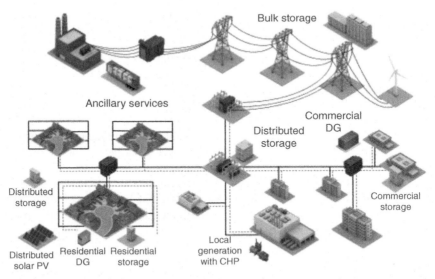

FIGURE 2.5 **Possible architecture for integrating controls to manage energy efficiency, demand response, photovoltaics and energy storage.** *(Source: EPRI, 2015 (unpublished).)*

system. The Integrated Grid offers a means by which distributed electric energy storage can be effectively integrated. Fig. 2.5 illustrates a possible Integrated Grid Architecture under which this could occur.

Distributed energy storage with utility control will have a substantial value proposition from several value streams. Incorporating distributed energy storage into utility planning and operations can increase reliability and flexibility. Dispatchable distributed energy storage can be used for grid control, reliability, and resiliency, thereby creating additional value for the consumer. Unlike distributed generation, the value of distributed storage is in control of the dimensions of capacity, voltage, frequency, and phase angle.

Consumer-sited storage has much of the same potential as utility-sided storage. The physics of power system behavior aren't really impacted by which side of the meter the storage is connected. The key differences come in operational objects: what are the benefits, and who gets them? For example, consumer-sited storage can be used to absorb energy from distributed generation in order to create a nonexporting solution. As a result, the consumer benefits from reduced bills and, possibly, from a simplified interconnection process. In the case of applying distributed storage to a distributed generation installation, the impacts of distributed generation on the grid may be less; however, there is also lost revenue for the utility, offset by the ability to utilize the asset.

In some jurisdictions, customer-sited energy storage systems are already using the Integrated Grid to bid into frequency regulation markets. In these instances, consumers benefit from market revenue, in addition to energy saving, and demand reduction.

Two examples of efforts to make greater use of distributed storage can be found in California and New York. In California, Southern California Edison is procuring over 135 MW of customer-sited storage to address the need for capacity. In New York, the Reforming the Energy Vision (REV) initiative, described elsewhere in the chapter, is undertaking proceedings which could require utilities to create markets for grid services provided for DER, including distributed storage.

3.3 Providing Ancillary Services

Fig. 2.6 illustrates how PV systems with advanced inverters can offer reactive compensation. Key to this ability is both the deployment of advance inverters, and the availability of a communications system and a distribution management system (DMS). In this example, the synergy between utility and consumer systems enables overall optimization of the system.[2]

3.4 Enabling Demand Response

The Integrated Grid will provide the connectivity and distribution management systems to facilitate consumers participating in a variety of demand response programs, including various customer notification schemes, interruptible tariffs, direct load control, real-time pricing, and critical peak pricing.

---

2. Depending on how the system is designed, consumers with local generation can provide the utility with various ancillary services, including operating reserves, regulation and load balancing, and voltage support. While the value of these kinds of services can be significant, it is highly variable (Electric Power Microgrids, 2006).

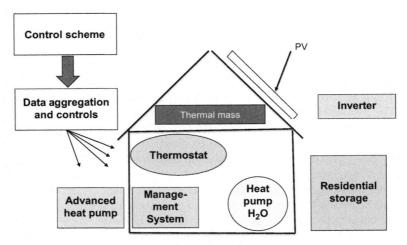

FIGURE 2.6 **Example integration of controls architecture.** *(Source: EPRI, 2015 (unpublished).)*

Examination of an event on the PJM system provides an excellent example of the value of DR and the ability of the Integrated Grid to enhance it.[3] PJM experienced record demand during Sep. 2013. In order to meet demand, the system operators deployed their demand response program on Sep. 9, 10, and 11. Fig. 2.7 illustrates the load shedding which occurred on Sep. 11. It illustrates how demand response (referred to as Emergency DR Event Period) was deployed on Sep. 11. Demand response was an essential, but insufficient resource to PJM during that period. The illustration does not reflect what the load would have been if the DR systems had not been activated. However, this event was reportedly the first time in ten years that load on the PJM had to be shed. However, in reviewing the event, several attributes are apparent.

Although overall demand response performed well during this period, restrictions such as the required 2-hour notification for invoking demand response, and the lack of granularity to utilize demand response on a more targeted subzonal basis, limited the operator's flexibility in using demand response as a tool in real time, in order to prevent some of these events. PJM has indicated that it will examine that issue, as well as other issues surrounding generator and transmission system performance, through its stakeholder process. The downside is that experience suggests that, if they make demand response resources more useful, more dispatchable, then participation drops.

The lack of divisibility of DR resources is usually because, by design, the stakeholders have never been able to agree on who would get called, if only some were needed. Some want to be called because they get paid on performance. Others who do not wish to be interrupted, however, are interrupted and get paid. Since Locational Marginal Pricing (LMP) during these periods is set using shortage

---

3. Refer to the chapter by Covina et al. on PJM.

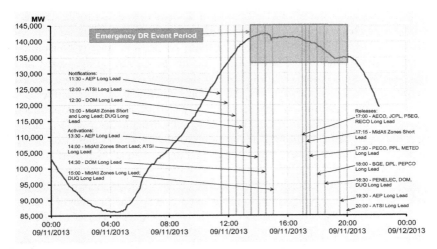

FIGURE 2.7 PJM geographic display of load shed events on Sep. 11, 2013. *(Source: www.pjm.com)*

pricing, it keeps a large load drop from DR from effecting. Should it not be affecting it? There is little enthusiasm for establishing a lot of dispatch granularity. However, this is a complex and controversial issue. DR advocates have prevented DR from being treated like a fungible resource—fact that reduces its value relative to a generator—yet DR usually gets the same level of compensation.

The integrated grid as described in this chapter would allow both the visibility at a more granular level of DR potential and local generation, as well as the ability to offer more flexible programs to engage customers.

3.5   Deferring Capacity Upgrades and Reducing System Losses

In many power systems, the installation of local resources may defer the need for capacity upgrades for generation, transmission, and/or distribution systems. In all cases where generation is increasingly distributed, rather than central, there will be a reduction in losses that would otherwise have occurred by the transmission and distribution of the displaced electricity.

4   ENABLING CONNECTIVITY

The integrated grid broadly enables connectivity. Connectivity of the electric power system refers to the increasingly widespread deployment of communicating equipment, providing access to data streams and functionality that can help inform decisions and behaviors all along the value chain, from the power plant to the end consumer and, potentially, through to their end-use equipment. As part of the integrated grid, connectivity will enable the power system to become more flexible, more resilient, and better able to capitalize on advanced digital functionality.

Cisco estimates there will be 21 billion devices connected to the internet by 2018 (see chapter by Cooper), three times the world population, up from 14 billion in 2013 (CISCO, 2013). Many of these are sensors, energy-consuming devices, energy resources, and other devices that play a role in the supply and use of electricity. As the electric power system transforms from a traditional one-way power flow network to one that enables intelligent and interactive two-way power and information flow, connectivity will be paramount. Continued investments in modernization to address aging infrastructure, respond to changing supply and demand profiles, incorporate renewables, and increase energy efficiency will introduce new technologies that depend on an increased level of connectivity.

Key drivers of connectivity include increased customer interest in connected products and services, and the rapid projected growth of the Internet of Things (IoT). However, connectivity presents a series of challenges. These include the large volume of data; proprietary legacy systems; the need for enhanced security; inconsistent lifecycle timescales of utility assets and connectivity technologies; the rapid pace of connectivity technology evolution; and the need for effective integration of connectivity technologies into the power system. Connectivity enables opportunities in all corners of the power system, from generation to transmission and distribution, and ultimately to the customer—opportunities that are described in this chapter.

One example of the connectivity enabled by the integrated grid is the ability to network consumer installations so as to engage them in collective demand response programs. In Charlotte, NC a not-for-profit organization, Envision Charlotte, teamed with Duke Energy and organized the owners of 61 of the 64 largest buildings to use the integrated grid in order to reduce energy usage by turning off lights and adjusting thermostats—thereby reducing energy use by 6.2% (Transitioning from Smart Buildings, 2015).

5 ENABLING HIGHER PENETRATION OF DER

Fig. 2.8 provides an example of where control schemes can also enable higher penetration of DER devices, such as PV, on distribution feeders, without necessary reconductoring or reinforcement. The illustration provides an example of how 20% PV penetration can impact primary voltage, with and without volt–VAR control. Better volt–VAR control allows more real power to flow on existing electrical circuits, thereby accommodating more PV installations.

Fig. 2.9 illustrates an analysis of a distribution system to determine the feeder hosting capacity, or the amount of local PV generation the feeders could support without substantial upgrading. As shown in the figure, the use of advanced inverters and control could substantially increase the feeders' capacity, or the number of PV systems installed. In this case, potential overvoltage situations can be mitigated eliminating the need for extensive reconfiguring. The "y" axis

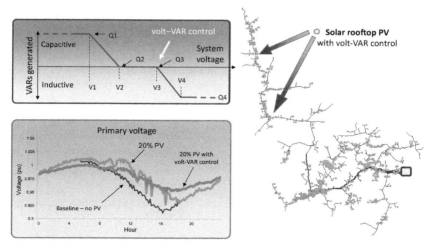

FIGURE 2.8   Volt–VAR control with high-penetration PV. *(Source: Stochastic Analysis, 2012)*

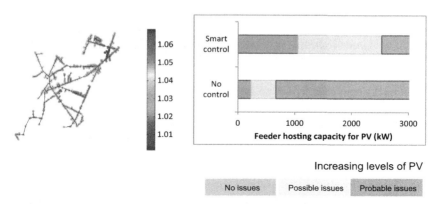

FIGURE 2.9   Smart inverters with communications and control can substantially increase the amount of PV a distribution system can support. *(Source: Stochastic Analysis, 2012)*

represents the voltage in "per unit," or PU values. A PU of 1.0 would translate into a service voltage of 120, on a 120-volt system.

Determining the capacity of a distribution circuit to accommodate various levels of DER is an essential part of understanding the value of the integrated grid. A circuit's ability to accommodate DER, without the need for mitigating cost or actions, is defined as hosting capacity, expressed as megawatts, or as a percentage of the total load served by the feeder.

An example of distribution capacity analysis is illustrated in Fig. 2.10. It shows the extent to which widely varying amounts of PV can be accommodated on distribution feeders with differing characteristics. (Distributed Photovoltaic

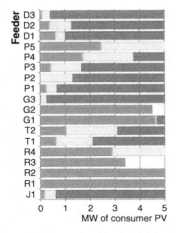

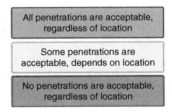

FIGURE 2.10   Sample distribution hosting capacity. *(Source: The Integrated Grid, 2015)*

Feeder Analysis, 2013). The feeders studied are listed on the vertical axis, and the megawatts of hosting capacity on the horizontal axis. The color coding for each feeder indicates the amount of capacity the feeder can handle, in three categories. Green indicates no problem accommodating the indicated level of PV. Yellow indicates that the location of PV on the feeder at that level becomes the determining factor, so that aspect must be considered. Red indicates the feeder load range over which no additional DER can be accommodated without increasing the possibility of adverse impacts to local loads, and possibly system loads.

Some feeders can host large amounts of PV (for example: R1, R2, R3, G1, and G2 in the figure). Others (G3, D1, D2, and D3) are limited in the amount of additional DER that can be accommodated or hosted without grid upgrades. Hosting capacity represents a consistent method for determining how much DER can be accommodated without grid upgrades, or aggressive demand-side management applied to the feeder.

5.1   Higher Penetration Enhances Resiliency

The presence of local generation also has the ability to increase the overall resiliency of the power system. In the event of an outage, local generation can help support portions of the distribution system that may be without power. It can also allow consumers to sustain part of their building services. The rising interest in the use of local generation has occurred, in part, due to heightened concerns about resiliency, improvements in generation technologies, power electronics, and the need for new capacity resources on the power system. A key potential benefit of local generation is the ability to improve the reliability of the power system by providing emergency power during interruptions of the bulk system supply. This benefit can only be realized if the local generation is operated in a configuration that facilitates islanded operation. The integrated grid allows for this type of operation, while also being able to capture all of the other benefits of local generation, such as waste-heat recovery, load reduction on the transmission and distribution system, and power quality improvements.

5.2 Higher Penetration Can Increase Environmental Benefits

Environmental benefits are among several externalities, or secondary impacts, of energy systems that are often highly uncertain, and not easily monetized. Examples of externalities are: emission impacts of local generation and of the central station power they may offset, economic activity resulting from expansion of local jobs, health impacts of increased or reduced emissions, national security impacts from shifting energy use, and a few other benefits. Often these impacts are assumed to be positive and are the motivation for policy enactment to encourage local generation.

6   THE FUTURE OF DISTRIBUTION UTILITIES

There is a growing sentiment, among industry observers, that the electric utility as we know it, as an institution, is nearing its end of life. For example, in response to the substantial changes in retail electricity markets, as summarized in this chapter, in Apr. 2014, the New York State Department of Public Service ("DPS") issued an order instituting a proceeding called "REV", aiming to reorient both the electric industry, and associated ratemaking paradigms, toward a consumer-centric approach that leverages both technology and markets. In the DPS order, DER will be integrated into the planning and operation of electric distribution systems, in order to achieve optimal system efficiencies, secure universal, affordable service, and enable the development of a resilient, climate-friendly energy system (see chapter by Jones et al.).

The goals of REV are expected to occur over a period of years, through the mutual efforts of industry, customers, nongovernmental advocates, and regulatory partners. The Apr., 2014 Order stated six objectives for the current initiative:

1. enhanced customer knowledge and tools that will support effective management of the total energy bill
2. market animation and leverage of customer contributions
3. system-wide efficiency
4. fuel and resource diversity
5. system reliability and resiliency
6. reduction of carbon emissions

The Commission's staff articulated the REV vision in two documents. The Apr., 2014 Staff Report and Proposal formed the basis for the initiation of the proceeding. The proceeding was separated into two tracks, with Track One focused on developing distributed resource markets, and Track Two focused on reforming utility ratemaking practices. In Aug., 2014, the Staff issued a Straw Proposal on REV.

REV intended to establish markets so that customers and third parties can be active participants, to achieve dynamic load management on a system-wide scale, resulting in a more efficient and secure electric system, including better utilization of bulk generation and transmission resources. As a result of this market animation, DER will become integral tools in the planning, management and operation of the electric system. The system values of distributed resources will be monetized in a market, placing DER on a competitive par with centralized options. Customers, by exercising choices within an improved electricity pricing structure and vibrant market, will create new value opportunities and, at the same time, drive system efficiencies, and help create a more cost-effective and secure integrated grid. The more efficient system will be designed and operated to make optimal use of cleaner and more efficient generation technologies. Enabling these markets will require modernization of infrastructure and operations, particularly communication and data management capabilities.

The reformed electric system will be driven by consumers and nonutility providers, and it will be enabled by utilities acting as Distributed System Platform (DSP) providers. Utilities are responsible for reliability, and the functions needed to enable distributed markets are integrally bound to the functions needed to ensure reliability. Technology innovators and third party aggregators will develop products and services that enable full customer engagement. The utilities, acting in concert, will constitute a statewide platform that will provide uniform market access to customers and DER providers. Each utility will serve as the platform for interface among its customers, aggregators, and the distribution system. Simultaneously, the utility will serve as a seamless interface between aggregated customers and the New York Independent System Operator (NYISO).

Reforming the Commission's ratemaking practices will be critical to the success of the REV vision. Utility earnings should depend more on creating value for customers and achieving policy objectives. Rather than simply building infrastructure, utilities could find earning opportunities in enhanced performance, and in transactional revenues.

6.1   The REV Market Framework

The Commission adopted the model of the DSP provider, in which DER providers will be viewed as customers and partners, rather than competitors of traditional grid service. The DSP will have the responsibility to offer services, whether in the form of information, interconnection, or dispatch services, at prices and under terms allowed by the Commission. DER providers and their customers are entitled to compensation from the DSP. The Commission staff has convened a stakeholder effort (Market Design and Platform Technology) to identify the necessary functional and business architecture for the DSP and DSP markets. To avoid overlapping jurisdiction over DSP activities, utilities will not purchase power that would constitute a sale for resale, under the Federal Power Act, except for purchases that are otherwise required by law.

6.2   Utilities as DSPs

Requiring utilities to serve as DSPs, under the Public Service Comission regulatory authority and supervision, is in the best interests of New York consumers. Having the utility expand its responsibility to include DSP functionality enhances the opportunity for integrated operation of the distribution system, and for realizing the economic value of DER investment (see chapter by Hanser and Van Horn). Utilities' behavior with respect to DER has been responsive to the regulatory structure under which they have operated. Reforming the regulatory model, and, by extension, utility behavior, is a critical component of the REV initiative. Unlike the wholesale market, the markets that will be enabled and potentially operated by the DSP will not establish commodity prices. Commodity prices, the prices for capacity, energy, and bulk ancillary services will be set by the NYISO. If DSPs are failing to meet the objectives of REV, the Commission will consider options to allow other entities to serve that function.

Could this signal the end of paying for electricity by the kilowatt-hour? The Integrated Grid would virtually prevent a telecom-like transition in several ways:

1. There is an absence of enabling technology, which could parallel the grid. The telecom transition resulted, in large part, in the establishment of a completely separate system—cell networks, which could displace land lines, while providing the added feature of portability. There are cases where a local generation system may be economical but overall from a societal perspective, an Integrated Grid is the lowest cost electric system.
2. The telecom system is also an electrical system—however, the energy levels used in sending and receiving bytes is nearly insignificant, compared to the massive levels in an electrical system. In communications, messages and digital files can be broken into groups, called packets. Electrical energy is transmitted not by the movement of electrons, but by electromagnetic vibrations. These vibrations require wires—substantial wires—to act as pathways. Electrical energy cannot be broken into packets and directed to take different pathways. However, an Integrated Grid can assure the optimal delivery and use of both central and distributed resources.
3. Substantially reducing power sector CO_2 emissions requires harvesting both bulk and distributed wind and solar power, from regions where there is a limited population and low electrical demand, and transmitting it to regions where electricity is needed. An Integrated Grid is needed to provide this transmission.

4. While the cost and overall impact of communicating a packet of digits is minimal, the equivalent delivery of kilowatt-hours has substantial cost, and associated environmental impact. Each kilowatt-hours is substantially more valuable than a packet, and must be metered so as to charge equitably for its production. In addition, studies have shown that when electricity remains unmetered (such as in some multifamily apartment buildings), consumers tend to be somewhat wasteful.

It is the author's view that most customers (and society) will be better off finding a way to maintain and improve the grid, while allowing DERs to flourish.

A successful example of a partial transformation in one industry is the adoption of the automated teller machine (ATM) network. Today, ATM machines have made banking completely distributed because of a secure, intelligent, and robust infrastructure that connects ATM terminals with central banks seamlessly. Much like the ATM "industry," a robust and intelligent grid can enable the most effective, low-cost use of resources.

The goals of REV are expected to occur over a period of years, through the mutual efforts of industry, customers, nongovernmental advocates, and regulatory partners.

6.3   The Integrated Grid Extends the Grid's Boundaries

Traditionally, the grid is viewed as the power delivery system, which connects bulk power generation with consumers. Its boundary starts at the central generator's bus bar, and proceeds through the transmission and distribution system, and ends typically at the meter. That model has been extended, with the interconnected approach facilitating adoption of local generation, storage, plug-in electric vehicles, and through the use of smart meters. In effect, the boundary of the grid now reaches "beyond the meter." The grid now extends all the way from the central generator's bus bar, through the transmission and distribution system, to the meter, and is now poised to go beyond, into buildings, as part of the Integrated Grid. This now enables a transactional relationship to be established between the bulk power system, local generation, storage, and end-use appliances and devices.

For example, as the grid begins to extended itself, so as to include inverters and the use of technologies like the intelligent universal transformer (IUT), the grid boundaries could be viewed as migrating further into the consumer's domain. Fig. 2.11 illustrates this evolution, and highlights the benefits it may bring to consumers.

7   OVERALL ATTRIBUTES AND VALUE OF AN INTEGRATED GRID

Table 2.2 lists a summary of the value of an integrated grid. On the left hand side of the table, it lists the improvements and benefits that result from the integrated grid, from the power system perspective. On the right hand side, it lists the

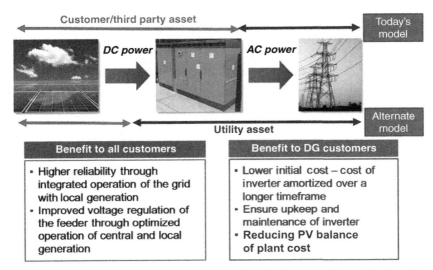

FIGURE 2.11 **Extending the distribution grid to include inverters.** *(Source: By the author (unpublished).)*

improvements and benefits from the consumers' perspective. Table 2.2 shows the various types of improvements that correspond to each of the critical attributes identified. A key aspect of the value estimation process, in general, is its consideration of improvements to the power delivery system (left column of Table 2.2), as well as improvements that consumers directly realize (right column of Table 2.2).

This was done to ensure that emerging and foreseen benefits to consumers, in the form of a broad range of value-added services addressed in the estimation of value. The "cost of energy" attribute is the total cost to deliver electricity to customers, including capital costs, operating and maintenance (O&M) costs, and the cost of line losses on the system. Therefore, the value of this attribute derives from any system improvement that lowers the direct cost of supplying this electricity. "SQRA" is the sum of the power security, quality, and reliability attributes because the availability part of SQRA is embedded in the power quality and reliability attributes. The quality of life attribute refers to the integration of access to multiple services, including electricity, the Internet, telephone, cable, and natural gas. This involves integration of the power delivery and knowledge networks into a single intelligent electric power/communications system, which sets the stage for a growing variety of products and services, designed around energy and communications.

Table 2.3 lists a preliminary analysis to quantify the benefits which could accrue from an Integrated Grid.

Table 2.3 shows avoided energy and capacity costs in order to represent the benefits that inure to all rate payers for PV energy generated. Avoided costs are equivalent to benefits because they reflect costs that would have been incurred,

TABLE 2.2 Attributes and Values – The Integrated-Grid Approach

Power delivery (improvements/benefits)	Attributes	Consumer (improvements/benefits)
O&M cost Capital cost of asset T&D losses	*Cost of energy (net delivered life-cycle cost of energy service)*	End-use energy efficiency Capital cost, end-use infrastructure O&M, end-use infrastructure Control/manage use
Increased power flow New infrastructure Demand responsive load	*Capacity*	Improved power factor Lower end user Infrastructure cost through economics of scale and system streamlining Expand opportunity for growth
Enhanced security Self-healing grid for quick recovery	*Security*	Enhanced security and ability to continue conducting business and everyday functions
Improved power quality and enhanced equipment operating window	*Quality*	Improved power quality and enhanced equipment operating window
Reduce frequency and duration of outages	*Reliability and availability*	Enhanced security Self-healing grid for quick recovery Availability included
EMF management Reduction in SF6 (sulfur hexafluoride) emissions Reduction in cleanup costs Reduction in power plant emissions	*Environment*	Improved esthetic value Reduced EMF Industrial ecology
Safer work environment for utility employees	*Safety*	Safer work environment for end-use electrical facilities
Value-added electric-related services	*Quality of life*	Comfort Convenience Accessibility
Increased productivity due to efficient operation of the power delivery infrastructure Real GDP	*Productivity*	Improved consumer productivity Read GDP

Source: Estimating the Costs (2011).

TABLE 2.3 The Costs and Benefits of an Integrated Grid

Benefit category	Avoided cost	Conditions
Energy	$36–44/MWh	
Generation capacity	$48–133/kW-yr	Promoted through DER visibility
T&D capacity	?	Highly site-specific
Ancillary services	?	Highly site-specific and enabled with smart inverters
Environment/RPS	?	
Grid resiliency	?	Requires standards and interoperability

Source: by the author (unpublished).

except for the PV energy produced. Avoided costs are the standard currency for valuing energy efficiency investments—it is appropriate as a general characterization of benefits. The avoided costs shown reflect estimates of 2016 capacity and energy costs, portrayed in a range of values to characterize the variety of circumstances of utilities across the United States.

To the extent that local generation, including PV, reduces the need for capacity to serve consumers because peak demand is decreased, then it theoretically reduces marginal capacity cost. In some cases, local generation may also impact ancillary service cost, depending on the penetration of local generation. As the penetration of local generation increases, the need for ancillary services will also increase.[4] Displacing energy reduces ancillary services costs, based on today's system, but to the extent that local energy production is highly variable, more ancillary services may be required in the future, and the net results may be positive or negative. The outcome will depend on the manner in which the mix of capacity portfolios evolve. Should there be rapid growth in resources, which have the flexibility to respond to the variability of a high percentage of DER, then the market price for these services will likely diminish.

The largest expected benefit from the Integrated Grid is in the categories of improved quality, security, reduction in overall capacity needed, increase in reliability, and reduction in the overall cost of energy. The actual costs incurred to both modernize the grid and upgrade the distribution system to accommodate

---

4. The avoided energy cost ($/MWh) reflects the California estimate of 2016 avoided energy costs, with a range EPRI researchers constructed using EIA estimates of natural gas price volatility (because gas is generally assumed to be at the margin). ISO-RTOs do not produce forward energy price estimates, so the California value was used as it is referenced.
The avoided capacity cost ($/kW-year) value is for 2016 PJM average forward capacity market price for 2016 (the low value), and the average California state avoided capacity (the high value). The difference is likely based on the particular circumstances of California (citing, environment compliance, land costs, etc.)

TABLE 2.4 Components of Marginal Energy Cost in California

Component	Description
Generation energy	Estimate of hourly marginal wholesale value of energy adjusted for losses between the point of the wholesale transaction and the point of delivery.
System capacity	The marginal cost of procuring resource adequacy resources in the near term. In the longer term, the additional payments (above energy and ancillary service market revenues) that a generation owner would require to build new generation capacity, in order to meet system peak loads.
Ancillary services	The marginal cost of providing system operations and reserves for electricity grid reliability.
T&D capacity	The costs of expanding transmission and distribution capacity to meet customer peak loads.
CO_2 emissions	The cost of carbon dioxide emissions (CO_2) associated with the marginal generating resource.
Avoided RPS	The cost reductions from being able to procure a lesser amount of renewable resources while meeting the Renewable Portfolio Standard (percentage of retail electricity usage).

Source: Technical Potential for Local Distributed Photovoltaics (2012).

substantial penetration of local generation will be made in the short-term, and are substantial commitments to be made, based on speculation that the widespread use of local generation and storage will indeed evolve. On the other hand, the projected benefits are even less certain and, for the most part, will not be realized until after the investments needed to enable the Integrated Grid have been realized.

7.1 Example of Avoided Costs

In Sep. 26, 2013, Energy and Environmental Economics, Inc. (E3) produced a report for the California Public Utilities Commission's (CPUC) Energy Division that evaluated California's net energy metering programs (Technical Potential for Local Distributed Photovoltaics, 2012). In the report, E3 updated an avoided cost framework which has been "developed in numerous proceedings at the CPUC since it was adopted in 2004." The framework is contained in Table 2.4.

7.2 Impacts on Average Residential Consumers

In a previous effort (Estimating the Costs, 2011), Electric Power Research Institute (EPRI) studied the macro level implications of grid modernization resulting

TABLE 2.5 Possible Consumer Implications of Smart Grid Costs

Class	Smart Grid Cost to Consumers—Allocated by Annual kWh[a]							
	$/Customer total cost[b]		$/Customer-year, 10-year Amortization[c]		$/Customer-month, 10-year amortization[d]		Percent increase in monthly bill, 10-year amortization[e]	
	Low	High	Low	High	Low	High	Low	High
	$/customer	$/customer	$/customer/ year	$/customer/ year	$/customer/ month	$/customer/ month		
Residential	1,159	1,679	116	168	10	14	14	14
Commercial	8,018	11,617	802	1,162	67	97	15	15
Industrial	489,545	175,310	48,955	17,531	4080	1,461	2	2

[a]Low refers to EPRI low estimate of $ total SG costs; high is the other SG costs. Customer numbers by class (residential, commercial, and industrial) are for 2009, from EIA. SG costs are allocated to customer classes based on kWh sales in 2009 (38% residential, 37% commercial, and 25% industrial).
[b]Total SG cost divided by customers for each segment (residential + commercial + industrial).
[c]Annual cost per customer per year for total SG cost spread out (amortized) equally over 10 years (nominal values).
[d]Annual cost per customer per month for total SG cost spread out (amortized) equally over 10 years (nominal values).
[e]Annual increase in monthly bill based on "d".
Source: Estimating the Costs, 2011.

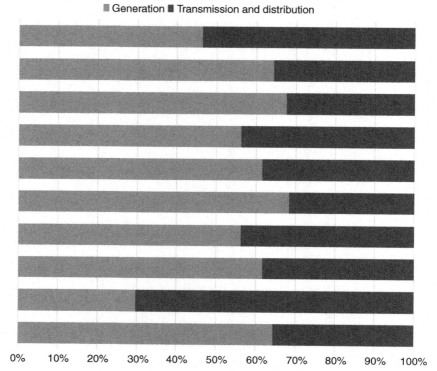

FIGURE 2.12 **Cost components of generation, transmission and distribution for a range of US utilities.** *(Source: Estimating the Costs, 2011)*

in an Integrated Grid. Table 2.5 summarizes these benefits, and apportions them to the average electricity customer. While these studies originated under the umbrella of the smart grid, they included most of all the attributes that integration would accomplish.

Extensive related work is underway to elucidate these costs and benefits further by the US Department of Energy, as part of their analysis of the smart grid demos conducted through funding of the American Recovery and Reinvestment Act of 2009.[5]

The costs and resultant benefits will vary widely among utilities. For example, Fig. 2.12 illustrates the cost components of generation, and the combination of transmission and distribution for a range of US utilities. These variations demonstrate that the relative impacts of integration will also vary, depending on the individual characteristics of the systems. Utility names are intentionally omitted from the graph.

---

5. Refer to the Department of Energy's website for more information: http://energy.gov

TABLE 2.6 Benefits to the Average Residential Consumer from Implementation of the Integrated Grid Approach

Net present worth ($B)		Residential amortized/month ($)	
Reliability	444	11.10	11
Security	152	3.80	4
Quality	86	2.15	2
Cost	475	11.88	12
Capacity	393	9.80	10

Source: Estimating the Costs (2011).

Table 2.6 elucidates the potential benefits to the average residential consumer from implementation of the Integrated Grid approach. It contains estimates of the benefits by major category of value.

7.3   EPRI's Integrated Grid Study

EPRI launched its Integrated Grid initiative with a concept paper, and the goal of aligning power system stakeholders on key issues. With widespread adoption of DER, potentially fundamental changes in the grid will require careful assessment of the benefits, costs, and opportunities of different technological or policy pathways. Four main areas requiring global collaboration were identified:

- interconnection rules and standards
- grid modernization
- strategies and tools for grid planning and operations
- enabling policy and regulation

Work on the following the three-phases of Integrated Grid initiative is intended to provide stakeholders with information and tools that are integral to the four areas aforementioned.

- Phase I. Stakeholder alignment, including the production of a concept paper, supporting documents, and related knowledge transfer efforts (The Integrated Grid, 2014).
- Phase II. Development of a benefit-cost framework, interconnection technical guidelines, and recommendations for grid operations and planning with DER (The Integrated Grid, 2015).
- Phase III. Global demonstrations and modeling to provide comprehensive data that stakeholders will need for transitioning to an integrated grid.

Preliminary results indicate that the key issue remains the ways in which DER interacts with the power system infrastructure. The formula for this answer has multiple dimensions. Beneficial and adverse circumstances can arise at differing levels of DER saturation. The interaction is dependent on the specific characteristics of the distribution circuits (design and equipment), existing loads, time variations of loads and generation, environmental conditions, and other local factors. Benefits and costs must be characterized at the local level, and the aggregated level of the overall power grid. These early findings reinforce the need for the industry to address the implications of DER systematically and thoroughly. This requires adopting planning protocols and operating procedures that see interconnected assets from end to end, and that operate the system in an integrated manner.

8 CONCLUSIONS

Changes in the retail electric business are the ordinary, not the exception. In order to provide society with affordable, reliable, and environmentally sustainable electricity, the Integrated Grid ensures the best use of scarce societal resources, when deployed in combination with the functionally essential technologies (local generation, smart distribution, etc.), and supported by enabling rate structures and interconnection guidelines. From a societal perspective, it is more sustainable and economic to interconnect local generation systems to the grid, rather than to have them operate independently, or simply connected. In addition, integrating local and central resources provides even greater opportunities to enhance the value of all of these resources.

The integrated grid will allow DER to be used in a virtual symphony with central resources in order to provide optimal security, reliability, quality, and availability to society, at the lowest reasonable cost, with the least environmental impact.

REFERENCES

California Net Energy Metering (NEM) Draft Cost-Effectiveness Evaluation, September 26, 2013. California Public Utilities Commission Energy Division, Energy and Environmental Economics, Inc.

Cisco Visual Networking Index: Forecast and Methodology, 2013–2018. http://www.cisco.com/c/en/us/solutions/collateral/service-provider/ip-ngn-ip-next-generation-network/white_paper_c11-481360.html

Distributed Photovoltaic Feeder Analysis: Preliminary Findings from Hosting Capacity Analysis of 18 Distribution Feeders, 2013. EPRI, Palo Alto, CA. Report No. 3002001245.

Distributed Photovoltaics: Utility Integration Issues and Opportunities, August 2008. EPRI Report No. 1018096.

Electric Power Micro-grids: Opportunities and Challenges for an Emerging Distributed Energy Architecture, May 2006. D.E. King, Dissertation, Carnegie Mellon University.

Estimating the Costs and Benefits of the Smart Grid, March 2011. EPRI Report No. 1022519.

Reforming the Energy Vision, New York State Department of Public Service (DPS) Staff Report and Proposal, 2014. Case 14-M-0101: Proceeding on Motion of the Commission in Regard to Reforming the Energy Vision.

Siemens, 2015. Deep Dive on Microgrid Technologies. FierceMarkets Custom Publishing. Washington, DC.

Stochastic Analysis to Determine Feeder Hosting Capacity for Distributed Solar PV, December 2012. EPRI Report No. 1026640.

Technical Potential for Local Distributed Photovoltaics in California: Preliminary Assessment, March 2012. E3 report for CPUC, pp. 55.

The Integrated Grid: Realizing the full Value of Central and Distributed Resources, February 2014. EPRI Report No. 3002002733.

The Integrated Grid: A benefit—Cost Framework, February 2015. EPRI Report No. 3002004878.

The Integrated Grid Phase II: Development of a Benefit—Cost Framework, May 2014. EPRI Report No. 3002004028.

Ton, D.T., 2014. Microgrids, Smart Grid & Our Energy Future. U.S. Green Building Council, Maryland.

Tighe, W.C., Apr. 2015. Transitioning from Smart Buildings to Smart Cities. NEMA Electroindustry, Rosslyn, VA.

Chapter 3

Microgrids: Finally Finding their Place

Chris Marnay

Microgrid Design of Mendocino, Comptche, CA, United States of America

1 INTRODUCTION

"Microgrids" has become a familiar buzzword in recent years, and current market reports claim dramatic growth in projects planned to around 12 GW total today worldwide, almost tripling the capacity estimate of a year ago (Navigant Research, 2015). The northeastern United States and Japan have embraced microgrids, following their twin disasters, the 2011 Great East Japan Earthquake and Tsunami (GEJE) and Hurricane Sandy in 2012. Consequently, fully 44% of the planned capacity lies in North America, and a further 15% in Japan. These traumatic events represented a turning point after which not only has the concept of microgrid and its perceived benefits shifted, but it has also begun to find a natural place in the regulatory and policy arena (Cuomo, 2015). This emergence will, in turn, bring new players into the sector, and have a major effect on electricity distribution utilities (DisCos), those entities operating the legacy distribution network. Indeed, some of the regulatory murmurings, especially in the northeastern United States, suggest a radical adjustment for the role of DisCos, although company responses vary widely to date.

In the future landscape that DisCos will face, microgrids will likely be a prominent feature. From a non-DisCo perspective, microgrids offer some interesting opportunities. In fact, the distinction between the domains where DisCos might be central players and where they might not represents one of the key fault lines described in this chapter. Since there will likely be microgrids of many forms, defining and classifying them, and then observing how their roles may differ, pose an early analysis task. Then, the significance of these actors' existence in the distribution network must be assessed, and the consequences gauged.

As will become clear, development of microgrids has followed a crooked path that has led to interest being currently high in the United States and Japan, whereas elsewhere there is some activity, but with much less dynamism and different drivers. A notable exception is China, currently estimated to have an additional 4 GW of microgrids in place within 5 years, fully one-third of the

Future of Utilities - Utilities of the Future. http://dx.doi.org/10.1016/B978-0-12-804249-6.00003-8

worldwide total. Microgrid drivers in China are somewhat different, mostly focused on enhancing overall renewable penetration in electricity supply. While China has successfully deployed considerable central station renewable capacity, it has lagged somewhat in distributed renewable deployment, and microgrids are seen as a possible solution. In other areas, notably Europe, there is only limited current activity in this space. Navigant Research estimates only 5% of worldwide planned capacity is here.

This chapter does not attempt a comprehensive survey or discussion of the many microgrid benefits that have been claimed. For a microgrid cost-benefit framework, see the Conseil International des Grandes Réseaux Électriques C6.22 Brochure (CIGRÉ, 2015) or Morris (2012). The major benefit categories listed are:

- Reduced electricity purchases from megagrid
- Infrastructure investment deferral
- Reduced emissions
- Ancillary service provision
- Increased reliability

Further, four case studies are covered in the CIGRÉ Brochure, one each in Canada, the United Kingdom, California, and Germany. The analysis of the United Kingdom, for example, the Holme Rd. project in Preston, is particularly interesting. It finds that the added benefits of forming a microgrid over standard deployment of similar distributed generation technology are an added £3.4–4.5 million ($5.3–7 million) over 20 years.

But the allure of microgrids cannot be fully appreciated in these terms alone. Rather, there is powerful appeal in the notion of local control itself. Trade-offs are evaluated and choices made among options for meeting energy service requirements on both the supply and demand sides, by any available fuel, including ground source, and locally or via public infrastructures. Local fuels, for example, solar, can be managed locally, and the microgrid can present itself to the megagrid as a controlled entity of a size and sophistication that allows it to fit within existing operating norms, thereby freeing the megagrid from the management burden of variable or tiny resources. The microgrid can be an aggregator of diverse and/or variable renewable resources, further described in some following chapters of this book such as the one by Athawale and Felder. Further, note that resilience does not appear in the list of benefits mentioned earlier.

This chapter's approach is qualitative, addressing the implications of microgrids' emergence, with a focus on just two technical capabilities that might deliver substantial benefits:

- Power supply resilience and reliability, which are the current principal drivers for microgrid deployment, especially in the northeastern United States, but also in Japan.

- Heterogeneous power quality and reliability (HePQR), that is, the possibility of delivering differentiated qualities of power to loads in order to match their individual requirements more closely than universal homogeneous service (HoPQR).

The questions at hand in this book are: how are the economic and technical forces of dispersed resources reshaping the industry; what are the consequences of the evolving structure for new and incumbent players, especially the DisCos; and what are the responses necessary for the policy and regulatory environment to catch up. More importantly, what role might microgrids play in the context of the rapidly evolving utility business models, as described in the companion chapters in this book.

Section 2 of this chapter is an introduction to the paradigm shift that microgrids represent. Section 3 addresses issues of reliability and power quality, which are at once both drivers and promises of microgrids. Section 4 offers some more formal microgrid definitions. Section 5 describes three notable demonstrations. Section 6 explores the policy and regulatory implications of microgrid's emergence, followed by this chapter's conclusions.

2   EVOLUTION OF MICROGRIDS

Perhaps surprisingly, microgrid history is fairly short, dating roughly from the turn of the last century. This is not to say power systems that meet modern definitions of a microgrid did not exist previously. Indeed they did. Rather, recent emerging technology has expanded their capabilities and the spectrum of opportunities greatly; at the same time, appreciation of the benefits of localized power systems has exploded. Many commentators have noted that the power industry began as numerous small isolated power systems, which over time have become increasingly interconnected and interdependent (Smith and MacGill, 2014). While the whole world is still not fully interconnected, major regions are. North America's Western Interconnection grid, for example, encompasses most of 11 western US states, plus 2 Canadian provinces, and a toehold in Mexico. The same applies to massive, interconnected megagrids in Europe, Australia, and elsewhere.

It is often noted that some legacy "microgrids" survive in remote locations, unconnected to the wider grid. This line of reasoning makes it easy to dismiss microgrids as nothing new, but rather a holdover or renaissance of the industry's roots. This perspective, however, misses key characteristics of the modern microgrid concept, conceived as a part of the whole legacy of electricity supply system, not as something separate from it, and yet it is indeed semiautonomous. The following section offers some formal definitions of the microgrid, but suffice to say it has two key characteristics:

- First, it's a locally controlled power system
- Second, it can function either grid connected, or as an electrical island

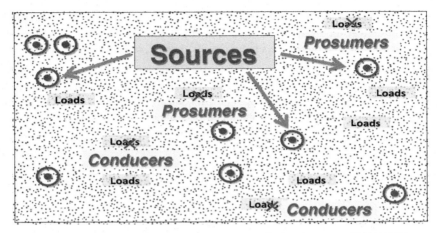

FIGURE 3.1 **The traditional power supply paradigm.** *(Source: Marnay and Lai (2012).)*

The corollary of the control requirement is that a microgrid can manage its profile with the legacy "megagrid," and can also establish economic relationships with it; for example, it can participate in grid operations and markets. The second requirement implies that a microgrid can connect and disconnect at will, whether for economic, reliability, or other reasons.

A simple way to visualize the change is shown by a comparison of the visuals in Figs. 3.1 and 3.2 (Marnay and Lai, 2012). Fig. 3.1 shows the traditional paradigm familiar to all: a small number of large remote generating stations represented by the sources, and the loads, represented by the gray dots spread widely. As shown, the disruptive force to the traditional model is typically characterized as the changing role of individual customers, notably becoming prosumers/conducers. That is, they are no longer fully dependent on their DisCo for electricity supply, and possibly becoming traders, meaning not dependent on the grid solely for the delivery of their purchased energy. This phenomenon is clearly happening, and its implications are discussed in the chapter by Gimon and others in this book.

Other chapters, including the one by Knieps, recognize the emergence of microgrids as a new entity in the power sector, and attempt to establish their place in a transactional framework. Fig. 3.2 shows a layered view of where the power sector might be heading. It envisions the emerging role of prosumers/conducers who deal directly with their DisCo, but additional new entities emerge that break the DisCo–customer bond, and permit multilateral trade. Let us assume that the traditional centralized meshed grid endures, that is, possibly upstream of the substation, called the *megagrid,* or Mgrid, in Fig. 3.2 and further in this chapter.

Three new types of power systems that comply with the microgrid definition are:

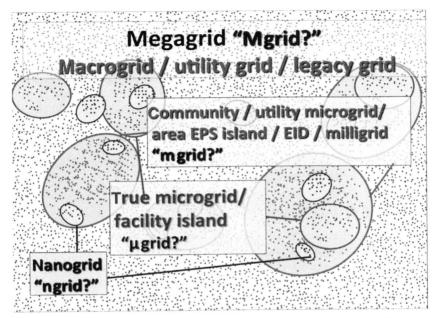

Megagrid "Mgrid?"

Macrogrid / utility grid / legacy grid

Community / utility microgrid/ area EPS island / EID / milligrid "mgrid?"

True microgrid/ facility island "μgrid?"

Nanogrid "ngrid?"

FIGURE 3.2   The emerging decentralized paradigm. *(Source: Marnay and Lai (2012).)*

- First, fragments of the traditional grid, called *milligrids* (mgrids), lying downstream of the substation, might function as islands in some circumstances. As seen in Fig. 3.2, many other names have been given to these systems, and an additional common descriptive one, not shown, is "utility distribution microgrid" (Asmus, 2014). An excellent example of such a system is San Diego Gas and Electric's (SDG&E) Borrego Springs project, further described in Section 5.
- Second, power systems or multiple ones closely colocated on sites that can function as electrical islands are called *true microgrids* (μgrids) in Fig. 3.2. These are the most familiar microgrids at present, and several examples will follow, including perhaps the world's most famous microgrid in Sendai, Japan.
- Finally, small local systems that match local sources and sinks, rather than conventional building circuits, merit a special mention. These circuits may be low-power local DC networks called *nanogrids* (ngrids).

While ngrids may not be as directly relevant to the topic of this book, they might play a significant role in future building power systems and should not be overlooked. They may be wired circuits operating at tens or hundreds of Volts, or may be just integrated into the increasingly ubiquitous data infrastructure using USB, Power over Ethernet, or other standards (Emerge Alliance, 2015). Technically, such systems have some compelling advantages for efficiency, reliability and power quality, and economy:

- Efficiency because so many loads—about a third of all commercial building usage—now involve DC, and the share is growing. Avoiding AC–DC conversions saves energy, and the effect is doubled if the source, for example, a PV array, fuel cell, or battery, is also DC.
- Reliability and power quality are good because low-voltage DC systems are simple and free of most AC's power quality and stability problems.
- Economy because simplicity and reduced need for skilled labor make installation, maintenance, and reconfiguration quick and cheap, even though low voltages do require bigger conductors.

Coexistence of AC and DC in a facility is an example of HePQR, discussed in more detail further in this chapter.

3 POWER QUALITY AND RELIABILITY

In some ways, electricity is a commodity, that is, a product with well-defined characteristics, such that kilowatt-hours are interchangeable. Indeed, they flow around the supply network according to physical laws, and no individual one can be traced, so how distinctive can they be? But the characteristics of electricity are much more complex. The huge differences in its value over time and space are well recognized, but there are other characteristics that also affect its value. This section seeks some understanding of these other dimensions of electricity service, namely, supply resilience, reliability, and power quality.

3.1 Resilience and Reliability

"Resilience" has emerged as the major driver for microgrids in the United States and Japan.[1] This focus comes in the wake of severe natural disasters the two countries have experienced and growing concern about the extreme weather events climate change will spawn. While the focus here is on the United States and Japan, these concerns are naturally shared globally, but have not generally quite yet been brought into such clear focus elsewhere (The Royal Society Science Policy Center, 2014). A widely accepted definition of resilience appears in US Presidential Policy Directive 21 (The White House, 2013): "… ability to prepare for and adapt to changing conditions and withstand and recover rapidly from disruptions." While this definition does not make explicit that resilience is measured relative to a major catastrophe, this attribute is quite clear in The Royal Society report.

The current interest in microgrids is motivated, in no small measure, by the promise that they have a better chance than the megagrid of delivering power

---

1. The difference between "resilience" and "resiliency" is subtle and they are used almost interchangeably. Resilience tends to refer to the quantifiable ability to withstand and recover from impacts, and will be used here. Resiliency is rather the quality of being resilient.

during a disaster, and/or they can recover faster. To achieve this, however, they must have a fuel source or adequate storage (Kwasinski et al., 2012). As described later, this has indeed been true in some notable cases. Another sector that needs power under highly adverse conditions is the military and, sure enough, it has shown great interest in microgrids. In the United States, particularly, a significant effort is being made to harden power supply to bases using microgrids, primarily under a program called Smart Power Infrastructure Demonstration for Energy Reliability and Security (SPIDERS), which is an excellent acronym/backronym.

By contrast, "reliability" focuses on the statistical expectation of power availability, and extreme events are often excluded from the calculation. Unlike resilience, reliability has been a paramount concern for power systems since their inception. The familiar dominant effect that the limited and costly storability of electricity has had on the economics of the commodity, that is, demand cannot usually be leveled by inventory at a utility scale, also has significant reliability implications. Backup generation resources and other measures, such as demand curtailment, must be primed and ready to be quickly called into service in case of supply shortfall. AC power systems are additionally vulnerable because of stability problems; that is, following a perturbation in system frequency, the system may not return to a stable steady state condition, leading to cascading failure.

Several widely used reliability metrics have been applied to power systems, but the two shown in Table 3.1 are the most used. System Average Interruption Duration Index (SAIDI) shows the average number of sustained interruption minutes per consumer during the year and System Average Interruption Frequency Index (SAIFI) shows the average number of interruptions per consumer (IEEE, 2012).

As is shown in Table 3.1, the United States generally has poor reliability performance, compared to other developed countries, although there is a wide range within the European Union. Japan was once viewed as having the most reliable electricity service in the industrialized world and, before the 2011 GEJE, it had been fairly consistently excellent for 2 decades.[2] Currently, about the best claimed performance in the world is Singapore. It reports a SAIDI of less than a minute per annum since 2008, and at such a high SAIFI, an average Singapore resident should experience an outage only once every 100 years. Nonetheless, some commentators propose pushing toward even higher levels of reliability, say 30 ms of outage per year (Galvin et al., 2009). The key question, of course, is: can the costs of providing the Singapore level of service, or even

---

2. Recently, Japan has performed similarly to the best in Europe, but as is clear in Table 3.1, the 2011 earthquake and tsunami temporarily upended Japan's long reliability run. Its system results have now returned to pretsunami levels. Bear in mind that the disaster not only caused significant direct damage, but also, all of Japan's nuclear capacity was idled as a result, particularly over the mid-2012 to mid-2015 period.

TABLE 3.1 Reliability Indices for Selected Countries, 2012 (Including Exceptional Events)

Country	Unplanned SAIDI (annual minutes of outage)	Unplanned SAIFI (annual number of occurrences)
USA	157[a]	1.4[a]
Poland	254	3.4
UK	68	0.7
France	63	0.9
Germany	17	0.3
Denmark	15	0.4
Luxembourg	10	0.2
Japan	14[b] 514[c] 16[d]	0.1[b] 0.9[c] 0.2[d]
Singapore	0.4	0.01

There are variations in the way these indices are estimated, hence data may not be fully comparable; for example, the exclusion of short outages, typically less than 5 min, is a notable source of inconsistency.
[a]2009.
Japanese fiscal years (Apr.–Mar.), [b]2009–10, [c]2010–11, and [d]2013–14.
Sources: Eto et al. (2012), CEER (2014), FEPC (2014), and Yoon (2015).

better, be economically justified? Microgrid thinking says no, mostly because only a small fraction of loads really demands this level of service, and these end uses are better served in other ways, such as securing backup power supplies, storage, or a self-managed microgrid employing HePQR.

While certainly expertise, technology, and available resources drive reliability, these divergent levels are, in large part, societal choices. Conditions in Singapore naturally favor highly reliable supply, that is, small compact system (as is Luxembourg's), all underground lines for aesthetic reasons, plus a relatively benign climate, with low seismic risk. Nonetheless, the country has striven to provide exceptionally reliable power, by, for example, maintaining a high generation reserve margin. Countries with the highest reliability generally have high electricity costs.

The United States, Japan, and Europe all deliver high levels of reliability, but they can never meet the requirements of the most demanding loads. For example, a nation wouldn't feel safe knowing its military would be paralyzed by a blackout, hence the interest on the part of the military to provide its own power sources. Codes often require alternative sources for critical civilian loads, such as backup generators for hospitals, banks, data centers, stock exchanges, grid operation centers, communication infrastructure, etc. Not surprisingly, it is exactly these market segments where microgrids have gained the most traction.

3.2 PQR Pyramid

Reliability can be thought of as one characteristic of electricity, among several that quantify the quality of power delivered, with voltage stability, harmonics, and others, being decreasingly discernible to the customer. In general, available data on these characteristics remain limited, although the arrival of advanced metering infrastructure (AMI) has dramatically increased the potential for collecting it. In Fig. 3.3, a conceptual schematic shows the overall power quality and reliability (PQR) of electricity on the y-axis. Demanding loads, such as military communications, appear high up, with discretionary loads at the base of the pyramid.

In practice, the grid's reliability cannot be perfect, hence a PQR cut-off must effectively be chosen, and loads above that level are on their own, in order to ensure procurement of the balance of gourmet power they crave. In other words, the effective cut-off at the pyramid's peak varies significantly across the developed world. Note that some end uses, such as pumping, have low or even negative PQR requirements. The growing electric vehicle charging load is particularly notable in this regard because it is potentially beneficial to the grid. Electric vehicles can provide storage and/or ancillary services

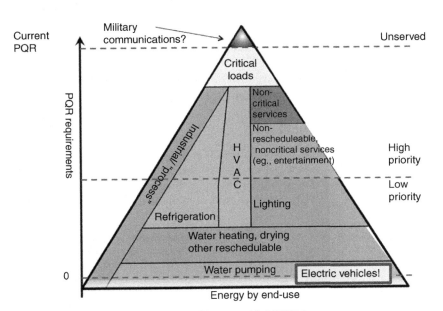

FIGURE 3.3   A PQR pyramid. *(Source: Marnay and Lai (2012).)*

by potentially discharging some of their stored energy back to the grid during emergencies. These topics are further described in chapters by Rowe et al. and Grozev et al.

This pyramid shape is inspired by the food pyramid, which entreats us to keep unhealthy foods at the top to a minimum, while nibbling freely on foods at the bottom. If electrical loads were classified in this way, it should be obvious there are demanding/costly ones at the top of the pyramid that we should try to keep to a minimum, while others should be pushed downward. Disaggregation of load can take place at four levels: the customer, the circuit, the device, or within equipment. While the final level might seem strange, consider that multiple devices, motors, controls, thermostats, etc., are often employed within a piece of equipment. Typically, the truly critical load, such as the thermostat in a refrigerator, is a negligible part of its total energy use, dominated in the refrigerator case by its compressor, which would appear down at the bottom of the pyramid.

Two aspects of this pyramid are noteworthy:

- First, to the extent the quality of power delivered can match the requirements, the more likely it is to be cost efficient; that is, don't squander gourmet power suitable for top-of-the-pyramid use on gourmand loads at the base.
- Second, this view of electricity is totally different from the way we see it today. Rather, the current utility paradigm delivers virtually the same high quality of power to *all* loads in *all* places, at *all* times; that is, it commoditizes electricity. This clearly is economically inefficient and expensive, since not all loads require a high level of reliability, as previously noted.

Microgrids offer an effective solution to the industry's traditional "one size must fit all" dilemma by allowing high value loads to secure much higher levels of PQR without raising, and potentially lowering, the costs to the more mundane loads that can survive with mediocre or low service levels.

3.3 Megagrid Treadmill

In fact, the drive to higher and higher levels of PQR is something of a treadmill for the megagrid, as shown in Fig. 3.4. Assume a similar index on the x-axis as seen on the y-axis of Fig. 3.3, for example, nines of availability, while cost appears on the y-axis.[3] First, consider the cost of unreliability curve, which falls

---

3. Nines of availability are the most common way of expressing reliability, in which 90% availability is equivalent to *one nine*, 99% to *two nines*, 99.9% to *three nines*, etc. In Table 3.1, for example, the average US SAIDI of 157 min lies between *three and four nines*, while Singapore exceeds *six nines*. The *Six Sigma* manufacturing methods, common in business literature, aim to achieve better than *six nines* of fault-free production. The Wikipedia entries are quite good for both nines and Six Sigma: https://en.wikipedia.org/wiki/High_availability and https://en.wikipedia.org/wiki/Six_Sigma

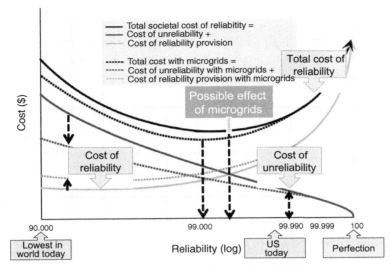

FIGURE 3.4   **The megagrid treadmill.** *(Source: Marnay and Lai (2012).)*

from the very high cost consequences of an unreliable grid to zero cost, if we had an infallible one. On the other hand, the cost of reliability shows the expenditure necessary to keep the megagrid to any level of performance. This curve must rise asymptotically as we approach grid perfection, although, of course, improving technology pulls the whole curve downward.

The cost of providing reliability actually has two components, one physical/direct/visible and one market/indirect/invisible. The physical part consists of undergrounding power lines, better quality equipment, more generation redundancy, etc.; we have a reasonable understanding of these costs. But the indirect part may well be more significant, coming from the conservatism with which the megagrid is run, driven by fear of service interruption. It comes primarily in the form of lost trade opportunities, precluded by overly tight standards. Nowadays, there is also carbon cost from resistance to accepting high penetrations of grid-hostile renewables into the mix. The sum of these two components, concrete and conservative, forms the societal cost of reliability, and the optimal level lies at its minimum. In this case, Fig. 3.4 is drawn provocatively with the optimal level below the three to four nines or so that the US megagrid achieves today.

The dashed curves show how the result might be changed by the existence of microgrids, which serve sensitive loads locally. They effectively lower the social cost of poor PQR, and let us assume there is some increase in the cost of providing reliability. As drawn, the sum of the dashed curves results in a shift of the optimal point, even further left to lower levels of PQR.

This exercise leads to a surprising conclusion. If microgrids exist in the distribution system and provide for sensitive loads locally, the PQR burden on the megagrid is lightened. In other words, the megagrid and its customers benefit, as well as the members of the microgrid. Further, the megagrid may capture a benefit far larger than captured within the microgrid. In other words, rather than being a gated community luxury, microgrids might deliver across-the-board power sector benefits.

The previous argument has become much more compelling because of the drive toward grid decarbonization. Much hand wringing results from the dread of a megagrid dominated by variable, nondispatchable sources. Similarly, volatile electricity markets are counter to a tightly controlled, highly reliable grid. Add to these concerns constraints on expansion of grid capacity because of costly rights of way, nimbyism,[4] etc., and it's clear the traditional high PQR paradigm is under threat. To the extent that microgrids provide high PQR local to loads, these problems are mitigated. Further, other power policy goals, such as decarbonization, become much more attainable as a result.

The role of microgrids in providing resilience and PQR where it is most needed is a complex story. To recap the main points:

- Resilience is currently the main driver for microgrid development. In other words, microgrids are seen as a way of delivering power during extreme conditions, and/or quickly restoring it afterward.
- While the requirements of various customers, loads, and even devices within pieces of equipment are wildly heterogeneous, the legacy megagrid operates on a universal PQR paradigm. It wastefully delivers unnecessarily high quality power to the majority of loads that simply don't need it, while the most demanding loads can never be served by any reasonable universal quality standard.
- If microgrids could serve sensitive loads in a local targeted manner, requirements on the megagrid could be relaxed, and costs reduced, potentially significantly.
- Importantly, but often overlooked, lowering PQR requirements on the megagrid would enhance its ability to pursue other pressing policy goals, notably decarbonization.
- Within a microgrid, emerging technology can supply HePQR, including DC, to various loads, in keeping with their requirements, generating further societal benefits. One notable example of modern technology being applied in this manner, the Sendai Microgrid, is described in Section 5.
- While resilience is now receiving considerable research and policy attention, not so PQR, and our understanding of the potential benefits of HePQR remain minimal. The HoPQR paradigm is deeply ingrained. The very concept of improving PQR is understood to be one of lifting all boats, regardless of whether the occupants want or can even benefit from the uplift. Scant attention is paid to the importance of choice.

---

4. NIMBY, not in my back yard.

4   WHAT IS A MICROGRID?

As is common in any young research field, universal definitions have been slow to emerge among the microgrid community, with the typical result that every analyst seems to have his or her own. Two formally established definitions are the following:

> *"A microgrid is a group of interconnected loads and distributed energy resources within clearly defined electrical boundaries that acts as a single controllable entity with respect to the grid. A microgrid can connect and disconnect from the grid to enable it to operate in both grid-connected or island-mode."*
>
> US Department of Energy Microgrid Exchange Group
> Ton and Smith (2012)

> *"Microgrids are electricity distribution systems containing loads and distributed energy resources (such as distributed generators, storage devices, or controllable loads) that can be operated in a controlled, coordinated way, either while connected to the main power network or while islanded."*
>
> CIGRÉ (2015)

Note that these definitions are quite simple and general, and both have the two distinguishing characteristics of a microgrid noted earlier:

- First, it's a locally controlled power system.
- Second, it can function either grid-connected or as an electrical island.

Despite the "micro" prefix, and the further emphasis noted in Fig. 3.2, the essence of microgrids is not scale. Notably, relatively small-scale power systems, such as remote communities, usually do not meet these requirements and are not microgrids as the term is defined here. Making this line fuzzier is the fact that new technologies for permanently islanded systems and microgrids are similar, and it is the emergence of these same technologies that is at the heart of the microgrid revolution taking place.

And make no mistake, this really is a revolution. While it may not sound that radical to nonelectrical engineers, the notion of control being distributed around the power grid actually represents a major paradigm shift. In the legacy grid, control is highly centralized, and the loads are mostly passive. Of course, there are many ways in which the smart grid is changing power delivery, and distributed control is only one. It can be thought of as one of three legs to the smart grid stool. The other two are customer interaction (AMI, etc.) and improved high voltage grid management (synchrophasors, etc.). So, these definitions provide more than just a classification for a microgrid—rather they actually say something profound about the evolution of electricity supply.

When presented with a definition, reflection on what it does not say is often as illustrative as what it does say. Please consider the omissions noted:

- There is no specification of size. Usually, analysts tend to assume microgrids are from a few hundred kilowatts to a few megawatts in scale, but there is no dividing line. Some large islandable systems, such as college campuses, are legitimate microgrids according to these definitions; at the opposite end of the size spectrum, so are single buildings able to function as islands.
- Both definitions clearly come from a power perspective; for example, an islandable boiler and heat distribution network probably cannot be called a microgrid. On the other hand, many current microgrids are built around combined heat and power (CHP) systems, but possibly with local renewables and other resources, notably some successful campus microgrids, such as Princeton University. The economic and resilience attractiveness of these installations make them highly promising early microgrids. In other words, microgrids may well be integrated energy systems of some kind. They do not have to be, but economics tends to favor them because, for example, heat storage is cheaper than electrical storage.
- Nor do the definitions specify any technology or application of technology. Microgrids must be locally controlled, grid connected, and able to island, but how control, disconnection, reconnection, etc., should be organized is not specified. These choices can have a major effect on the performance and cost effectiveness of microgrids. While the microgrids we've seen to date tend to apply modern or even highly sophisticated technology, the door is open for low-tech microgrids, too. For example, these might be in relatively undeveloped regions, or may employ low-grade resources, such as biogas, that would not be considered good enough for a utility application. Further, the definitions make no mention of renewables, storage, load control, AMI, etc., which many analysts tend to assume will be found in microgrids.

5   MICROGRID EXAMPLES

Three notable microgrid examples are presented in this section. Since US microgrid thinking is so driven by the Hurricane Sandy experience, it's fitting to mention briefly, first, one of Sandy's microgrid stars, the Princeton University campus. Subsequently, examples of two main microgrid types are presented.

The Princeton University campus in central New Jersey islanded for a day and a half after Hurricane Sandy hit. In addition to the 15 MW CHP plant shown in Fig. 3.5, the campus has a 5 MW PV array and a comprehensive strategy for load prioritization during islanding, characteristics making it a μgrid. As a 2014 news article on the system clearly shows, Sandy was a true turning point in the recognition of the benefits of microgrid systems such as Princeton's. It has actually had these capabilities for some time, but they were never showcased, and their benefits not fully appreciated (Kelly, 2014). Another Sandy star campus CHP system was the New York University campus around Washington Square in Lower Manhattan.

FIGURE 3.5 **The combined heat and power plant that served Princeton University campus for 36 h following Hurricane Sandy.** *(Source: Chris Marnay.)*

Sections 3 and 4 have established definitions and terms for classifying microgrids. Two terms were introduced to describe the two major types of microgrid of most immediate relevance, and an example of each is described.

- A mgrid (milligrid) is a fragment of the legacy-regulated distribution network that might island when economic, reliability, resilience, or other conditions warrant. The Borrego Springs project is chosen as an example mgrid.
- A μgrid is akin to a current customer, that is, downstream of a single (or at least very few) meter(s). For example, the widely known Sendai Microgrid, is a μgrid close to the GEJE epicenter with a strong HePQR component, which is one focus of this chapter.

The fact that the Borrego Springs demonstration is being led by a DisCo is significant. This type of microgrid involves legacy rate–based utility assets, which has significant implications. Borrego Springs lies in the far northeastern corner of SDG&E's service territory, at a small remote community surrounded by the Anza-Borrego Desert State Park, as shown in Fig. 3.6. This is a hot location that has recorded a 50°C temperature, and the long-run average temperature for Jul. and Aug. is over 40°C.

The first phase of this project was funded under the US Department of Energy's Renewable and Distributed Systems Integration (RDSI) program (SDG&E, 2014). The goal for RDSI's projects was peak feeder load reduction by 15%. Nonetheless, SDG&E chose the Borrego Springs site largely because of the reliability problems experienced there. The 3500-strong village lies in a valley bottom, at the end of a vulnerable 32 km, 28 MW feeder. Two 1.8 MW mobile gensets were installed at the substation as shown in Fig. 3.7, as well as a 1.5 MWh, 0.5 MW Li-ion battery.

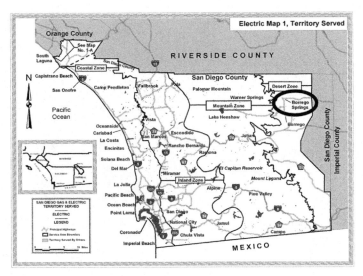

FIGURE 3.6 **The Borrego Springs microgrid is in the far northeast of SDG&E service territory.** *(Source: California Public Utilities Commission.)*

FIGURE 3.7 **The two 1.8 MW Borrego Springs microgrid diesel gensets.** *(Source: SDG&E.)*

Note in Fig. 3.7 the exceptional spill control measures taken. This is an environmentally sensitive area, and additional requirements were imposed accordingly. During the RSDI demonstration, several islanding events occurred, both planned and unplanned. The most notable occurred on Sept. 6, 2013, when a severe thunderstorm hit the area. The 69 kV tie line was lost at 14:20, and the community went black for 3 h. After some manual reconfiguration of the substation, the engines were started and some power restored. It was less than 1 MW at first, but rose to 2.5 MW for the final 8 h of the 24 h blackout.

Not surprisingly, this area has an excellent solar resource, and there are two merchant PV arrays in the area, one 26 MW and a second of 5 MW. Note that the tie line is problematic in both directions. Its vulnerability represents a demand-side problem, and also its limited ability to export all the output of the two solar installations creates supply-side congestion. SDG&E has started

FIGURE 3.8 Original Energy Center at the Sendai Microgrid. (A) Photovoltaic panels, 50 kW; (B) gas engine genset, 350 kW × 2; (C) molten carbonate fuel cells, 250 kW; (D) dynamic voltage restorer #1, 600 kVA; (E) dynamic voltage restorer #2, 200 kVA; (F) Building #1, integrated power supply; and (G) building #2, back-to-back voltage source converters (for test equipment). *(Source: NTT Facilities.)*

a 4-year second phase to the project. Three additional batteries have been installed, and much more sophisticated segmentation and reconnection of the local distribution feeders are planned. The goal is to provide service to a few critical centers during an outage.

The Sendai Microgrid shown in Fig. 3.8 was one of four New Energy and Industrial Technology Development Organization (NEDO) microgrid demonstration projects conducted between 2006 and 2008. This project was intended to demonstrate the delivery of multiple power qualities to various circuits on the small Tohoku Fukushi University campus and nearby municipal facilities, a school, and a water treatment plant. The campus trains caregivers, especially for geriatric care, and has both a senior residence home and a hospital. An Energy Center was constructed, as shown in Fig. 3.8. A control center was installed and techniques for practical local heterogeneous power supply systems developed. This was the only one of the four NEDO demonstrations to continue beyond the 2008 end of the program. CHP was added to the campus, rectifying one of the key limitations of the NEDO program design, and the city buildings were dropped. These changes were sufficient for the University to find it economic to continue operation, and it was still functioning in this mode at the time of the GEJE.

Parts of Sendai were severely affected by the GEJE, although damage was minimal in the immediate area of the microgrid. Further, the city was well prepared for earthquakes in many ways. Particularly, it has a hardened natural gas supply infrastructure, which enabled import of natural gas from Niigata Prefecture. Together with the good design, excellent skill, and creative improvisation of onsite technicians, this asset enabled the Sendai Microgrid to perform well during GEJE's aftermath. This seminal experience is described in more detail elsewhere, but it spoke loudly for the advantages of robust localized power systems, and microgrids in general (Marnay et al., 2015). When the earthquake

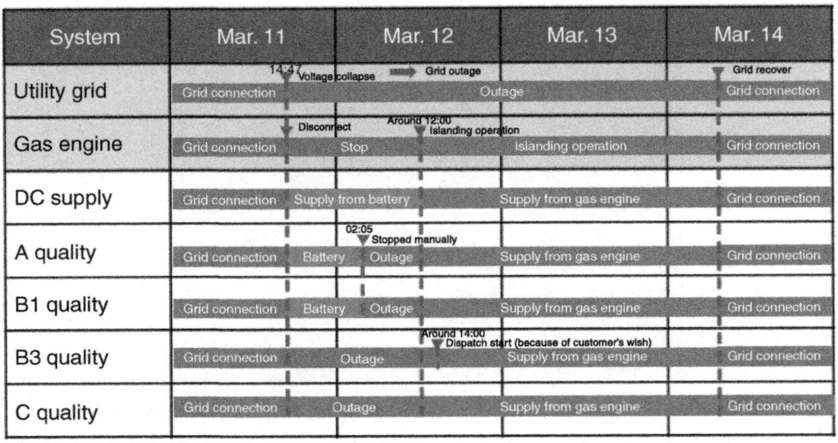

System	Mar. 11	Mar. 12	Mar. 13	Mar. 14
Utility grid	Grid connection	14:47 Voltage collapse ➡ Grid outage — Outage	Outage	Grid recover — Grid connection
Gas engine	Grid connection	Disconnect — Around 12:00 Islanding operation — Stop	Islanding operation	Grid connection
DC supply	Grid connection	Supply from battery	Supply from gas engine	Grid connection
A quality	Grid connection	Battery — 02:05 Stopped manually — Outage	Supply from gas engine	Grid connection
B1 quality	Grid connection	Battery — Outage	Supply from gas engine	Grid connection
B3 quality	Grid connection	Outage	Around 14:00 Dispatch start (because of customer's wish) — Supply from gas engine	Grid connection
C quality	Grid connection	Outage	Supply from gas engine	Grid connection

FIGURE 3.9 Operation of the Sendai Microgrid during the 2011 Japanese earthquake and tsunami. *(Source: NTT Facilities.)*

occurred, Tohoku Power Co. (ToPo) stopped supplying power to the area surrounding the campus, resulting in a 3-day outage. Nevertheless, the microgrid was able to supply power continuously to some small critical loads within the campus, and provided full heat and power service for almost 2 days.

The microgrid operations timeline is presented in Fig. 3.9. Following the GEJE, megagrid service was lost for almost 3 days. During the blackout, various qualities of service were provided for differently, which is the strength of this microgrid. The DC circuit was never dead, since it has battery backup. The A and B1 circuits were also served by battery for several hours, but then went black. A natural gas genset was restarted about a day into the blackout, when all circuits were restored. Thus the University's hospital, rest home, and other facilities were partially served and were available with near full functionality. Surprisingly, this microgrid maintains no onsite backup fuel, so its resilience rests on its lifeline. Natural gas service to the microgrid was not interrupted as supply was available from the nearby undamaged Saiwaicho storage facility until access to import from another prefecture was established.

6 THE EMERGENCE OF MICROGRIDS CALLS FOR NEW THINKING ON REGULATION AND POLICY

Many of the regulatory and policy challenges posed by the emergence of microgrids are quite familiar to analysts of distributed self-generation by current utility customers, such as rules and pricing for interconnection and power export. These issues are not discussed in this chapter, but this does not suggest they are unimportant, just well covered elsewhere. Other less discussed issues, such as the benefit microgrids can provide by managing difficult resources locally, are also not addressed here.

The evaporation of revenue generating load to efficiency and self-generation, interfuel competition, and other difficulties confront incumbent utilities and their regulators. The combined effect constitutes the death spiral described in the chapters by Sioshansi and Athawale and Felder, with the relentless decay of the traditional DisCo business model well covered in Sioshansi's chapter, as well as others. The emergence of microgrids is covered, using the example of the University of California San Diego campus, which is able to function as an island, albeit with much reduced service levels. Indeed, because microgrids provide multiple benefits tending to make electricity self-supply more attractive, they are indeed an enabling technology for further leakage of electricity sales powering the spiral (Bronski et al, 2014, Glick et al 2014, Asmus, 2015).

The recent traumatic disasters will likely move the regulatory needle toward greater investment in resilience and socialization of the resulting costs in rates, but it seems unlikely policy will move significantly in that direction. In other words, the traditional regulatory pact is unlikely to deliver significant improvement in resilience, while microgrids demonstrably can. And importantly, on the cost side of the scales, the benefit of investment in resilience and reliability is highly uneven. While some services are critical, such as first responders, not only are others less so, but there is, in fact, a whole spectrum of load requirements. Hardening the megagrid until all end uses, regardless of importance, are served with the same resilience and reliability as the most critical ones is clearly economic folly. Currently, developed societies implicitly determine the societal level of service now, and clearly otherwise similar countries appear to come to quite different conclusions.

But the implications of microgrid emergence have other dimensions. The excellent performance of some microgrids, for example, Princeton and New York Universities in the United States, and Sendai and Roppongi Hills in Japan, during the twin catastrophes, opened the eyes of regulators to a key fundamental limitation of the legacy megagrid, its inherent vulnerability. While much can be done to improve the resilience of the megagrid, evaluating investments to mitigate statistically rare events is no easy challenge. Simple economics doesn't work as well when these considerations enter the picture. We don't have good intuition for evaluating the costs and benefits of mitigating the effects of rare catastrophic events, and even normal probabilistic reliability indicators are not easily applied. Feelings about resilience and reliability are much more visceral and heavily influenced by recent experience. Hence the seminal importance of the GEJE and Hurricane Sandy, versus the quite different perspectives toward microgrids seen around the developed world.

Solving this problem requires two steps:

- The first is to separate the high voltage meshed grid, roughly upstream of the substation, from the network extremities below. Upstream, the basic heterogeneous model still holds.

- The second is the type of reform of the distribution network driving New York's Reforming the Energy Vision process, further described in the chapter by Jones et al.

Countries with fully developed high voltage grids, which include many middle-income countries and not just the rich ones, are unlikely to abandon them, or change their operating approach radically. In many ways, life in this sector may remain relatively unchanged, although the death spiral of falling throughput and the technical challenges of the "duck curve" remain (CAISO, 2013). The big regulatory pill that must be swallowed in the meshed grid is acceptance that reliability simply isn't an absolute, as indeed few goods are. Reliability is a normal good, in the economic sense; that is, the more of it we have, the happier we should be—if it weren't for the costs. The economic problem is rather one of budget and priorities. What is the optimal level of reliability to provide to the substation, given that current technology allows us considerable leeway to establish varied levels of service that matches the actual end-use requirements, and to escape the tyranny of universal service?

Downstream of the substation, the winds of change are blowing the hardest and the coldest. On the one hand, HePQR promises both enhanced service to sensitive loads, and a more economic paradigm than the homogeneous one DisCos and regulators are locked into. True microgrids, μgrids, are well positioned to take advantage of the HePQR opportunity, but many barriers stand in the way of milligrids, mgrids, doing so. The notion of universal service is deeply ingrained in regulation, technical standards, and by sheer inertia. While legacy rate-based assets could conceivably be devolved into a milligrid, the path looks very steep. Here is where, perhaps, a new entity may emerge. For example, could a community power supplier take over the distribution assets of a "utility" and run them along HePQR lines?

While much can be done to improve megagrid reliability, as Singapore has clearly demonstrated, the key questions are "at what cost," "what fraction of loads really needs it," and "who should pay for it?" Sadly, the status of research and policy debate on these questions remains thin, which keeps utilities on the treadmill; that is, only improvements in PQR can be considered. And, in fact, estimating the benefit of lowering PQR is no trivial task. There are both equipment and operational considerations, and we really do not have an established method for measuring the effects of PQR variation.

There are additional complications. For example, when an entity steps in to run a local microgrid, should there be a penalty or a subsidy? On the one hand, as long as the microgrid remains grid connected, leaning on it for backup and/or other service beyond the level it finds attractive to self-provide, a penalty seems appropriate. In this sense, a microgrid is no different from any other prosumer, and contributes to the death spiral equally. That is, infrastructure being provided by the DisCo, as well as upstream in the high voltage grid, cannot be sustained without adequate sales to recoup fixed costs. Without the necessary net sales, investments become stranded assets burdening other ratepayers. Conversely, if the microgrid is providing a valuable service, such as providing high quality service to a sensitive load like a hospital, is it lowering the DisCo's burden, for example, by releasing otherwise sacrosanct neighborhoods to rolling blackouts, and thereby merits a subsidy?

What is quite clear is that the current push of DisCos to recover more of their costs through fixed charges is likely to be counterproductive, encouraging disconnections, and disadvantaging both parties. Simple economics says microgrid disconnection is unlikely, simply because self-supplying a load unleveled by the law of large numbers benefiting the megagrid is an expensive proposition. However, if the megagrid makes itself an unnecessarily expensive luxury, as in the Australian case described in the chapter by Nelson and McNeil, the death spiral will be accelerated.

As further argued in the preceding chapter by Gellings, the megagrid offers many services that are potentially beneficial to microgrids. If these services, such as backup or wheeling among microgrids, can be isolated and offered individually or in packages, they may well be attractive to prosumers and microgrids. Or, as described by Gimon, providing extra reliability or supply security could be offered as a service to consumers who are willing to pay a premium for it. Is this not the essence of the New York's Reforming the Energy Vision initiative, further described by Jones et al.?

Knieps offers some ideas regarding the way in which services provided by microgrids can or should be valued and priced transactionally, as do others, including Covina et al., who look at the intersection of wholesale and retail, transmission and distribution, generation and storage, both centralized and distributed. In the same vein, Cavanagh and Levin suggest that electricity service is not a one-dimensional "commodity" to the exclusion of its other attributes. Clearly, there is a lot more to the microgrid story than can be covered in this chapter.

What is the correct regulatory treatment for a milligrid, such as Borrego Springs? On one hand, a higher level of service is being provided to local customers, and they should be responsible for its incremental costs. On the other, hardening this remote vulnerable node of the distribution system frees SDG&E assets for emergency restoration, thereby benefiting other customers. Restoration following the Sep. 2013 flood involved 200 utility staff!

7   CONCLUSIONS

Microgrids are rapidly emerging entities that currently lack a seat at the regulatory and policy table, although this is changing. Regions that have experienced natural disasters in which the resilience of microgrids was convincingly demonstrated are the first to provide one. The important consideration that might easily be missed is that the effect of microgrids will be felt not only by its members/loads, but also by the Mgrid (megagrid). While the well-recognized death spiral caused by departing load has been much discussed, the potential benefit to the

Mgrid from microgrid control of relatively small local resources and PQR of service, potentially at the level of loads or even below, is underappreciated.

Providing a quality of power that matches the requirements of loads is relatively little discussed and yet potentially a big win. While our familiar "one size fits all" HoPQR service provides obvious economies, when the range of comfortable sizes is so wide, the benefits of tailoring the quality of power delivered are surely commensurately large. The concept of HeQR merits more policy research, both to assess its direct benefits, but also to determine whether reducing PQR requirements on the megagrid can deliver the significant cost savings intuitively available. This is particularly needed, given our strong desire to achieve ambitious policy goals for renewable penetration, competitive markets, etc. which are impaired by the megagrid PQR treadmill. In other words, the gains from getting off the treadmill are growing.

It should be added that, while the current HoPQR paradigm is one of uniform service provided at socialized costs, the reality does drift away noticeably from this norm. Service quality on Wall Street actually differs significantly from the experience of a residential customer at the end of a long rural feeder. Unfortunately, rather than having a rational process for establishing an appropriate level of service and corresponding cost, we currently have a squeaky wheel system. She/he who complains the most gets the best PQR and its cost socialized. At a minimum, a more equitable system wherein the highest PQR comes at the highest cost should be pursued. Even better would be an economically selected universal standard level of service, with subsidies for "customers," including microgrids that accept a lower level of service and drive down the cost of universal service to the benefit of all. Those loads that desire a higher level need to join a microgrid that provides it, or, even better, provides heterogeneous service options.

GLOSSARY

µgrids A *true microgrid* on a single customer site
AMI Advanced metering infrastructure
CAISO California Independent System Operator
CEER Council of European Energy Regulators
CHP Combined heat and power
CIGRÉ Conseil International des Grandes Réseaux Électriques
DisCos Electricity Distribution Companies, such as SDG&E
FEPC Federation of Electric Power Companies of Japan
GEJE Great East Japan Earthquake and Tsunami
HePQR Heterogeneous power quality and reliability
HoPQR Homogeneous power quality and reliability
Mgrid The legacy centrally controlled hierarchical *megagrid*
mgrid A *milligrid* or islandable segment of the legacy regulated distribution network
NEDO New Energy and Industrial Technology Development Organization, energy research and development arm of the Japanese economy ministry
ngrid A *nanogrid* or distinct (e.g. DC) local network of sources and sinks

PQR Power quality and reliability

RDSI Renewable and Distributed Systems Integration

SAIDI System Average Interruption Duration Index, average duration of interruptions per consumer during a year

SAIFI System Average Interruption Frequency Index, average number of sustained interruptions per consumer during a year

SDG&E San Diego Gas and Electric

SPIDERS Smart Power Infrastructure Demonstration for Energy Reliability and Security, the US military program for microgrid development on bases

ToPo Tohoku Power Company, the local vertically integrated supplier in the Sendai area

ACKNOWLEDGMENTS

Several recent research collaborations have contributed to this chapter. My understanding of the situation in Japan, as well as the nature of resilience, was greatly assisted by coauthors on a recent Power and Energy article, Hirohisa Aki, Keiichi Hirose, Alexis Kwasinski, Saori Ogura, and Takao Shinji. Work in this area has been supported by the National Science Foundation, the US Department of Energy Offices of Electricity Delivery and Energy Reliability and Building Technology, the US State Department, and the US Trade Development Agency. Many of the ideas and their graphical representations are derived from my earlier work with Judy Lai.

REFERENCES

Asmus, P., 2014. Utility Distribution Microgrids: Investor-Owned and Public Power Utility Grid-Tied and Remote Microgrids: Global Market Analysis and Forecasts.

Asmus, P., 2015. Microgrids: Friend or Foe for Utilities? Public Utilities Fortnightly, Feb. 2015. https://www.navigantresearch.com/research/utility-distribution-microgrids

Bronski, P., Creyts, J., Guccione, L., Madrazo, M., Mandel, J., Rader, B., Seif, D., Lilienthal, P., Glassmire, J., Abromowitz, J., Crowdis, M., Richardson, J., Schmitt, E., Tocco, H., 2014. The Economics of Grid Defection. Rocky Mountain Institute, Boulder, CO, http://www.rmi.org/PDF_economics_of_grid_defection_full_report.

California Independent System Operator (CAISO), 2013. What the Duck Curve Tells Us About Managing a Green Grid. http://www.caiso.com/Documents/FlexibleResourcesHelpRenewables_FastFacts.pdf

CIGRÉ, 2015. Microgrids 1: Engineering, Economics, and Experience, Working Group C6.22 Microgrids Evolution Roadmap, October.

Council of European Energy Regulators (CEER), 2014. Benchmarking Report 5.1 on the Continuity of Electricity Supply: Data update. http://ceer.eu/portal/page/portal/EER_HOME/EER_PUBLICATIONS/CEER_PAPERS/Electricity/Tab3/C13-EQS-57-03_BR5.1_19-Dec-2013_updated-Feb-2014.pdf

Cuomo, A., 2015. Governor Cuomo Announces Launch of $40 Million NY Prize Microgrid Competition. Available from: http://www.governor.ny.gov/news/governor-cuomo-announces-launch-40-million-ny-prize-microgrid-competition

Emerge Alliance, 2015. http://emergealliance.org/Standards/OurStandards.aspx

Eto, J.H., Hamachi LaCommare, K., Larsen, P., Todd, A., Fisher, E., 2012. An Examination of Temporal Trends in Electricity Reliability Based on Reports from U.S. Electric Utilities. Lawrence Berkeley National Laboratory, Berkeley, CA.

Federation of Electric Power Companies of Japan (FEPC), 2014. FEPC Infobase. http://www.fepc.or.jp/library/data/infobase/pdf/infobase2014.pdf

Galvin, R., Yaeger, K., Stuller, J., 2009. Perfect Power: How the Microgrid Revolution Will Unleash Cleaner, Greener, and More Abundant Energy. McGraw Hill, New York.

Glick, D., Lehrman, M., Smith, O., 2014. Rate Design for the Distribution Edge. Rocky Mountain Institute, Boulder, CO.

IEEE, 2012. Guide for Electric Power Distribution Reliability Indices. Institute of Electrical and Electronic Engineers Standard 1366-2012.

Kelly, M., 2014. Two Years After Hurricane Sandy, Recognition of Princeton's Microgrid Still Surges. News at Princeton, posted October 23, 2014; 02:00 p.m. https://www.princeton.edu/main/news/archive/S41/40/10C78/

Kwasinski, A., Krishnamurthy, V., Song, J., Sharma, R., 2012. Availability evaluation of microgrids for resilient power supply during natural disasters. IEEE Trans. Smart Grid 3 (4), 2007–2018.

Marnay, C., Lai, J., 2012. Serving electricity and heat requirements efficiently and with appropriate energy quality via microgrids. Electricity J. 25 (8), 7–15.

Marnay, C., Aki, H., Hirose, K., Kwasinski, A., Ogura, S., Shinji, T., 2015. How Two Microgrids Fared After the 2011 Earthquake. IEEE Power and Energy Magazine, May/June.

Morris, G.B.C.W.Y., 2012. On the Benefits and Costs of Microgrids. M.S. thesis, Department of Electrical Engineering, McGill University, Montreal, December 10, 2012.

Navigant Research, 2015. Identified Microgrid Capacity has Tripled in the Last Year, press release available at https://www.navigantresearch.com/newsroom/identified-microgrid-capacity-has-tripled-in-the-last-year

San Diego Gas and Electric, 2014. Borrego Springs Microgrid Demonstration Project. Final Technical Report.

Smith, R., MacGill, I., 2014. Revolution, Evolution, or Back to the Future? Lessons from the Electricity Supply Industry's Formative Days. In: Sioshansi, F.P. (Ed.), Distributed Generation and Its Implications for the Utility Industry. Academic Press, Oxford, UK.

The Royal Society Science Policy Center, 2014. Resilience to Extreme Weather. https://royalsociety.org/policy/projects/resilience-extreme-weather/

The White House, 2013. Presidential Policy Directive—Critical Infrastructure Security and Resilience. https://www.whitehouse.gov/the-press-office/2013/02/12/presidential-policy-directive-critical-infrastructure-security-and-resil

Ton, D., Smith, M., 2012. The U.S. Department of Energy's microgrid initiative. Electricity J. 25 (8), 84–94.

Yoon, K.T., 2015. Powering the Nation: Smart Energy. http://www.i2r.a-star.edu.sg/horizons14/pdf/Powering%20the%20Nation%20(Smart%20Energy).pdf

Chapter 4

Customer-Centric View of Electricity Service

Eric G. Gimon

Energy Innovation LLC, San Francisco, CA, United States of America

1 INTRODUCTION

The book explores the implications for corporate strategy and public policy of electricity consumers becoming more active behind their own meters, in generation, storage, and demand management, that is, distributed energy resources (DERs). The rapidly falling costs of self-generation and smart systems, plus economic regulations that encourage decentralization, enable growing numbers of consumers to reduce their electricity bills. Through their decisions, consumers are often able to lower the system's avoidable costs. However, they may also avoid paying for the system's historic costs, an issue that raises the question of who will cover those expenses. As explored in other chapters in the book, this is an important question for the future of the grid, but also reflects a common mode of analysis where distributed resources are principally evaluated on the basis of their physical and financial impacts on the bulk grid, and the entities that manage it.

This chapter flips the traditional analysis on its head by looking instead at what the grid has to offer to customers empowered by DERs to keep them as customers. The most relevant context for examining this new analysis is one where DER costs are low enough that they make the case not only for customers to employ some mix of behind the meter resources with grid-supplied ones, but for customers to defect[1] from the grid altogether, if grid-supplied resources are unattractively priced. Today, this situation only holds true in a few rare localities, such as Australia[2] but is likely to become increasingly prevalent. Given that context, the structure of this chapter is organized around the following question: in the new world of DERs, *what are the services customers might be interested in getting from the grid, and*

---

1. Two terms of art now exist: "load defection" (Bronski et al., 2015), when customers choose to supply some fraction of their own electricity, and "grid defection" (Bronski et al., 2014), when they disconnect from the grid entirely. The context of interest here is one where load defection is more likely, but grid defection is possible.
2. See chapter by Nelson and McNeil about specific strategies of various major incumbents in Australia with the highest penetration of residential solar PV.

Future of Utilities - Utilities of the Future. http://dx.doi.org/10.1016/B978-0-12-804249-6.00004-X

Box 4.1

Our customers have opportunities to understand what great customer service is every day of their lives as they interact with Amazon or they go to Nordstrom or ... anywhere where they have the opportunity to experience another service provider's product or service.

They want flexibility. They want different tariffs. They want their bill options to be different. They want products and services. They might want rooftop solar. They want communication in the form that they want to be communicated with, and in a language that they understand.

—Lynn Good, CEO Duke

CS Week Keynote, May 27th, 2015

how should they be priced? As can be seen from Lynn Good's statement (Box 4.1), this is becoming a pressing issue.

To start answering this question, Section 2 looks at all the services or benefits that customers get from the grid. These are typically bundled into a few line items on a monthly bill or connection charge. The first task will be to unbundle them, and then reexamine them from the point of view of DER-empowered customers. Section 3 will tease out what services customers need in the new world of DERs. After considering some of the implications that flow from these new sets of services in Section 4, this chapter will then discuss in Section 5 different ways all these services might be priced, so as to address efficiently the needs of a variety of customers, while providing an optimal network for all, a "social good," as in the spirit of Gellings in the chapter Value of an Integrated Grid.

2 THE TRADITIONAL MODEL

Historically, customer desire has commonly been seen as electricity delivered dependently when they want, in any quantity, at a reasonable and predictable price. Traditionally, customers or ratepayers were seen only as passive consumers of power, disconnected from market dynamics, except as aggregate load, and free from worries about externalities. A customer's main lever to influence the service provider was choosing how much to consume, and electing a voice via his or her elected representatives in the legislative and regulatory arena. The only other alternatives to the traditional grid model were expensive diesel generators and lead-acid batteries for small-scale backup, or fairly inflexible generators for larger industrial customers looking to harvest heat for onsite consumption, as well.[3] The grid has had a lot to offer in terms of power costs, quality, and convenience. To understand how these play out today, it is worth unbundling this large set of traditional advantages. They can be loosely clustered around three themes (Fig. 4.1):

---

3. Here, we will focus on commercial and residential customers, but similar concerns apply for large consumers.

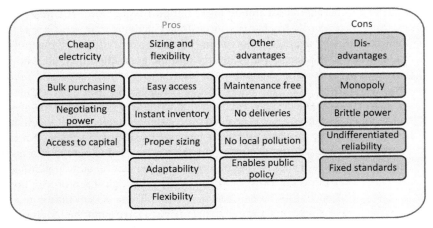

FIGURE 4.1   **Pros and cons of a traditional grid.**

- Cheap kilowatt-hour unit costs. The pooled purchasing of power allows for economies of scale from big plants, high-capacity transmission lines, and markets. Up until the end of the last century, the steady increase in power plant size brought lower cost electricity to customers. The grid also allowed a purveyor of electricity, like a utility, to purchase large amounts of cheap electricity through direct access to competitive wholesale markets, as well as through aggregate contract negotiations. Utilities then turned around and offered these low prices to their customers. Additionally, utilities could leverage their size and monopoly franchise to take on debt on behalf of customers, accessing long-term financing on more attractive terms than individual customers can.
- Sizing and flexibility. By aggregating customer demand, the grid is able to accommodate huge variations in customer behavior. That way, it can provide easy access, instant inventory, proper sizing, adaptability, and flexibility. Access means that, except for some rare cases, when customers first hook up to the grid they can immediately access the amounts of electricity they want.[4] For most existing properties, utility customers don't have to install new equipment—inventory—or take it with them when they sell their property or end their lease. A grid connection can be sized fairly easily and cheaply for whatever maximum power one might draw, even if that is much bigger than daily peak consumption. The grid adapts automatically for daily, seasonal, or annual variations in consumption; volumetric charges go down if use goes down due to natural factors (efficiency upgrades, kids move away, less demand at the restaurant, vacation, etc.).

---

4. There is also the concept of universal service, that is, that all customers who want to be served will be served. This is an important social good captured by the concept of social solidarity later.

Finally, and perhaps most importantly, the grid is flexible[5]: it adjusts in fractions of a second to changes in customer demand. The grid accomplishes all this by averaging fluctuating demand over millions of customers, and throwing some peak flexibility into the generation mix, all while maintaining (in developed countries at least) a very reliable product.[6]

- Other advantages. There are many other advantages the grid delivers, and services it makes unnecessary. A connection to the grid is relatively maintenance- and hassle-free; there is no need for consumers to clean their junction box, and the utility provides all maintenance on its side of the meter. When customers are exclusively using the grid, there is no need to purchase or arrange for delivery of fuel. While power generation can be quite polluting, it mostly takes place far away from customers (although it's probably not without consequences for human health, overall). There is very little equipment that can be stolen, and the meter and breakers have a minimal footprint on the customer's property. Finally, although the grid can push externalities unfairly onto others, it is also a powerful tool for promoting social solidarity and other public policy goals, and is seen as a benefit for the poor or otherwise disadvantaged (see chapter by Gellings).

This considerable package of services explains the dominance of the existing utility model for all but the most impractical off-grid applications. Yet, there are also several traditional downsides that should be noted (Fig. 4.1), clustered as follows:

- Monopolies. Most grid users have to deal with a monopoly, be it an investor-owned utility, a public-owned one, or some other arrangement like a rural coop. In some places, competition exists at the wholesale level or at the level of the particular retailer selling the wholesale power to the customer, but there still remains the natural monopoly over the poles and wires—the distribution grid—that delivers the electricity to customers. Regulation mitigates some of the effects of a monopoly regime, yet connecting to the meter is largely a "take it or leave it" proposition where many choices have already been set in stone. For example, most utilities don't provide many avenues for interested customers to market their load flexibility or trade against differences between high marginal costs and average costs.
- Brittle power. This nomenclature, coined by the Lovins' classic 1982 book of the same name (Lovins and Lovins, 1982), refers to the problem that, due to accidents, natural disasters, or malevolent intent, our power grids have a tendency to go down all at once. What might otherwise be a mere

---

5. Adaptability and flexibility are very similar ways to describe the same thing. Here, the two distinguish between shorter and longer time periods. Flexibility can economically be provided DER, adaptability less so, without major oversizing, and the ability to offload extra capacity.

6. At the transmission level, the grid typically aims for 1 h in 10,000 downtime. Some distribution circuits experience much more frequent outages, due to events like trees falling on distribution lines in storms and other weather related events.

inconvenience—the loss of power for several hours to several days—becomes a crisis when it happens to everyone else all around. Equipped, as most people are, to only draw power from the grid, most can't easily use alternative energy infrastructure, like natural gas or stores of energy from wood or diesel, to supply their electricity needs.

- Undifferentiated reliability. Grid-connected customers are unable to differentiate the level of reliability desired, or designate critical circuits that the grid would continue to serve in emergencies. This would be useful for a wide range of customers. For example, microfluctuations in voltage that happen over a handful of cycles are not always well understood, or managed, by the utility provider, yet can do a lot of damage to electronic equipment. Customers can purchase onsite battery-based systems that "clean" their power,[7] and provide backup. They can also purchase generators for longer outages, but all these have been fairly expensive. In part, this is because, in a reliable grid with one-way flow of services, these backup systems are redundant almost all of the time. On the flip side, the grid can't supply any kind of alternate rudimentary service during blackouts; even customers with photovoltaic (PV) solar systems cannot ride through these blackouts and continue to provide power without a battery and energy management system. People looking for more reliability than the grid has to offer can at least buy DERs to obtain it. Those who might want less reliability in order to pay less[8]—or because their existing DERs already provides a fair amount of reliability—are out of luck.

- Other disadvantages. Obviously, to use the grid, one needs to be near it and this is not possible for remote customers. There are other less obvious disadvantages of depending on the bulk grid. For example, customers are stuck with the AC voltage and frequency the grid provides, even if they want to feed DC appliances, or other AC formats. In general, customers lack direct agency, and have varying, generally insignificant, degrees of influence on what the utility does. Hence, if poor planning results in bad decisions, customers are stuck with them.

Altogether, the grid provides a remarkable panoply of useful services and advantages. These benefits more than justified living with the disadvantages of the traditional model. With the rise of cheap DERs, however, comes the need for some reexamination of what services customers might want, and of how they might be the best way to charge them for it.

---

7. Here, "clean" is electrical engineering parlance for electricity provided with an AC waveform free of harmonic distortions and abrupt changes in waveform, and doesn't refer to any pollution mitigated in its production.

8. The idea that electricity could be provided at lower reliability in return for lower cost is socially controversial in electricity circles: a minimum standard of reliability is seen as the only humane way to provide universal service. Paradoxically, some customers might effectively get more reliable service, on average, if they paid for a less reliable service. If the discount was significant enough, they would avoid facing disconnection and connection charges when they couldn't afford to pay their bills and, hence, have fewer service interruptions.

3 THE PROMISE OF A NEW WORLD

Imagine a new set of customers with access to attractively priced DERs like PV, batteries, fuel cells, smart load controls, or smart-charging electric vehicles, all behind the meter at their homes or small businesses. These resources are all integrated by an intelligent energy system manager or a microgrid (see Marnay's chapter on Microgrids), which can act as the customer's agent, or a group of customers' agent, with respect to the grid. Looking at the advantages and disadvantages of the centralized grid listed in Section 2, one might see how investment in such a system of DERs can compete with the grid in terms of advantages offered, especially regarding price, while mitigating some of the disadvantages of solely relying on the grid.

It is important to consider the *residual services* and *complementary services* that a utility or other party could offer via a grid connection in an integrated grid. Residual services are services that require no change in grid functionality—essentially what it has always offered and charged for. An example might be the provision of bulk electricity to fill in where DERs are not sufficient. Complementary services are new services that can be offered to customers with DERs that enhance the economic value of DERs by reducing their associated investment, or extracting more value out of them. An example might be an outside balancing service for a DER-centric home energy system.

Looking at the unbundled functionalities discussed in Section 2, a variety of residual and complementary services will emerge from what were interrelated traditional services. The exact difference between residual and complementary service is not particularly important; it is mainly a useful way to capture the idea that the grid will provide some of the same old services—although perhaps in new ways—while also providing entirely new services. This section will analyze all these services in functional terms before addressing their pricing in Section 5.

3.1 Low Cost Electricity from Everywhere

The rise of DERs provides a brand new source of cheap electricity at the service of customers. This will drive competition for bulk electricity services with the traditional power plants (Fig. 4.2). Distributed resources, such as fuel cells and PV panels, also benefit from scaling through mass production of the basic hardware, and mass adoption. Third parties can also give customers efficient access to capital markets by aggregating power purchase agreements or loan agreements.

As the grid loses market share in kilowatt-hour sales of electricity, its main residual service in this category is to provide electricity to those who can't afford DERs or whose energy needs aren't entirely covered by DERs. Customers may be underserved by DERs because they don't have access to capital, they find the next incremental purchase of DERs uneconomical, or they may

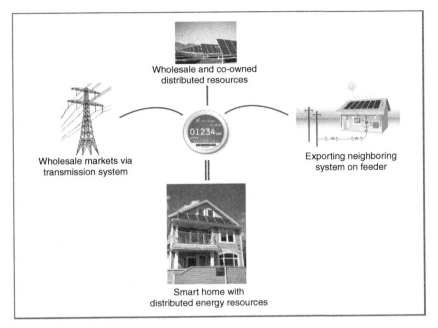

FIGURE 4.2   Electricity from everywhere.

just have insufficient local resources[9] to provide such an increment. The grid thus serves as a benchmark against which DERs compete to provide the most optimal electricity services.

At least two possible paths exist for how complementary services might emerge from the existing grid:

- First, the grid provides a conduit for DER owners to export their power for sale, and this helps make DERs more economical.
- Second, the grid can offer a variety of new channels for obtaining electricity through the meter, allowing customers to shop around for the best options. For example, customers could purchase electricity from nearby DER systems that export through the local distribution grid. The grid can also allow customers to purchase electricity through arrangements like solar gardens, or offer them generation from DERs owned in bulk by the utility or a third party in the same substation territory, that is, wholesale distributed power.

As a new complementary service in a highly distributed future, the grid would either allow customers to switch seamlessly to the cheapest available power, or just facilitate bilateral power purchase agreements.

---

9. For example, only a fraction, albeit a significant one, of homes and businesses have sufficient sunlit area onsite to provide for all their energy needs from solar.

3.2   The Grid as a Balancing Resource

The ability of the grid to provide adaptive and flexible service is where it really shines as an option for DER consumers, since a single grid tie, typically sized much bigger than any customer demand, allows customers to access any amount of energy they need, at any time. Even with storage and load control, distributed generation is likely to either underproduce or overproduce at some time during the year. The grid is so adaptable in terms of sizing of service, and offers so much flexibility in terms of power output, that it will likely almost always make sense to stay connected, if these size and flexibility services remain attractively priced.[10]

Residual balancing services are hard to describe or price, as size and flexibility benefits to consumers are typically paid for in a bundled manner, through electric bills. Their relatively low cost is effectively covered through fixed charges, demand charges, time-of-use charges, and volumetric charges. Demand charges seem the closest proxy for charging a customer for their balancing needs, but they tend to be more tied to cost incurred from peak system demand, than any direct measure of distribution constraints incurred by a particular customer. Variable sizing of a customer's connection to the grid, and its flexibility in meeting demand, is an important residual service that is likely to become more prominent, when compared to various options for self-balancing.

The main complementary balancing service that will likely emerge is an extension from on demand servicing of variable demand, to a more unbundled balancing service. To see how this plays out, first imagine a DER owner who never overproduces—feeds back to the grid—either because they only produce a minimal amount of power, or because they use storage or load shifting to ensure self-consumption of all they produce. The grid works much as it did before, balancing by servicing residual demand. However, once customers overproduce and feed into the grid, an ideal balancing service includes compensation for excess power. Customers can overbuild systems to ensure they almost always have enough energy on hand to remain self-sufficient (especially if they have storage), and this yields reliability/resilience benefits. By so doing, they also limit exposure to overpricing. The more the grid will pay for excess production, relative to the cost of production, the more attractive overgeneration becomes, as a way to manage variable demand (Box 4.2). An important part of any balancing service is, thus, the ability to suck up a customer's extra power.

As balancing services demanded by customers become more significant than total electricity supplied on average, the grid may experience more volatile flows in aggregate, as distributed energy resources continue to proliferate. The level of investment in grid infrastructure will still be driven by the size of peak power draws, irrespective of lower total flows; wear and tear in items like tap-transformers corresponds to the frequency of adjustments in settings due to volatility (see chapter by Daniel Rowe et al.). If DER penetration

---

10. This is an important premise for Tim Nelson and Judith McNeill in their chapter.

> **Box 4.2 Example of a New Service**
> A whole new kind of service the grid could offer to distribution customers would
> be to mediate exchanges or swaps of electricity production. For example, week-
> day overproducers like residential customers could trade power with weekend and
> holiday overproducers like commercial customers behind the same.

accelerates chaotically, some kind of charge tied to the extent and quantity of volatility that more closely tracks the costs incurred by the grid may be prudent. DER owners could then invest in more load shifting and storage, so as to reduce their volatility and balancing costs, or decide that the grid proposition is more attractive.

3.3 Reliability and Resilience

Reliability is related to flexibility[11] and sizing, much in the way contingency reserves on the transmission grid, held for unexpected outages, are related to regular resources that provide operational and planning reserve margins. In the traditional grid supply framework, like flexibility and sizing, reliability is expected, and not charged for. Yet, as described in Section 2, the reliability one gets under the traditional model is not an all-positive proposition. While generally reliable on a day-by-day basis, it is also brittle and undifferentiated. In the new world, the grid may need to shift what it offers as a reliability service, maintaining advantages while mitigating disadvantages it used to offer.

One interesting feature of reliability is the multiplicative nature of the probabilities that define it. For example, if the grid promises end-users at most one hour of downtime every ten thousand hours, and the end-user has a self-generating DER package—either a small battery system or a full microgrid, as in the chapter by Marnay—that offers the same reliability standard, then, by hooking up to the grid, that customer would see his or her reliability move up to one hour of downtime in every hundred million hours, roughly *three seconds every ten years*! Even for the most demanding customers, this may be too much reliability for the costs. Customers could reduce these costs by spending less on reliability from DERs, but at some point they might rather just pay for less reliability from the grid, a new kind of service.

Differentiated reliability allows a hybrid between a residual and a complementary reliability service to emerge, where the grid and DERs can share responsibility for a customer's reliability, and support overall grid reliability together. The grid can support DERs in providing day-to-day reliability, and provide insurance against unusually disruptive weather, and other year-by-year variations in DER output, by supplying extra electricity as needed to maintain reliable service on the customers' premises. Meanwhile, DERs can provide

---

11. See chapter by Luis Boscán and Rahmatallah Poudineh, later in the book.

critical services during periods of system stress, for example by providing demand responsive contingency reserves for the grid. The logic for DERs and the grid to work together on reliability is very similar to the logic that supports wider balancing areas and reserve sharing on the grid in order to lower the cost of integration of variable generation and overall reliability.

4 PUTTING IT ALL TOGETHER

Section 3 covered how the grid can modify existing services and offer whole new ones to customers empowered by DERs. Before moving on to how utilities providing these grid services can charge their customers in a manner best aligned with customer needs, this section will briefly cover a few key ingredients that must also be considered: the role of monopolies, the importance of automation and computerized agents, and analogies to the financial world.

4.1 The Monopoly Goes Away…

For most of the twentieth century, the grid was mainly operated through vertically integrated, regulated monopolies that controlled every aspect of electricity delivery. The advent of independent power producers, wholesale deregulated markets, and retail competition in some jurisdictions have eroded some elements of the grid monopoly, yet pretty much everywhere a "natural monopoly" still remains, as far as the last mile of poles and wires that connect customers to the grid, that is, the distribution grid. Connecting to the grid today is still an all or nothing proposition for most customers.

As the cost of DERs drop, various services will gradually become more competitive, that is, they will give the customer competing choices, in different ways, at different times. The story starts when wholesale power provided through the grid competes with that from DERs. Unit delivery prices will have to drop, and equilibrium should emerge. Even today, it is striking how solar PPA prices track average utility rates in the major distributed solar markets. Still, for almost all customers today, going off-grid is not an economical option, so utilities still have a competitive advantage in selling their whole portfolio of services that warrants careful regulation and scrutiny.

As the need to modify existing services and offer new ones arises, new pricing models will have to emerge, and incremental changes like net metering, despite their appealing simplicity, may not suffice. New rate design, however, should not become a conduit for utilities to stifle competition under the excuse that DER customers must "pay their way" under the traditional utility world view that mixes cost-of-service with historical costs incurred, and concentrates on short-term cost causation, instead of the medium- or long-term financial impact of customer choices. Instead, the utilities that manage the grid should think very carefully about exactly what value they can offer to customers with DERs and align that with the prices they charge. This way, they will be fully prepared for an era of full competition with DERs.

4.2   The Role of Automation Increases

A lot of the potential services the grid can offer, as discussed in Section 3, will operate at their best efficiency if the interaction between the grid and the customer is automated. In an automated future, an intelligent energy manager[12] acts as the customer's agent in deciding how to optimize the use of onsite DERs with the offerings of the grid for energy services, balancing services, etc. As the grid-customer interface gets more sophisticated, the rate of interaction is likely to go up, and involve a lot of different adjustments throughout the day. Customers are likely to want to set their preferences ahead of time, with the ability to override them, of course, and will gravitate to the simplest and most intuitive interfaces that provide the most value. Much like the seamless handshakes our ubiquitous communications devices carry behind the scenes, the details of the customer–grid interaction should fade into the background, as customers will have plenty of other things on their minds.

Pilot projects currently carried out already demonstrate the advantageous role automation plays in delivering better integration with the grid, while also providing savings to customers.[13]

4.3   Financial Ways of Thinking about Grid Services

Because in the new world of DERs the nature of services customers are interested in purchasing from the grid will change, in ways still unforeseen, it is useful to look to other areas of commerce for guidance. Given the commonality of electricity with other commodities, and even currencies, the use of financial analogies offers many models for innovation. For example, similar creative transaction possibilities and derivative contracts present in financial markets are possible in energy markets.

---

12. An intelligent energy manager is a sophisticated form of what is commonly called a Home Energy Management System (HEMS) (see, for example, http://www.greentechmedia.com/articles/read/home-energy-management-systems-redefined), able to act as an agent for the home. It takes in various forms of external information, like price signals, synthesizes it with information about onsite resources, and then dispatches these accordingly. One of the principal distribution circuit modeling tools, Gridlab-D from the Pacific Northwest National Lab, incorporates this kind of behavior through an agent-based modeling approach.

13. For example, a study (Herter and Okuneva, 2014) by Herter Energy analyzed 2 years of Sacramento Municipal Utility District (SMUD) pilots looking to elicit customer response to demand reduction incentives during peak periods. Their study concluded that "voluntary Time-of-Use/Critical Peal Pricing (TOU/CPP) rates with customer-controlled event automation outperformed information-only and load control programs, having higher energy savings, non-event peak demand savings, and event peak demand savings." And "for customers on the TOU/CPP rate the use of a programmable communicating thermostat (PCT) for event automation more than doubled event savings during nonevent and event peak periods"—as compared to a pilot with similar rates, but no PCT. As a result, the study recommended that SMUD implement time-variable rates matched to system needs, coupled with automation.

One caveat is that even well regulated financial markets tend to see rent-seeking behavior—many may remember the foibles of Enron—and electricity is impractical to store for long periods in bulk, absent much cheaper storage. At least initially, storage is likely to play the role of a shock absorber rather than that of a vault, or facilitate the role of a market maker.

When thinking of financial analogies for the services discussed in Section 3, at least two distinct classes of services emerge:

- A *banking or marketing* service where the grid allows customers to either to sell extra electricity or store locally generated electricity offsite, and get it back later with daily or seasonal time constants.
- An *insurance and backup* service, characterized by its probabilistic nature. Because power needs may fluctuate, or distributed generation can fail because of conditions like broken equipment or extended cloudiness, the grid can offer customers insurance against a sudden need, even if this may happen just once in a while. The insurance policy can cover different kinds of conditions, trigger some event-based payment, and should work with differentiated reliability, for example, by covering only a few key circuits or a fixed amount of power. In contrast to the finance world, the insurance agreement could run both ways, with a distributed pool of DERs agreeing to be a supplier of last resort to the grid—insurers, rather than insurees—in a manner akin to emergency demand response.

Note that, in this financial analogy, the bulk transmission grid also mirrors the global financial system. It provides services similar to interbank lending, derivative contracts and reinsurance that most consumers aren't ever aware of, unless a large failure occurs.

5 PRICING IT ALL: HOW DOES THIS ALL PLAY OUT?

There is no silver bullet rate restructuring that will automatically allow an efficient transition from a traditional grid model to one that efficiently integrates with DERs. The modern electricity grid, often referred to as the largest machine ever built, is a complex system, and is likely to become even more so with the rise of DERs. The rapidly changing landscape surveyed previously will see the unbundling of residual services, the emergence of complementary services driving real competition for utilities, more automated transactions, the evolution of energy management systems, and new energy market products that will evolve in unpredictable ways. Given these uncertainties, there are three broad directions one could anticipate emerging for compellingly pricing grid services toward customers.[14]

---

14. Other chapters covering pricing and tariffs include Biggar and Reeves, and Knieps cover (pricing schemes), as well as Jones et al.

5.1   Real-Time Pricing

The most obvious way to deal with a complex new marketplace is to move to nodal[15], bidirectional, real-time with very granular resolution, perhaps down to a minute or even less, if latency times for energy management systems (EMS) allow it. This enables the market to expose customers to an independently discovered *value* of electricity, as opposed to just offering them prices based on someone else's estimate of the *cost* to provide service. Market-based real-time pricing also automatically adapts to changing conditions on the ground, and is no longer tied to rate cases that may not occur often enough to keep up with technological changes. Transmission and distribution tariff adders would be automatically built into offer prices, giving distributed resources the possibility of competing with the grid on real-time service. Parties anxious about being at the mercy of the markets can manage risk via derivative contracts, in a manner that is not distortive to value discovery, and much like the solution to ensuring capacity in European wholesale markets, proposed in the chapter: European Utilities: Strategic Choices and Cultural Prerequisites for the Future by Christoph Burger and Jens Weinmann.

There are several potential downsides to real-time pricing. First, if value discovery becomes automated and drives dispatch, someone still must take responsibility for system failure. As actors become more and more disintermediated from the impacts of their decisions, some guardrails need to be put in place to ensure smooth operation. Second, markets can become very opaque and subject to manipulation, if not properly regulated and/or there is insufficient competition; after all, plenty of rent-taking behavior can be seen in financial markets. Finally, markets also become inefficient if there is too much volatility; they need market-makers[16] and others guarantors of stability. None of these are necessarily deal breakers, but it is sure to be a challenge charting a smooth incremental path from today's traditional model to a full real-time market.

5.2   Rearrange the Bill

Another broad direction is the incremental evolution of pricing for future grid services marketed to DER-empowered consumers. New rates specifically targeted to them would gradually bring tariffs into closer alignment with the services customers ultimately want. This process entails unbundling residual services, blending them with new complementary services, and then repackaging them. Taking residential rates as an example, the simplest first step might be to move to a three-part tariff which simply adds demand charges to fixed charges and

---

15. See the chapters by Knieps, and by Biggar and Reeves for more on nodal pricing.

16. In stock markets, market-makers are there to ensure there exists a buyer for every sell order, and a seller for every buy order, at any time, in order to maintain liquidity. On the distribution grid, the distribution utility or platform provider would likely be the primary market-maker, leveraging local storage or ties to the transmission grid to perform this function.

volumetric rates. This prices the size of the connection, and brackets the volatility DERs can export to the grid. In time, this type of pricing scheme could evolve into a virtual bank with wheeling/brokering charges for trading energy back and forth, with other actors on the grid, all backstopped by an insurance service.

5.3 Scheduled Load

Ultimately, prices and rate design may be optimized more for machines than for people, leaving many customers to interact mostly with their own energy management system, and ultimately just looking to the bottom line to see if their DER investments are delivering as promised. This leaves room for entirely new kinds of pricing models. One such example would be scheduling load.

A basic description of scheduling load would work as follows. It starts with three kinds of electricity blocks:

- energy blocks
- power supply blocks
- reserve blocks

The customer's EMS would bid in a day-ahead market for straightforward energy blocks of imported (consumed) or exported (generated) power in kilowatt-hour units or fractions thereof.[17] The EMS would also bid for power supply blocks—either independent or subordinate—in each hour, defined by a fixed power range (eg, 0–2 kW). The independent supply blocks would vary as a function of customer needs, with the grid expected to match the demand. Subordinate supply blocks would match a signal from the grid by adjusting demand, perhaps with a minimum consumption floor.[18]

Because the subordinate supply blocks relinquish control over demand to a central operator, electricity under such arrangements would be significantly cheaper than an independent supply block, and may perhaps even receive compensation. Typical times to bid a subordinate block would be during periods when the customer's batteries were charging or discharging, far from empty or full, or when a large, controllable load—like a charging electric vehicle—is available. All other times would use an independent supply block. Customers could also bid reserve blocks of exportable power, or cancelled load that they would rather not supply, but could provide to the grid for a premium on a contingency basis, with sharp penalties for noncompliance. Pricing doesn't necessarily have to be connected to a market, but could be subject to a prenegotiated price schedule.

---

17. Barrager and Cazalet (2014) in Transactive Energy, identify the need for customers to self-reveal their energy/capacity needs in advance and contract for it. More from the GridWise Architecture Council on Transactive Energy can be found at: http://www.gridwiseac.org/about/transactive_energy.aspx
18. In a July 2015 filing to the CA PUC, San Diego Gas & Electric propose a voluntary tariff where they would take control of a customer's storage system; this a more primitive and less abstract version of the subordinate block concept.

Under this scheme, certain aspects of managing the grid, reserve mechanisms, and real-time matching of supply with demand would be more directly under the control of grid operators, making their lives much easier, and providing some certainty to the system. On the other hand, it is dependent on smart machine interacting, and represents a significant step change from existing practice.

All of the price schemes aforementioned are not mutually exclusive. For example, even in a grid completely managed by real-time pricing, a third party could offer to trade the real-time stream of payments and credits incurred by a customer for a simpler cash flow, tied to the customer's behavior. This is akin to what electricity retailers already do when they buy power from the wholesale markets, and then market it to their retail customers. Whichever price schemes are contemplated, decision-makers should look to game-theoretic concepts and agent-based modeling, coupled to empirical customer responses and pilot programs to seek the best path forward.

6 CONCLUSIONS

Early pioneer states like New York (with Reforming the Energy Vision) and California, with its Distribution Resource Planning framework, are taking steps to explore a future where customers are more engaged, either directly or through autonomous devices and third party intermediaries. These evolving regulatory regimes will be closely watched by all utilities wanting a glimpse of the future. To some extent, legacy regulatory practice is blocking innovation, but pressure from many sources is building up; nobody knows exactly sure what will happen when the dyke protecting established incumbents breaks. Established utilities most directly exposed to the leading edge of change, like PG&E, will struggle to overcome a sprawling internal corporate culture not known for nimbleness, and suspicious of factors that might drive revenue erosion. Their key to survival will be to focus on creating new revenues by providing value-added services, to accept a crowded business environment with new competitors, while partnering with innovative upstarts, and to develop a relentless appetite for serving their customers better, the way they want.

Utilities, with the help of their regulators, need to move toward better customer relationships. This means understanding better what the grid really has to offer, and pricing that delivers real value to customers, in a way they will find equitable and fair. This may entail a role less as a top-down conduit from large generators to customers, and more as an intermediary providing a much richer variety of services. Regulators and utilities must accept that Pandora's Box has already been opened. With customers becoming more and more in control of their electricity service, incumbent actors should focus on the hope that an integrated grid can emerge for the benefit of all. Some will learn to thrive under the new regime, others will gradually be ground down or bought out, but all will have to change.

BIBLIOGRAPHY

Barrager, S., Cazalet, E., 2014. Transactive Energy. Baker Street Publishing, San Francisco, CA.

Bronski, P., Creyts, J., Guccione, L., Madrazo, M., Mandel, J., Rader, B., Tocco, H., 2014. The Economics of Grid Defection: When and Where Distributed Solar Generation Plus Storage Competes with Traditional Utility Service. Rocky Mountain Institute, Boulder, CO.

Herter, K., Okuneva, Y., 2014. SMUD's Residential Summer Solutions Study: 2011–2012.

Lovins, A.B., Lovins, H.L., 1982. Brittle Power. Brick House Publishing Company, Baltimore, MD.

Bronski, P., Creyts, J., Crowdis, M., Doig, S., Glassmire, J., Guccione, L., Lilienthal, P., Mandel, J., Rader, B., Seif, D., Tocco, H., Touati, H., 2015. The Economics of Load Defection: How Grid Connected Solar-Plus-Battery Will Compete with Traditional Electric Service, Why it Matters and Possible Paths Forward. Rocky Mountain Institute, Boulder, CO.

Chapter 5

The Innovation Platform Enables the Internet of Things

John Cooper[1]

Siemens PTI, Business Transformation & Solution Engineering, Austin, TX, United States of America

1 INTRODUCTION

The need for a new vision in electricity is driven by new competing alternatives to grid electricity, made possible by new technologies and business models. For the first time in an impressive utility history, competitive grid alternatives have arisen that challenge the status quo, and beg the question of a Second Act. What is next for the electricity sector? How do you top what has been described as the singular engineering feat of the 20th century?[2]

As is well documented in other chapters in this book, the electric grid was the fundamental infrastructure of the 20th century, enabling all the other infrastructures that drove social and economic progress. Arguably, anything that follows would first seek to leverage such singular success. We'll continue to need the grid for a long time, after all. And therein lies the key to formulating a sustaining vision that will take this sector on to a similar foundational role in the 21st century: *"Reimagine foundational infrastructure, the core competency of utilities the world over, for a new purpose."*

When it comes to the next big thing in infrastructure, it's hard to find a better vision than the Internet of Things (IoT), a doubling down of the revolutionary Internet that sprang up to drive disruptive changes over the past two decades. The IoT, is an expansive concept of billions of intelligent devices, networked to trade data in the background, and to execute commands intelligently, providing unimagined new levels of personalization and value, the currencies of our 21st century tech-driven economy and society. This chapter argues that, in order to achieve long-term sustainability as the foundational infrastructure for the IoT, utilities must first transform themselves into sustainable platforms that enable transition, innovation, retail market integration, and transactive interaction, in that order.

---

1. This chapter represents the opinion of the author and does not reflect the views of Siemens Industry, Inc. or its affiliates.
2. http://www.greatachievements.org

Future of Utilities - Utilities of the Future. http://dx.doi.org/10.1016/B978-0-12-804249-6.00005-1

An obstacle to realizing such a vision is the challenge of change. Change, relative in both scope and speed, looms with potentially disruptive leaps and bounds. What we called fast yesterday, literally, will be considered slow tomorrow. Where change used to happen over generations, decades, or years, now it just takes months. For incumbent companies, accelerating change requires considerable adaptation and innovation to survive, lest their offers become uncompetitive and irrelevant in the face of new solutions that provide new value (Kotter, 2014). For historic companies in our essential electric utility industry, accelerating change begs a new vision to recognize these dramatically new conditions, and to lead progress into a new era. The vision recommended in this chapter is relatively straightforward: *to become the sustainable foundation of the IoT.*

Electric utilities have an opportunity to become the vital platform providing the needed power, communications, and intelligence to enable compounding future value, including the IoT. This chapter explores the implications of accelerating change, highlights the challenges of innovation, delves into the nature of crafting a compelling vision, and connects the dots between the grid of the 20th century and the IoT of the 21st century, between the emerging *Personal Energy* service economy, and the foundational infrastructure to support it.

Addressing these challenges will lead utilities down new paths. Understanding what we might call the *Infrastructure to Services Transition* is a first step in understanding these challenges. New business models that rely on innovation to provide greater value must be considered and evaluated. Finally, utilities will need to open up to collaboration with third parties who have what they lack, and to consumers who are maturing into new *prosumer*[3] roles, with new capabilities, but also new needs, and new expectations. In summary, today's situation presents a bold new proposition that underlies any vision utilities may devise.

Underlying Proposition: Infrastructure for commodity delivery naturally evolves to support greater personalization and value. As utilities seek to become the enabling foundation for innovative services, their evolution ultimately leads to utility and sector transformation. Transformed for new purposes, utilities enable new infrastructure, new platforms, and new apps (services).

In short, what brought past success to utilities will still be *required* in the future, but it will no longer be *sufficient*. Utilities will need to change significantly, adopting new roles. Utilities will need to rise to the occasion, or be swept along by historical forces. Utilities have an historic opportunity to redefine themselves for another century-long run of success as *the essential foundation of what is to come.* Such redefinition starts with a vision, followed by a transition plan.

The transition will involve moving from a current focus on *continuity* and *stability*, where both regulators and utilities emphasize holding on to the traditional role of commodity distribution. Today, few doubt the continuing need for

---

3. As explained in Gunelius (2010).

a stable utility to manage the grid. But it will not be sufficient to evolve slowly from ensuring affordable, highly reliable electricity distribution, to enabling the integration of distributed energy resources (DER) and renewables, with minimal disruption to business as usual.

Instead, utilities need a bolder, more comprehensive vision. They must design a transition with a greater end in mind than simply responding to new technologies, or reacting defensively to slow change down to fit their revenue and operational requirements. A new focus on *adaptability* and *flexibility* must be blended with the traditional focus on continuity and stability. If the electric utility of today is to grow into a new vision of supporting the emerging IoT, it must put itself on a transition path, taking the necessary steps to become a platform for innovation—this is the main focus of this chapter. This chapter connects the dots between new business models and a new purpose for the electric grid.

The rest of the chapter is structured as follows: Section 2 examines the challenge of accelerating change and its consequences, disintermediation, and disruption. Section 3 focuses on the rise of platforms, a compelling business model enabled by, and designed for, the 21st century economy, driven increasingly by technology change and innovation. Section 4 applies this analysis to the critical infrastructures of the IoT, the telecom and information network, and the electric grid, followed by the chapter's conclusions.

2   ACCELERATING CHANGE, DISINTERMEDIATION, AND DISRUPTION

With all the new benefits that technology brings, it also brings a new headache: accelerating change. Electrification may have taken decades to unfold in modern societies, but, as it took hold, it drove the 20th century economic miracle, lifting living standards, and enabling no end of wonders. Change continued to accelerate in the 20th century, with each generation experiencing a different pace of life. Digital technology (only possible with electricity) entered the picture in the 1960s, with the transistor and the integrated circuit, and the pace of change found a new driver. High Tech sped everything up, as better, more powerful, less costly chips were embedded in more and more devices. Then, networking compounded the pace of change with mobile telephony and the Internet.

And *business model innovation* accelerated each technology innovation, leveraging new capabilities to challenge old ways of doing things. The postal service became "snail mail," and then, e-mail became the way that parents communicated. Younger generations shifted to texting, then Facebook, then Pinterest and SnapChat, rapidly streaming through ever more ways to stay in touch.

Now, the IoT is emerging with retail applications like Uber,[4] which challenges the traditional transportation paradigm. Popping up like mushrooms,

---

4. https://www.uber.com/

mobile apps riding on the emerging IoT show up daily to delight us and change our lives, a little bit at a time. Vivino[5] places a wine steward at hand at the point of sale, replacing confusion with seamless wine recommendations based on your particular sense of value. Shot By Shot[6] provides a normal golfer the experienced counsel of a local caddie—for free.

Each new retail approach, enabled by advances in digital technologies and new business models, leverages the foundational networks of power, communications, and information to provide greater personalization and value. Less apparent, but perhaps more impactful, is the industrial vision of the IoT, which like these retail apps, depends upon sensors and data gathering devices supported by power, communication, information, and logistical networks in order to automate today's more manual processes, and generate new value. Predictive analytics make for more efficient factories, and more reliable performance from machines and appliances. Autonomous vehicles, for instance, will depend upon the IoT. And underlying all this new value will be a more reliable and automated modern electric grid.

Beyond this still hazy future, though, two factors in particular act on the electricity sector today to challenge the status quo, and introduce radical changes:

- First, energy technologies such as LED lighting, solar PV onsite generation, electric vehicles and energy storage—collectively, distributed energy resources or DER—combine to offer a compelling complement and, for some, alternative to the grid.
- Second, recognition of climate change and the imperative to address pollution from fossil fuel combustion challenge the existing central resource system of fossil–fuel-based power production and distribution.

Disruptive change is materially different in the age of platforms and rapid technological and business model innovation, leading to the term *Big Bang Disruption* (Downes and Nunes, 2014). Where incumbent companies previously had time to plan and execute marketing plans, in today's world, they must start planning differently.

> *...big bang disruptions differ from more-traditional innovations not just in degree but in kind. Besides being cheaper than established offerings, they're also more inventive and better integrated with other products and services. And today many of them exploit consumers' growing access to product information and ability to contribute to and share it. (Downes and Nunes, 2013)*

As a change concept, big bang disruption observes that the *platform*, a business model designed to match the constant innovation flowing from digital technology, will have a devastating impact on incumbents. Incumbent businesses that lack platform benefits are less nimble, adaptive, and competitive than platforms that focus on particular segments with compelling offers leveraging new

---

5. https://www.vivino.com/
6. http://www.shotbyshot.com/

technology, adapted infrastructure, and innovative business models. The challenge for the utility is repositioning itself to address such a threat by leveraging their valuable core competencies, and history of market success. The emergence of platforms points to a new strategy for incumbent electric utilities, described thus by the authors:

> ... to survive them, incumbents need to develop new tools to detect radical change in the offing, new strategies to slow down disrupters, new ways to leverage existing assets in other markets, and a more diversified approach to investment. (Downes and Nunes, 2013)

In Germany, solar PV, a new energy technology embraced and promoted with Feed In Tariffs (FITs), turned the entire sector upside down.[7] In the United States and Canada, rebates and net metering, rather than FITs, have led to slower growth. But the introduction of power purchase agreements (PPAs) stimulated market acceptance; and now, North American electric utilities in high penetration zones such as Hawaii, Arizona, and California must address that question: "PV: threat or opportunity?" At least one recent poll of utility executives[8] suggests that 2014 saw an attitudinal shift from threat to growth opportunity. Utilities in Arizona have even gained regulatory approval to own solar PV generation on rooftops and in community gardens, and community solar business models in particular are gaining traction.[9]

But, in a less predictable future, utility executive decisions will rely on less information, will envision far more renewable energy, and will accommodate the impacts of affordable energy storage. Disruption and disintermediation pose significant risks to traditional rate-based revenue, either through sudden changes, or, more likely, through long-term revenue erosion. Utilities will face an ongoing disruptive environment, requiring a more permanent solution, one that is more in keeping with new market realities and, more specifically, one that acknowledges and embraces the *personal service revolution*—namely, utilities have a tremendous opportunity to transform themselves by becoming platforms to deliver value beyond reliable and affordable kilowatt-hours.

3   THE INNOVATION PLATFORM

Long ago, Charles Darwin famously studied Nature and recognized that a standard framework could result in elegant, highly complex outcomes. With emergent behavior, and a few simple rules (trial and error, survival of the fittest, closed feedback loops, etc.), Nature has managed to organize immense complexity over thousands of centuries, all without any supervisory management.

---

7. For more on Germany, see Chapters by Burger & Weinmann; Lobbe & Jochum.
8. http://www.utilitydive.com/library/the-state-of-the-electric-utility-2015/
9. https://www.solarelectricpower.org/utility-solar-blog/2015/january/arizonas-utility-owned-solar-programs-new-price-models,-grid-integration-and-collaboration.aspx

Further study over the past two centuries since Darwin has revealed Nature to be a worthy teacher for human affairs. For instance, Nature is a great model for understanding the relationship between *networks* and *emergence*, two phenomena that are critical to understanding the potential of the IoT.

Popular author Steven Johnson uses examples from Nature to describe the concept of emergence and self-organization, showing how the distinct activity of individuals emerges with the help of network connectivity (Johnson, 2002). Emergence and networks together present a compelling alternative to highly engineered and structured hierarchies based on top-down control, which characterize much of the man-made business models from the 20th century. Witness the innovations that have unleashed value, once the Internet matured and access became widespread, first with connected PCs and laptops, then with mobile smart phones and tablets.

Futurist Jeremy Rifkin connects the dots between infrastructure and innovation as well, describing the IoT as the emerging synergy of three pervasive, powerful networks (Rifkin, 2014). First, wired and wireless communication networks move data rapidly between millions, then billions of points. Second, energy networks comprised of the traditional grid and newer interconnected DER move energy to where it is needed, in the most optimal fashion. Finally, smart logistics networks move goods and people from point to point, safely and efficiently.

Whether seen as the evolutionary path of machine-to-machine and smart grid, the synergy of three fundamental forms of infrastructure, or a complement to Big Data and the Cloud, the IoT promises to unleash new innovation and greater value than seen heretofore, notwithstanding the incredible twin revolutions of power and telecommunications in the 20th century, or the more recent Internet revolution over the past two decades.

Imagine the changes possible when utilities deliver the IoT. Power outages become a quaint concept, when reliability is redefined as a choice between different levels of service, much as seat belts, air conditioning, and power windows started off as options in luxury automobiles, but became standard offers in time. Metered-based pricing may be replaced by flat rates and free weekends, as with cell phone plans today. Smart meters and inverters become more like the routers of the energy Internet, pushing power back and forth in response to price signals in the transactive energy economy. Positive energy buildings become distributed power plants, as nanogrid operating systems proliferate, and REITs finance new power investments. While these ideas may seem farfetched and whimsical to the practical energy executive today, so did blogs and apps when they were first introduced. Then we got used to them. Then we all needed them. To realize this potential, the current electric grid must transform its operating model to become more automated, digital—more modern—and it must evolve from its 20th century business model to ensure long-term sustainability.

A central premise of this chapter is that a third platform,[10] an emerging IT and business model, offers the most rational transition path for such business

---

10. http://en.wikipedia.org/wiki/Third_platform

and sector transformation. This third platform[11] recognizes the technology advances that are steadily transforming IT organizations and business itself. In its predictions for 2014 and 2015, industry analyst IDC is breathless in its descriptions about the potential disruptive impact of moving to this new technology platform, claiming that disruption will be widespread across industries.[12] And what about the electricity sector?

One way to consider the new utility-as-platform is to imagine two interconnected halves. The first, an operations platform, will transform system operations with its focus on automating, optimizing, and interconnecting the functionality of power systems and smart grid functions, like distribution automation and demand response, with emerging site-based power systems and functions. So far, this type of platform has received the most attention in forums like the New York PSC Reforming the Energy Vision (REV),[13] but the other half may prove as transformational.[14]

The external focused market platform, interconnected with, and complementary to, the operations platform, will be necessary to interface with the emerging complexity of commercial activity from thousands to millions of third-party energy systems, and hundreds of millions of energy consumers/prosumers. The market platform will enable the emerging energy services that will bring tremendous value and personalization to maturing energy prosumers. Working in tandem, these two halves will realize the vision of an energy Internet[15] imagined, but not heretofore possible.

Platform business models enable the ultimate in personal value and customization. To explain platforms[16] and contrast their new business model with the more traditional grid operations by electric utilities, let us start with a definition/description, and move on to examples. For our purposes, a platform may be described as "an extremely valuable and powerful ecosystem that quickly and easily scales, morphs, and incorporates new features, users, customers, vendors, and partners." As summarized in Table 5.1,[17] platforms (1) embrace third-party collaboration; (2) foster symbiotic, mutually beneficial relationships; (3) move beyond selling to promote customer utility and communications; and (4) enable rapid adaptation.

---

11. Mainframes, then PCs/client/server/internet were the first and second platforms, respectively.

12. http://www.idc.com/getdoc.jsp?containerId=prUS25285614

13. http://www3.dps.ny.gov/W/PSCWeb.nsf/a8333dcc1f8dfec0852579bf005600b1/26be8a93967e 604785257cc40066b91a/$FILE/REV%20factsheet%208%2020%2014%20(2).pdf

14. See the chapter by Jones et al. in this book.

15. http://energy.gov/national-clean-energy-business-plan-competition-2014/energy-internet (see also, Bob Metcalfe's *enernet* or Jeremy Rifkin's *intergrid* and the vision of transactive energy promoted by Ed Cazalet and others).

16. I refer the reader once more to an earlier citation of the "third platform," which comes closest to the platform I discuss herein. http://en.wikipedia.org/wiki/Third_platform. See also Hyatt (2012) and Rossman (2014).

17. Material in the table is derived from Simon (2011),

TABLE 5.1 Platform Business Rules for Success

Rule of success	Description
1. First mover/first to mass market	Early experiments give rise to a platform that gets the details and value prop right enough to encourage rapid, mass adoption
2. Customers to affiliates/leverage community	Create community to encourage frequent visits. Marketing is the key to create community/sense of belonging
3. Personalization and value	Focusing on customer benefits drivers greater value, and the platform gets more valuable over time
4. Leverage innovation	The platform enables steady improvement with low-risk experimentation (new tech, services, business models, etc.) until it becomes institutionalized as a core competency and value
5. Delay is death	Irrelevance, obsolescence, and death for dawdlers - ironically, waiting for risk to go away becomes the riskiest path of all
6. Open, partnerships, collaboration	More rapid customer acceptance and better innovation become possible when the ecosystem, not the company, is the driver - sharing a huge pie is better than owning a smaller pie
7. Scale and investment	Up front investment enables rapid scaling (e.g., Cloud)

Described thus, platforms represent a unique 21st century business model inconceivable before the Internet. A threat to incumbents because they acknowledge and embrace ubiquitous competition, platforms leverage Internet and technology to combine falling costs with rising value. Platforms create unrelenting margin pressure, and consistently build a brand, moving beyond adapting to change in order to become its driver. Businesses that implement a platform successfully harness disruption to their advantage, shifting from defense to offense.

As introduced previously, these business concepts converge on what is now referred to in IT as the emerging third platform era that follows the first platform era (mainframe computing) and the second platform era (client/server computing with PCs, laptops, and the Internet). Advancing technology capabilities have led us to this jump to a new platform characterized by four maturing technologies: (1) the cloud that facilitates data access from any device from anywhere; (2) mobile devices and apps that become primary means to access the Internet and communicate; (3) social networks that bind stakeholders in an ecosystem; and (4) big data and data analytics that feed the cloud, platforms, and apps with incredible amounts of data to enable pattern detection, and new insights. Understanding this new perspective is critical to understanding the forces that are driving change in the electric industry.

The *Cloud* reflects the emerging collection of servers that comprise the Internet, managing increasing amounts of data. The Cloud describes a remote data access and data processing capability, either public (ie, as a service from a vendor like Amazon); or private, with dedicated servers. The public cloud has opened up tremendous business opportunities for smaller companies that now may avoid large investments in IT infrastructure.

Mobile devices and *apps* represent the most visible aspects of the emerging third platform. Smart phones and their apps are powerful technology that is already edging out laptops and desktops as the prevailing means to access the Internet (and each other). Mobile apps now implicitly access the Cloud to generate value and win attention.

Social networks have emerged as a powerful marketing tool. People map their lives on these platforms, and share valuable data. When linked with apps, social networks drive increasing personalization and value.

Big Data and *data analytics* promise tremendous value, as sensors proliferate, data piles up, and patterns begin to reveal themselves, flowing back into the Cloud, Mobile Devices and Apps, and Social Networks, as value-added personalization and new services.

Finally, the IoT, the manifestation of all these technological advances, depends upon the third platform and its four converged technologies. The IoT converges these IT trends with trends in power, communications, and logistics infrastructure. While such convergence may appear academic—part of a distant future—we are already seeing its manifestation in New York, in 2015, where the REV proceeding is opening up new market potential and driving utilities to adapt.

Many of the NY Prize microgrid demonstration projects supported by NYSERDA[18] use the Power Analytics EnergyNet platform,[19] the closest example to date of the Innovation Platform described in this section. Pioneer markets in California, Texas, and Hawaii are likely to converge on a platform approach, as well. And, given the attention to the NY REV activity paid by utilities, vendors, and policy makers worldwide, platform convergence is well on its way to global market adoption—other chapters in this book document similar activity in such pioneer markets as Germany,[20] Japan,[21] Australia, and New Zealand,[22] among others.

Facing such a changing landscape, electric utilities have an opportunity to prepare internally in order to be able to leverage these trends as they emerge. The goal of such internal development, as the following analysis shows, is to become the infrastructure that supports value-added personalization. This path is the path of Personal Energy, offered as a service.

---

18. "[The NY Prize is] a first-in-the nation $40 million competition to help communities create microgrids - standalone energy systems that can operate independently in the event of a power outage" (83 communities were selected by NYSERDA in the first round of NY Prize demonstration projects). See http://www.nyserda.ny.gov/ny-prize
19. See http://www.poweranalytics.com/energynet/
20. See the chapter by Loebbe and Jochum.
21. See the chapter by Marnay on microgrids.
22. See the chapters by Sioshansi, and Nillesen and Pollitt.

TABLE 5.2 The PEAAS Model for Personal Energy Evolution

Metric	Telecom/internet	Electricity
Market - Evolution to competitive services	National monopoly service grant, cables, long distance, wireless, CLECs, ISPs, satellite companies, web, platforms	Regional/local monopoly grants, deregulated retail service, DER disintermediation
1. Commodity *infrastructure* POTS/POES	1890–1982: Monopoly grants, universal service, capital/construction wires, local/long distance	1890-present: Monopoly grants, universal service, capital/construction wires, light/power, reliability/cost
2. Enhanced *infrastructure* post POTS/post POES	1980s: Call waiting/ forwarding, messages, long distance/international calling, faxing, paging, dial up access	2000s: Outage notification/ mitigation, flexible billing, bill reduction, peak avoidance, energy efficiency/demand response, energy management systems, net metering
3. Basic *services* pre PTAAS/pre PEAAS	1990s: Cell phones (mobility), broadband (data access)	2012+: DER: (DG, EV), smart thermostats, rooftop PV, LED, electric vehicles
4. Advanced platform *services/ AAS* PTAAS/PEAAS	2000s: Web 2.0, smart phones, social networks, platforms and apps, cloud, big data	Emergent: Energy storage, EV charging, microgrids, nanogrids, platforms and apps, integrated platforms

4   NEW BUSINESS MODELS: TELECOM AND ELECTRIC UTILITY TRANSFORMATION

To understand better the relationship between infrastructure and innovation, the historic shift in telecom services is examined, from infrastructure delivery of commodity dial tone voice telephony in the 20th century, to an amazing array of personalized services driven by the march of increasingly flexible digital technology in the 21st. Starting in roughly 1980, telecom began its long slow unwind, adapting to technologies that provided greater personalization and value. Evaluating changes in the telecom industry in this way provides valuable lessons to help explore fundamental changes in the electricity industry in the recent past and present, and prospects for fundamental changes in the near term.

Table 5.2 shows the gradual evolution from basic commodity services delivered by infrastructure companies (ie, utilities) to advanced personalized services delivered by competitive private sector companies, in four distinct stages: (1) Commodity Infrastructure Service; (2) Enhanced Commodity Services; (3) New Basic Services; and (4) Advanced Platform Services. The transition from one stage to another is driven by new value unleashed by technological advances, often spurred by changes in governmental regulation and/or innovation in business models. At the infrastructure level, transformation is constrained because progress cannot occur at the expense of stability. But for those businesses that leverage improvements in infrastructure, transformation is stimulated and accelerated. After all, the innovation of the web and the app economy wouldn't have emerged without the Internet, fiber optics, and cell towers. The dynamic tension of stability constraints and technology drivers define this transformation.

This emerging platform economy leverages existing and new infrastructure to offer new levels of personalization and value to consumers. Ultimately, the platform's business innovation depends upon synergies between infrastructure, commerce, and fashion. A platform exists to offer innovation opportunities as widely as possible, stimulating tremendous increases in real and perceived value, while allowing the ultimate in personalization. Apps demonstrate the dramatic changes possible in this fourth stage.

Charts in this section map progress in telecom through these four stages. Viewed together, they suggest a corollary emerging in the electricity industry. Telecom monopolies built infrastructure to deliver *universally available, high quality dial tone connectivity*—the ability to make voice phone calls, which within the industry, came to be called Plain Old Telecom Service, or POTS. Likewise, electricity monopolies built infrastructure to deliver *high quality, affordable, highly reliable power*—following the telecom example, this could be labeled Plain Old Electric Service or POES.

In the next "post" stage, third parties and incumbents enhance the basic commodity service—dial tone or power—in a first wave of innovation beyond the "plain old" service offered by monopoly infrastructure companies.

As personalization increases, thanks to new technologies and innovation, a new service paradigm emerges, which I've called either Personal Telecommunication as a Service (PTAAS), or Personal Energy as a Service (PEAAS). But, before that final era can take hold, we experience a third stage of transformation—call it Pre PTAAS or Pre PEAAS—where personalization and value begin to depart from the core commodity, first complementing, then offering a substitute for the commodity's core value. When the PTAAS or PEAAS stages emerge, infrastructure-enabled competitive platforms create a new marketplace of innovation and disruption, delivering the ultimate in personalization and ongoing, increasing value, where the consumer benefits from new technologies, new business models, and changing tastes.

4.1 Telecom Evolution

The four stages of evolution in telecom flowed naturally from (1) constructing the poles, wires, and switches to provide a commodity—this is the POTS stage; (2) enhancing dial tone with long distance and equipment innovations—the Post POTS stage; (3) introducing mobility and data access, innovative new services made possible by new technologies—the Pre PTAAS stage; and (4) introducing advanced services, enhanced by new Internet business models, mobility and platforms—the PTAAS stage. Noteworthy, initial market focus looks backwards, where powerful incumbents are the frame of reference. As innovators shift focus to prepare for the future, infrastructure incumbents must adapt their assets to new market realities.

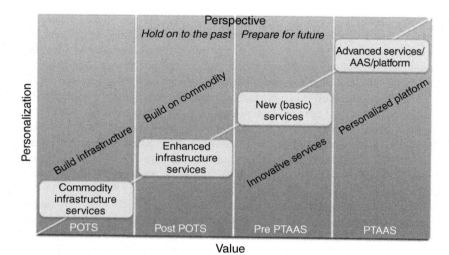

FIGURE 5.1   **Personalization and value: telecom.**

In telecom, that meant consolidation, network expansion and repurposing, and a decline in POTS revenue. Fig. 5.1 shows this change in four stages.

As shown in Fig. 5.2, new technologies lead change from stage to stage. For instance, deregulation of telephone monopolies was driven by two catalysts: first, advances and choices in handsets that introduced greater value and personalization; and second, new long distance options from MCI and others

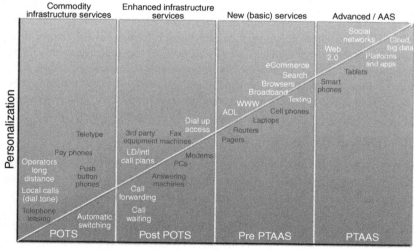

FIGURE 5.2   **Technology drives change: telecom.**

that spearheaded lower costs and greater value. Over time, cellular telephones unshackled callers from the wall, desk, or phone booth, and new equipment, such as pagers, routers, and laptops brought new levels of value to customers. As the Internet matured, new business models accelerated these changes, ultimately leading to the third platform family of technologies, including mobile platforms and apps, social networks, and data analytics leveraging Cloud services.

This historic progression is marked by the rise and fall of major brands in each stage. Notably, in the Post POTS Stage, infrastructure company brands predominate. In the Pre PTAAS Stage, we see early cell phone leaders, PC manufacturers, and Internet pioneers. It is only in the PTAAS Stage that we see the rise of such giants as Google, Amazon, Facebook, and Apple, the four horsemen of the Internet Apocalypse. By creating platforms based on ever increasing personalization and value, these companies created altogether new markets in Search (Google), eCommerce (Amazon), Social Networks (Facebook), and Mobility (Apple), where they each could dominate and drive evolution according to their own visions. In so doing, they provide a model for analysis of changes underway in the electricity sector, detailed in the following section.

4.2 Electricity Evolution

In Fig. 5.3, the telecom graphic is adapted to show the four evolutionary stages in electricity, revealing change that evolves more naturally to support a new kind of energy; what may be called PEAAS™. This transformation has gone from (1) building power plants and the grid to provide POES; to (2) enhancing low rates and reliability—the Post POES Stage; to (3) introducing innovative

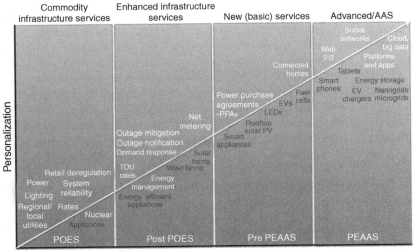

FIGURE 5.3 **Technology drives change: electricity.**

services made possible by new DER technologies and new business models—the pre PEAAS stage; and finally, to (4) introducing advanced services enhanced by new business models and platforms—the PEAAS Stage. As with telecom, initial market focus has built upon the core service value of monopoly electric service, enhancing the low costs and high reliability of POES. As DER technology progresses, innovators reach beyond the grid, challenging utilities to adapt or decline. Implementing a Personalized Energy platform represents the core challenge of electric grid operators today.

From stage-to-stage, new energy technologies drive change. In the electric sector today, most market activity remains focused on commodity electricity in the first stage. In the next stage, new Internet and Smart Grid technologies already enhance the low cost and reliability of POES through energy efficiency and demand response innovations. The third stage, emerging only recently, features the steady decline in price of DER, and fosters new business models. To date, the PEAAS stage remains mostly a concept, with only nascent energy platforms that promise to enhance value and personalization significantly.

4.3 Applying the PEAAS Model

To ensure system reliability, grid operators have traditionally sought top-down control. But taken together, the rise of alternatives to grid electricity, the introduction of third parties, and decentralized, value added, personalized services present a fundamental challenge to this approach. Most utilities have so far been more oriented to slow change down, than they have been to charge ahead with new ideas. Occupied as they are with technology-based grid optimization, few if any utilities have yet embraced new business models.

Just as modern telecom/IT services evolved into the leading platforms of today—Google, Amazon, Apple, and Facebook—emerging DER technologies can be expected to stimulate the rise of PEAAS. The undifferentiated market of incumbent utilities, utility ratepayers, and passive loads will be replaced by extensive market segmentation, energy prosumers, and interactive resources and loads.

In contrast to mass market POES, the emerging PEAAS business model will offer intelligent prosumers an array of options. Value-oriented service companies, whether transformed electric utilities or new third parties, will segment customers based on maturing perspectives on new energy options, perceived needs, explicit and implicit desires and other dimensions. With unique combinations of rates, DER equipment, services and new business models, competitive service companies (including some evolved utilities) will offer new personal energy services (i.e., PEAAS) designed to meet specific needs, much like the online design of a Dell computer at the point of sale.

4.4 Getting to PEAAS: Platforms Lead to IoT

The emerging personal energy services market will drive utilities to become the essential platform for the emerging IoT. Energy services and the IoT go

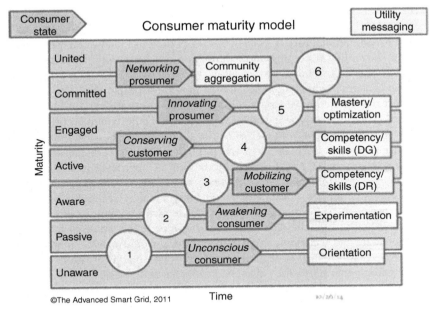

FIGURE 5.4   Consumers mature on a continuum. *(Source: Cooper, A., Carvallo, J., 2015. The Advanced Smart Grid: Edge Power Driving Sustainability, second ed.)*

hand in hand. Just as the World Wide Web drove expansion of the Internet as the emerging norm in energy, PEAAS will accelerate the transformation to the IoT. Becoming a modern service-oriented platform will provide the necessary revenue for the electric utility to upgrade the grid with millions of sensors interconnected to the cloud and to data analytics engines.

But even those utilities ready and willing to become New Age energy service platforms will depend on the emergence of a more mature energy services market. Emerging consumer behavior, described in the Consumer Maturity Model (CMM) in Fig. 5.4, tracks six stages, intertwined with technology advancement.

Most energy customers in today's POES era are still relatively unconscious ratepayers, still passive, still unaware (or uninterested) in the opportunities for better value and personalization unfolding before them. A customer vision limited to lower prices in a slowly emerging market constrains both vendors and utilities that seek to benefit from the new value promised by new energy technologies.

As the CMM shows, before new business value may be realized, a marketing strategy based on orientation and experimentation messaging is needed to awaken and activate customers. As customers begin to mobilize, they seek to acquire new skills associated with smart consumption (ie, conservation) and assume a more innovative role as *prosumers* (ie, both producers and consumers). Energy companies embracing a bold future will lead customers along this maturity path with greater intensity and more deliberate activity, disrupting incumbents along the way.

Prosumers represent an emerging market ready to acquire new value added energy-related services. Today's utilities may foster a nascent personal energy market by driving maturity in consumer energy outlooks, and experience with telecom shows that becoming an enabling platform is the optimal means for utilities to accomplish this transformation.

Such changes could come much sooner than most people think. In contrast to the organic evolution we saw in telecom and IT, the maturing third platform economy promises to both stimulate market forces and relieve market constraints. But for utilities to benefit from such transformation, they must evolve from active resistance to passive acceptance to active promotion of customer maturity.

What would such a transformation look like? A high level process would allow market creation, starting with a platform in place. With a virtual space common across utilities, consumers would enjoy a readily accessible means to learn, experiment, and acquire the skills associated with becoming more mature consumers and prosumers. And utilities and vendors alike would benefit from the efficiencies and acceleration that such a platform would provide. In this way, the PEAAS platform may be seen as a catalyst to drive market creation, helping market participants across the value chain embrace new roles.

The elements of the platform that utilities adopt to serve the prosumer—cloud computing, data analytics, social networks, and mobile communications and applications—are also essential to the IoT. Shifting the business model while investing in grid modernization is the key to long-term sustainability for electric utilities and the grid.

The transition to a new business model starts with a compelling vision, then follows several logical steps that include planning, innovation, and transformation.

- Preliminary Planning
  - Traditional and new forms of system planning (eg, IRP, DRP, Risk Mitigation)
  - New resources and requirements (i.e., DER technologies)
  - SWOT Analysis to identify gaps
  - Alignment with Vision
  - Innovation Experimentation: Managed Trials
- Advanced Planning
  - Industry Sector Best Practices Review
  - Market Needs Assessment
  - Business Model Evaluation (including platforms)
  - Organizational Change Readiness Assessment
  - Regulatory Alignment
  - Innovation Experimentation: Phase 0 Projects
- Transition Implementation
  - Scenario Development
  - Product/Services Alignment, Partnering
  - Transition Plan (Transformation Roadmap)
  - Program Management Office, Value Management Office

- Platform Creation
- Partner Alignment

With such business model transitions well underway, the focus of electric utilities may shift to inculcating innovation into every aspect of their new business models. This focus on innovation not only harnesses the inherent creativity of all utility employees inside the organization, but also opens up the conversation by engaging the creativity of the marketplace, including both third-party energy companies, and emerging energy prosumers.

As innovation matures, valuable energy services will meet blossoming needs in the rapidly expanding Personal Energy marketplace, for consumers and prosumers, for third-party service providers, and for transitioning electric utilities. As services and service providers proliferate, as consumers mature, as utilities transition, and as DER energy resources proliferate, the conditions for Personal Energy platforms will emerge. We can expect market entry by new platform providers during this transformation.

With the innovation platform and a maturing IoT, converged power, communications, and logistics networks will create the conditions that enable widespread adoption of transactive energy, a new paradigm that envisions massive numbers of energy transactions between buyers and sellers, including peer-to-peer transactions of the type that we see in the maturing Internet economy. Both Personal Energy and Transactive Energy may be seen as manifestations of the Third Platform, and natural consequences of an emerging IoT.

5 CONCLUSIONS

This chapter has described a vision for electric utilities close to their historic role as grid operators, but more in alignment with emerging trends that promise a more sustainable footing. If a core task of leaders is to craft and explain a vision that inspires others to follow, emerging electric industry leaders must step up with visions that promise more than regulatory compliance, steady incremental change, or operational efficiency, all essential functions of grid operators, but not close to the inspiring visions of our forefathers in the late 1800s. From Edison to Tesla to Westinghouse to Insull, pioneer visions laid the foundations for today's grid and utility sector creatively and persistently and, consequently, to enable our 20th century economic miracle. Transformation requires today's leaders to step up and craft new visions for a new century.

An inspiring vision for the grid must be rooted in the traditional role that an infrastructure provider plays in a society and economy, but must also point to a logical future that is both within traditional core competencies, and in alignment with trending opportunities, regardless of current capabilities. These three conditions are critical to any compelling vision for the Utility of the Future.

- Tie *Smart Grid* to the *IoT*. Modernizing core grid infrastructure so that it gains the flexibility and intelligence to converge with other infrastructures to support the IoT.

- Embrace *Personal Energy* as Opportunity, not Threat. Developing new sources of service revenue to replace the erosion of rate-based revenue and to ensure financial stability.
- Develop into the *PEAAS Platform*. Defining the electric grid and utilities as the enabling infrastructure of the emerging platform-based economic paradigm.

Far into the future, we'll need the grid and all it has to offer as the connecting and supporting infrastructure of tremendous energy value, but we won't want or need the negatives we have long had to endure. The Ubers and Vivinos of electricity will be closely associated with the elimination of outages, high utility bills, ongoing rate increases, and lingering monopoly attitudes. In the United States, ATT and Verizon represent success stories of infrastructure-to-services transformation for electric utilities to emulate. If leaders emerge to provide compelling, inspiring visions of robust futures that match or exceed the glorious past of this proud industry, a truly remarkable energy future is in store for all of us. The Age of Energy Abundance beckons.

REFERENCES

Downes, L., Nunes, P., 2013. Big bang disruption. Harvard Bus. Rev. 3, 46.

Downes, L., Nunes, P., 2014. Big Bang Disruption: Strategy in the Age of Devastating Innovation. Portfolio Penguin, New York, NY, p. 47, http://www.amazon.com/Big-Bang-Disruption-Devastating-Innovation-ebook/dp/B00DMCUWW4/ref=sr_1_1?s=books&ie=UTF8&qid=1440455332&sr=1-1&keywords=big+bang+disruption+strategy+in+the+age+of+devastating+innovation.

Gunelius, S., 2010. The Shift from CONsumers to PROsumers, Forbes, Jul. 3. http://www.forbes.com/sites/work-in-progress/2010/07/03/the-shift-from-consumers-to-prosumers/

Hyatt, M., 2012. Platform: Get Noticed in a Noisy World. Thomas Nelson, Nashville, TN.

Johnson, S., 2002. Emergence: The Connected Lives of Ants, Brains, Cities and Software. Penguin Books, London, http://www.amazon.com/Emergence-Connected-Brains-Cities-Software/dp/0684868768/ref=sr_1_2?s=books&ie=UTF8&qid=1431811088&sr=1-2&keywords=Emergence.

Kotter, J., 2014. Accelerate: Building Strategic Agility for a Faster-Moving World. Harvard Business Review Press, Boston, MA, http://www.amazon.com/Accelerate-Building-Strategic-Agility-Faster-Moving/dp/1625271743.

Rifkin, J., 2014. The Zero Marginal Cost Society: The Internet of Things, the Collaborative Commons, and the Eclipse of Capitalism. Palgrave Macmillan, London, http://www.amazon.com/Zero-Marginal-Cost-Society-Collaborative/dp/1137278463/ref=tmm_hrd_title_0?ie=UTF8&qid=1431812955&sr=1-1.

Rossman, J., 2014. The Amazon Way: 14 Leadership Principles Behind the World's Most Disruptive Company. CreateSpace, North Charleston, SC, http://www.amazon.com/The-Amazon-Way-Leadership-Principles/dp/1499296770/ref=tmm_pap_title_0?ie=UTF8&qid=1431873214&sr=1-1.

Simon, P., 2011. The Age of the Platform: How Amazon, Apple, Facebook, and Google Have Redefined Business. Motion Publishing, Henderson, NV, http://www.amazon.com/The-Age-Platform-Facebook-Redefined/dp/0982930259/ref=tmm_pap_title_0?ie=UTF8&qid=1431872972&sr=1-1.

Chapter 6

Role of Utility and Pricing in the Transition

Tim Nelson*, Judith McNeill[†]
*AGL Energy, North Sydney, NSW, Australia; [†]Behavioral, Cognitive and Social Sciences, Institute for Rural Futures, University of New England, Armidale, NSW, Australia

1 INTRODUCTION

In many ways, Australia is at the forefront of the wave of disruption currently affecting the global utilities industry. Since 2008, the cost of installing embedded solar photovoltaics (PV) has declined by around 80% (Simshauser, 2014). In South East Queensland, around one in five residential properties is now operating its own distributed generation, placing the jurisdiction among the highest in the world for solar PV penetration. Like the other countries discussed in this book, and many other countries in the developed world, Australian electricity demand has declined: by around 5% over the past 5 years (Saddler, 2013). Utilities have increased prices markedly in recent years to recover costs associated with significant upgrading of transmission and distribution infrastructure. This has resulted in a near doubling of prices and record low levels of consumer trust (AMR, 2015). Further background on the uniqueness of the Australian situation is explored in chapters by Grozev et.al and Smith and MacGill in this book. The significant increase in electricity prices relative to other countries is shown in Fig. 6.1.

The response by Australian utilities to this perfect storm of disruption has been relatively slow and cumbersome. Declining grid-based electricity demand, and the emergence of a nongrid supply substitute, caught many participants in the Australian market off guard. Most businesses failed to embrace new technologies in a meaningful way. While policy uncertainty[1] and oscillating feed-in tariffs (FiT) inhibited the adoption of successful long-term business models, it would be reasonable to state also that the sector was not prepared culturally to adapt to rapid change. After all, the production and consumption of electricity in Australia had not changed in any meaningful way for decades.

The Australian utility of 2015 is now beginning to look very different to the utility of the past. The first significant market offering for battery services has

---

1. See Nelson et al. (2015) for a discussion on renewable energy policy in Australia.

Future of Utilities - Utilities of the Future. http://dx.doi.org/10.1016/B978-0-12-804249-6.00006-3

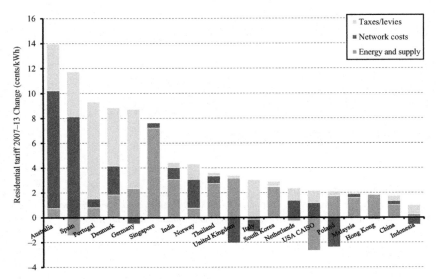

FIGURE 6.1 **Electricity price increase by selected countries, 2007–13.** *(Source: UBS data presented in Simshauser (2014).)*

been launched by an Australian utility; metering and behind the meter services are becoming a ubiquitous service offering; and new embedded solar PV financing models are being rolled out by many providers. At the same time, utilities are adjusting the way in which they price energy, with different strategies being adopted by different participants.

This chapter outlines how utilities are responding to disruption within the Australian market, as well as the role pricing strategies are playing in the transition to a more decentralized energy future. Section 2 provides a brief backdrop of Australian energy market developments for context. In Section 3, the commercial response of utilities now and that foreshadowed for the future is examined. The economics of coexistence, rather than complete grid substitution are explored in Section 4 with different pricing strategies based upon this coexistence considered in Section 5, followed by the chapter's conclusions.

2 SOME CONTEXT: RECENT AUSTRALIAN ENERGY MARKET DEVELOPMENTS

2.1 A "Perfect Storm" of Conditions in Electricity Markets

To understand the commercial response of utilities to technological disruption, it is necessary to consider the "perfect storm" of difficult conditions within which Australian electricity markets have been operating. Capital expenditure on networks increased from around $5 billion between 2005 and 2010 to around $30 billion in the period 2011–15. This expenditure was predicated on

TABLE 6.1 Average Annual Growth in Peak and Average Demand From FY06 to FY14

	New South Wales	Queensland	Southern Australia	Victoria	Tasmania
Peak demand (%)	−0.6	2.1	1.8	2.2	−0.4
Average demand (%)	−0.8	0.1	−0.4	−0.5	−1.4

Source: esaa Annual Reports (www.esaa.com.au).

significant increases in electricity demand, which, in the event, did not materialize. While peak demand has increased, or decreased only slightly, underlying demand has decreased (Table 6.1).

Australian residential tariffs are largely structured as two-part tariffs with a small fixed component and a primary "average cost" throughput charge. Persistence with these tariffs has resulted in system capacity factors falling substantially as customers economize on consumption at times most convenient to them, not at times of system peak demand. Lower demand has resulted in higher tariffs, and these, in turn, have prompted further declines in consumption. In some ways, a "death spiral" could be said to be underway (see chapter by Felder). The "grinding down" of underlying electricity demand is shown in Fig. 6.2 and annual customer bill increases and declining system capacity utilization are presented in Fig. 6.3.

Given the near doubling of grid-connected electricity prices in the past 5 years in Australia, it is unsurprising that Australian households have been among the highest adopters of solar PV as a partial grid-substitute. Electricity consumers have voted with their feet and installed PV to avoid some of the increases in grid-connected bills. Many in the industry expected installation rates to decline after a peak in 2012 as premium FiT were wound back by governments due to concerns about their regressive social impacts (Nelson et al., 2012). While installation rates have fallen, nearly 700 MW of embedded generation was installed across Australia in 2014 (around 1.5% of total Australian generation capacity). Solar remains very popular for a range of reasons.[2] The growth rate of solar PV and cumulative installation is shown in Fig. 6.4.

The circumstances facing Australian utilities are not unique to Australia, but they are among the more extreme. Prices have increased markedly, relative to other nations, and demand has declined, but not in a way that facilitated reduced capital

2. For example, 85% of homeowners believe solar panels increase the value of their home, with 78% believing it adds up to $A10,000 to the price (Adelaide Advertiser, 2015).

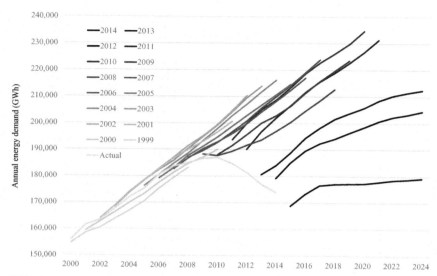

FIGURE 6.2   Grinding down of actual and projected electricity demand. *(Source: Frontier Economics analysis based upon data from AEMO (2014).)*

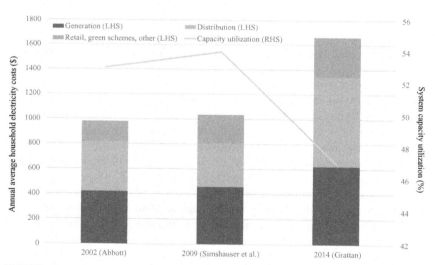

FIGURE 6.3   Annual household electricity cost and system capacity utilization. *(Source: Abbott (2002), Simshauser (2010), and Grattan Institute (2014).)*

expenditure requirements. System capacity utilization has therefore plunged and persistence with "average cost" pricing has resulted in customers reacting to higher bills by shifting to a new substitute for grid-connection, solar PV. With such events having occurred, it is not surprising that consumers are expressing a lack of trust in their energy utility (see Wood et al., 2015 for further discussion).

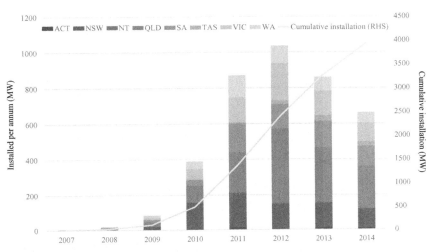

FIGURE 6.4   Installation rates of solar PV. *(Source: Clean Energy Council data.)*

2.2   Commercial Response of Utilities

Australian electricity markets are largely restructured with clear separation between network operators and retailing. Given that networks are regulated, an assessment of the commercial response of utilities necessitates a focus on electricity retailers. Within Australia, there are three large retailers with integrated generation portfolios: AGL Energy, Energy Australia, and Origin Energy. Beyond these three "gentailers," there are dozens of smaller retailers and brokers operating in the retail market. Furthermore, providers of new technologies and energy services are also competing actively for electricity customers. Such arrangements are not unique to Australia, but there are differences in the way in which industrial structure has impacted on the response of incumbent utilities (see chapter for evidence of the response of European utilities).

It could be argued that, until recently, the commercial response to new technology from the incumbent retailing sector has been almost nonexistent. However, all three of the large "gentailers" have launched significant product offerings in 2015 to compete with new technology providers. There seems to be universal acceptance within the industry that some decline in grid-connected business is inevitable and business growth is dependent upon the adoption of new products and services.

Official market forecasts for the continued slowdown of residential (and non-residential) electricity demand and the implications of continued uptake of embedded solar PV are shown in Fig. 6.5. Average household consumption is projected to fall, albeit at slower rates than between 2011 and 2015. This is due to grid-connected electricity prices falling and some exhaustion of noncapital intensive energy efficiency opportunities. However, average grid consumption

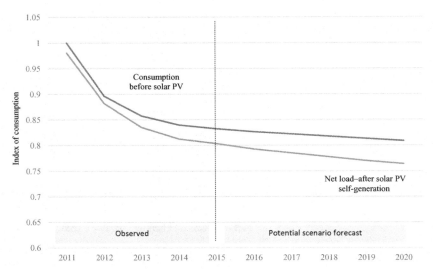

FIGURE 6.5 **Residential consumption projection scenario.** *(Source: Based upon forecasts by AEMO (2014).)*

by households is projected to fall further due to continued deployment of embedded generation. While this represents one potential future, it is indicative of the industry shifting its expectations from grid-connected consumption growth to expectations of decline. This necessitates the development of alternative business models.

Australian utilities are also cognizant of a growing list of international companies with significant balance sheet strength developing new products and services. Apple, Google, and Samsung have all expressed some interest in solar, security, telecom, and energy. Their emerging interest appears built around partnerships to develop "internet-of-things ecosystems" under an umbrella of home energy management solutions (See chapter by Cooper in this book, and Vergetis Lundin, 2015). There has been extensive media interest in the launch of Tesla's battery products in Australia starting in 2016 (Francis, 2015).

2.2.1 AGL Energy

AGL Energy has announced through a company Greenhouse Gas Policy that it intends to transition its business away from coal-fired generation gradually, in recognition that, "the electricity sector is likely to undergo significant change over the coming decades" (AGL Energy, 2015a). This corporate positioning is being supported by AGL being the first—of the "gentailers"—into the market with a solar Power Purchase Agreement (PPA) offering for residential consumers, a smartphone app to manage energy consumption in real time, and battery storage products. The business has created a *New Energy* division, which has been given license to compete directly with the traditional retail arm of the business. The head of the division has stated, "AGL has a history of re-inventing energy for over 175 years in order to meet customer needs and harness new technologies. Our new energy strategy continues that trend. Our aim is to have a presence in one million Australian homes and businesses by 2020" (AGL Energy, 2014).

Importantly, AGL Energy has also stated that, by the late 2020s, up to one-third of Australian households will be partially or fully off-grid. This is significant as it represents a shift in focus for utilities away from being solely reliant on grids, to servicing customers with new products and services. In the case of AGL, it would appear that metering is seen as a key facilitator of other value-added services, such as embedded generation, microgrids, and storage. The acceptance of off-grid solutions is a key reason why the business has already begun marketing a 6 kWh battery offering to complement solar installations of between 3 and 4 kW (AGL Energy, 2015b).

2.2.2 Energy Australia

Energy Australia has increased significantly offerings related to downstream services, such as end-to-end heating, cooling, hot water, and solar PV. The company has shifted focus from selling systems to being focused on the lifecycle of the product and offers repairs and service over the life of the asset. Interestingly, the business appears increasingly focused on customer service and guarantees to ensure it arrives within 2 hours of any scheduled appointment time. Much of this appears focused on rebuilding trust with consumers.

2.2.3 Origin Energy

Origin Energy estimates that they currently have around 30% of the market of "solar customers," and are seeking to grow this share. The business has appointed a Group Manager for "Connected Customers," and is seeking to capitalize on its early position in solar PV by bringing together smart metering technologies, batteries, electric vehicle charging, and solar into a single business proposition (Giarious, 2015). The company views this offering as providing customers with the products and services they *want* (energy management), rather than those that they *need* (a simple energy connection).

Origin has calculated that there are still 5.3 million Australian homes and businesses that are untapped solar energy sources and could produce electricity worth an extra $4.4 billion (Keane, 2015, p. 48). The business also sees the deployment of new technologies as a key way to reengage customers and earn higher levels of consumer trust, stating, "Building customer loyalty and trust is the most powerful mitigant to the impacts of a highly competitive market" (King, 2015, p. 4). This is an important development as it indicates that the business sees itself akin to a "traditional retailer," competing not just with other homogenous providers of the same product, but also companies with different, heterogeneous, or substitute products.

2.2.4 Other Responses

Other retailers are pursuing a range of commercial strategies aimed at engendering consumer trust through the provision of new products and services. At the same time, electricity provision has become acutely political, given the shift toward less greenhouse gas intensive energy sources. An active political group within Australia called GetUp! has partnered with Greenpeace to launch a campaign for customers to switch away from the "Dirty Three" integrated retailers to Powershop, a company focused on renewable energy and energy management.[3] This is in addition to the successful "One Big Switch" campaign aimed at using residential consumer bulk buying power to drive electricity costs lower.[4] Evidentially, such activity is not inconsistent with global trends. Burger and Weinmann discuss the importance of *alliances* in the development of the future utility model, Utility 2.0.

3 COEXISTENCE RATHER THAN COMPLETE SUBSTITUTION

As outlined in the previous section, Australian utilities themselves are now predicting declining grid-connected average household demand and increased uptake of embedded generation and associated services. AGL Energy has forecast that, in the late 2020s, one-third of consumers may be partially or fully off-grid. This implies that coexistence, rather than complete product substitution, is likely to occur.

Analyzing the way households use energy is critical to understanding the likely challenges and opportunities facing utilities as customers install more distributed generation and storage. For example, during an average day, consumption (in kilowatts) shows a morning and evening peak. Fig. 6.6 outlines weekday household consumption by appliance end-use in Queensland, averaged for a year. The data was produced by the CSIRO (2013), in a study involving Queensland households that had individual real-time consumption meters installed at various points in the home, behind the network connection point. On an average day, it is clear that households have a morning and evening peak, with increased use of lighting and power being the strongest contributors to peak demand.

Peak demand days show a distinctly different pattern. Fig. 6.7 presents the data for the same households, but during the critical peak demand days where spatial cooling consumption increases rapidly. Also shown in Fig. 6.7 is the output of a typical solar PV system, measured through the CSIRO study. Note that

---

3. Further information about the GetUp! campaign is available at: https://www.getup.org.au/campaigns/renewable-energy/switch/join-the-switch-to-save-renewables, accessed online on May 11, 2015.
4. Further information about the One Big Switch campaign is available at: https://www.onebigswitch.com.au/campaigns/snapshots#.VVBIaPmqpuA, accessed online on May 11, 2015. Around 250,000 consumers became members of the campaign.

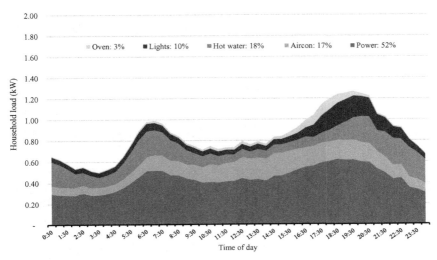

FIGURE 6.6   **Average weekday household consumption by appliance.** *(Source: Simshauser (2014) and CSIRO (2013).)*

the peak kilowatt consumption is around two-thirds higher than it is in Fig. 6.6, but it occurs well after the solar PV system has produced its maximum output.

Utilities have an interesting dilemma in this scenario. Introducing demand tariffs—that is, tariffs based upon the household's measured peak demand in kilowatts—would be one way of ensuring that households contribute to the cost

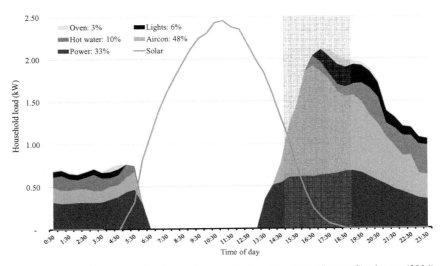

FIGURE 6.7   **Peak consumption by appliance and solar PV output.** *(Source: Simshauser (2014) and CSIRO (2013).)*

of building the network to meet peak demand. Given the network is largely fixed capital costs, such pricing could be said to be economically efficient. However, demand tariffs are likely to incentivize the deployment of batteries as households attempt to store peak solar generation for later use at peak consumption times. This may be a welcome development, if network augmentation could be avoided at a lower cost through installing distributed generation and storage. Alternatively, it may also reduce the utilization of the existing infrastructure necessitating an asset write-down. Such consideration is not unique to Australia, with US utilities also pondering this dilemma (see Athawale and Felder).

A further critical understanding utilities need as they design tariffs and product offerings for batteries and embedded generation relates to whether partial or full grid substitution is possible, or desired, by their customer cohorts. Full grid substitution, or "going off-grid" implies a different commercial response from partial grid substitution, where customers provide some of their power requirements from self-generation and utilize battery technology to shift power production from the middle of the day to the evening peak.

To answer this question, it is necessary to examine the proportion of customers with varying levels of consumption and contrast this to the output of typical embedded generation solutions. Fig. 6.8 presents the proportion of customers by their consumption band. Average (both mean and median) annual consumption in the Australian NEM is between 5 and 7 MWh, depending upon the

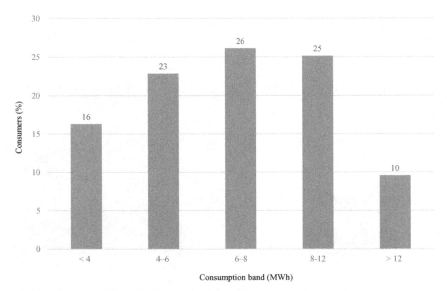

FIGURE 6.8 Proportion of Energex customers by consumption band. *(Source: Energex (2010) and Nelson et al. (2012).)*

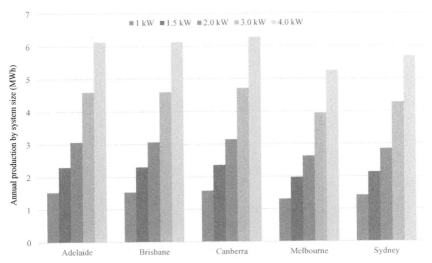

FIGURE 6.9   Annual output of solar PV by system size and city location. *(Source: Clean Energy Council (2014).)*

jurisdiction—areas with higher penetration of natural gas, such as Victoria, tend to have lower average electricity consumption.

The annual output of solar PV by system size and city location is presented in Fig. 6.9. While there is some regional variation, the annual production is relatively similar, for each city and system size.

Fig. 6.10 combines the data presented in Figs. 6.8 and 6.9 to provide a snapshot of the proportion of customers that could generate enough energy to satisfy their own electricity consumption (self-satisfy). The nonweighted average solar system output[5] by system size is plotted (LHS) against the proportion of customers that would be able to use such a system, combined with battery storage to manage variable consumption, to effectively disconnect from the grid (RHS). Around one in six customer connections would be able to "disconnect" from the grid by installing up to 2 kW.

A 3 or 4 kW system would provide enough power *on average* to supply the requirements of around 40% of households, over an annual cycle. However, such analysis ignores the seasonal and diurnal nature of electricity demand. Several days of below average solar production due to overcast conditions, combined with above average consumption due to colder weather necessitating spatial heating, would be likely to result in an inability to produce and store enough energy to ensure self-consumption. Many of these households are likely to want to maintain a grid connection for security of supply purposes.

---

5. A limitation of this analysis is that a nonweighted average is used, which would be likely to overestimate solar PV production. There are higher populations in Sydney and Melbourne, and these cities have lower average solar PV output.

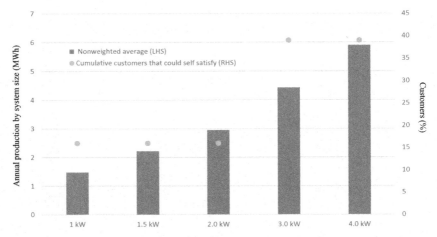

FIGURE 6.10 Solar system output and percentage of customers who could "self-satisfy." *(Source: Adapted from data in Nelson et al. (2012) and Clean Energy Council (2014).)*

This analysis also ignores the significant physical constraints related to the existing housing stock and roof space. A 4 kW system currently requires the installation of 16 panels (see http://www.originenergy.com.au/content/dam/origin/residential/docs/solar/current-models/Origin_4000d_solar_panel_brochure.pdf) and many households do not have rooftops that would allow 50 m² of solar PV panels to be deployed in a way that would enable optimum output and aesthetics to be maintained. Furthermore, around one-quarter of Australia's population does not live in a free-standing house (ABS, 2013). There would be significant additional energy requirements if households adopted electric vehicles.

In addition to analyzing solar output patterns to see how many consumers might look to go partially or fully off-grid, the financial feasibility of either option will also be critical. Fig. 6.11 shows the results of a simple financial model of an average household that installs an integrated solar PV and battery system.[6] At current pricing, installation of such a system today would result in the household's investment breaking even in 2035. With a 25 and 50% reduction in capital costs, the investment would be breakeven in 2029 and 2023, respectively.

This analysis indicates that capital costs would need to fall significantly before customers would want to go "off-grid," even if the physical energy production and consumption limitations outlined earlier could be successfully addressed. The CSIRO has estimated that because of these types of limitations,

---

6. The assumptions used: $A24,760 capital cost for 5 kW continuous power system; inflation 2.5%, cost of capital 5%, 15 kWh per day consumption, and grid-connected electricity tariff of $A0.29 per kWh.

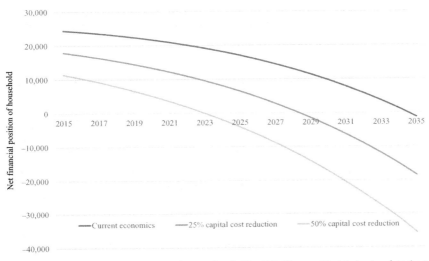

FIGURE 6.11   Net financial position from going "off-grid." *(Source: Modeled using data from Offgrid Australia. Information obtained from http://www.offgridenergy.com.au/. Accessed online on May 12, 2015.)*

the medium cost of household disconnection is $0.47 per kWh, with a low and high range of $0.23 and $0.93 per kWh, respectively (Graham et al., 2013). The analysis suggests strongly that coexistence, rather than complete substitution, is the likely reality facing incumbent utilities. This finding is supported by Gellings, in this book.

4   DEVELOPMENT OF ALTERNATIVE PRICING MODELS

There are a range of ways in which utilities could price their products. Biggar and Reeves, as well as Knieps, discuss the role of demand tariffs and nodal pricing, among others. Athawale and Felder explain the role of rate design and the increasingly referred to "death spiral" concept. However, a precondition of such discussion is the realization that Section 3 of this chapter suggests that solar systems and batteries are not able to facilitate mass grid-disconnection. Even so, utilities facing a future of coexistence with distributed energy solutions, will need to reassess how they price their products. Within Australia, this pressure is even more acute, given historically low levels of consumer trust of electricity retailers. Furthermore, while most customers are unlikely to go "off-grid," they may utilize new "virtual trading" models using advanced information technology. It is not beyond the realms of possibility to imagine how Uber or AirBnB platforms could be transferred to energy, facilitating trading between households or microgrid development. For example, households producing energy during the day may wish to sell it to households with substantial evening peak consumption

(Cooper provides further discussion on platforms for energy dispersion within communities). As such, pricing is likely to be a key factor, which utilities must consider to meet emerging customer needs.

Currently, residential electricity tariffs in Australia are conventional two-part tariffs, with energy throughput measured in kilowatt-hour, utilizing either average cost, or inclining block structures. For example, Queensland residential tariffs are 20.1 cents/kWh, with a fixed charge, irrespective of individual household connection peak demand requirements, of 79 cents/day (Simshauser, 2014). For a household consuming an average 7000 kWh p. a., 83% of the total annual bill of $1695 is based on energy throughput, with the remaining 17% effectively a smeared average cost connection charge.

There are significant equity issues associated with such an approach (Athawale and Felder in this book discuss the importance of equity issues). For Australia, Simshauser (2014, p. 25) estimates that households without solar PV are paying around $70 p.a. more in electricity tariffs than their use of capacity costs would dictate. Similarly, households with air conditioning are up to $200 p.a. better off due to persistence with average cost tariffs. To address such inequity and provide greater cost reflectivity, Australian utilities are considering a number of tariff reform options, driven by a recent determination by the Australian Energy Market Commission (AEMC) (2014) that all tariffs be "cost-reflective." These options include: significantly increasing the "fixed" component of the "average cost" tariff; introducing a peak demand or kVA charge; re-introducing declining block tariffs; and ToU pricing, including dynamic pricing. An assessment of these tariff designs can be found in Nelson et al. (2014).

In many ways, Australia is emerging as a "test case" for the rest of the world. There is significant penetration of solar PV, and battery storage and advanced metering together with different tariff structures are now being offered by utilities in various Australian jurisdictions. In the coming years, it will be possible to determine which pricing structures have been the most effective for meeting customer demand. Table 6.2 summarizes the revised tariff proposals of the different networks operating within the NEM.

Table 6.2 shows that networks in Queensland, South Australia, and Victoria are shifting away from "average cost" tariffs to demand-based tariffs. At face value, it would appear that introducing kilovolt-amps demand-based fixed charges would be ideal for ensuring allocative efficiency. There will be a need to be cautious here, however. As noted earlier, the profile of solar production in Australia, appears to do little to reduce peak demand because "peak" solar output and peak demand do not coincide. Introducing demand tariffs would ensure that a household would pay network charges to reflect their use of the network, irrespective of the *output* produced by their solar PV system. One problem is that there are now over one million Australian households with solar PV on their roofs. Each of these households is likely to view their investment in solar as a "sunk cost."

TABLE 6.2 Shifts in Network Pricing Structures

Jurisdiction/ network	Regulatory period	Regulatory status (at time of writing)	Proposed network pricing changes	Application date
Queensland				
Ergon	2015–16 to 2019–20	Preliminary decision made on May 1, 2015	Introducing "Seasonal demand" pricing for all small customers.	Jul. 1, 2016
Energex	2015–16 to 2019–20	Preliminary decision made on May 1, 2015	Introducing a voluntary (opt-in) demand tariff for residential customers with interval metering enabled. Energex plan to make demand tariff mandatory from 2020.	Jul. 1, 2016
New South Wales				
Ausgrid	2014–15 to 2018–19	Final decision	Converting any small customer inclining block tariffs to a flat (or possibly declining block) structure, over the next 4 years. No plans to introduce demand-based tariffs.	Jul. 1, 2015
Essential	2014–15 to 2018–19	Final decision		
Endeavour	2014–15 to 2018–19	Final decision		
ActewAGL	2014–15 to 2018–19	Final decision	No indication of a move away from variable and TOU pricing as yet.	Jul. 1, 2015
Victoria				
Citipower	2016–20	Proposal made on Apr. 30, 2015	Citipower and Powercor have indicated they will explore demand pricing using LRMC as a basis but have made no firm proposals as yet. Preparing to consult.	Jan. 1, 2016
Powercor	2016–20	Proposal made on Apr. 30, 2015		

(Continued)

TABLE 6.2 Shifts in Network Pricing Structures (cont.)

Jurisdiction/ network	Regulatory period	Regulatory status (at time of writing)	Proposed network pricing changes	Application date
United Energy	2016–20	Proposal made on Apr. 30, 2015	Introducing new seasonal demand-time of use tariff for residential customers. Will use actual maximum demand (kilowatt) between 3:00 pm and 9:00 pm weekdays, with the demand rate increasing in summer.	Jan. 1, 2016
Jemena	2016–20	Proposal made on Apr. 30, 2015	Introducing a demand charge to all residential network tariffs. The proposal is for a monthly maximum demand charge based on the maximum demand (kilowatt) in each month, recorded between 10:00 am and 8:00 pm weekdays, with no minimum chargeable demand level.	Jan. 1, 2016
SP Ausnet	2016–20	Proposal made on Apr. 30, 2015	Considering including a demand component in network tariffs for small customers, based on the LRMC of supply. Options include charging customers based on anytime maximum demand, maximum demand in peak period, or average demand in peak period. Also considering adopting geographic differentiation for demand charges across the AusNet network region.	Jan. 1, 2017
Southern Australia				
SAPN	2015–16 to 2019–20	Preliminary decision made on May 1, 2015	In 2014 introduced an opt-in monthly demand/capacity tariff for residential and small business customers with interval metering. Plan to roll out this tariff as mandatory to all customers that install interval meters. Also flattening current inclining block tariff.	Jul. 1, 2015

Source: Australian Energy Regulator determinations, company reports.

In other words, the investment case for any battery technology that allows the consumer to shift the output of their solar PV system to the time of their peak demand is not likely to include the cost of the existing solar PV system.

The clear commercial driver for networks is to introduce cost reflectivity in pricing, thereby providing customers with incentives to shift *demand* and *output* into periods that alleviate network congestion. In this context, demand tariffs are being structured differently, with subtle but important distinctions. Some networks are moving toward "anytime maximum demand" tariffs, while others are adopting "maximum demand" tariffs within a particular time period. These differences will provide varying incentives for individual customers. It is evident that utilities are beginning to adopt different pricing strategies for optimizing coexistence between the grid and distributed generation and storage.

Interestingly, New South Wales networks are restructuring tariffs in a very different way. Rather than shifting to demand tariffs, these businesses are proposing to reintroduce "declining block tariffs." Such tariffs provide customers with additional incentive to use more energy from the grid as prices decline with greater volumetric consumption. This is particularly problematic in an era where energy efficiency is a public policy objective in its own right (Cavanagh and Levin explain this further in this book). These businesses have effectively argued that with declining demand, there is very little network augmentation required and reducing unit costs for consumers depends upon increasing throughput in the network. In many ways, such an approach represents the typical response of a business with surplus stock, reducing prices to clear inventories. This may conflict with broader environmental policy however, with a need to reduce coal-fired generation utilization due to long-term climate change policy.

Irrespective of the design of tariffs, electricity prices in Australia are likely to fall in real terms as regulators significantly reduce revenue caps (the total revenue allowed to be collected by regulated networks from customers) between 2015 and 2020. Recent determinations by the Australian Energy Regulator (AER) indicate that allowable revenue is likely to be up to 30% lower than the previous regulatory period. Consequently, Simshauser and Nelson (2013) estimate that electricity bills in 2020 may be 10% lower in real terms, than 2013. This presents an opportunity for networks and retailers to price grid-based electricity more effectively as a complement to embedded generation and battery storage, with a focus on its reliability benefits.

Figs. 6.6 and 6.7 and the analysis of the previous section, showed that the vast majority of Australian households are likely to continue to use grid-connected electricity for reliability purposes. Put simply, their energy consumption exceeds their self-generation and storage capacity. Therefore Australian energy businesses can begin to consider how to adjust their tariffs to reflect the value of reliability to the consumer.

The analysis in this chapter supports the findings of previous studies. McIntosh (2014) states that if electricity utilities offer competitive tariffs, the majority of customers will remain connected to take advantage of the cost-effective reliability offered by the grid. Riesz et al. (2014) go further by stating that businesses may adopt shadow pricing so as to remain just cheaper than the disconnection option, to ensure they don't lose customers.

One of the key challenges facing utilities during the transition is the speed of tariff adjustment. The Bonbright (1961) tariff principles state that "stability and predictability" are important elements of tariff design. Given the speed at which distributed generation is entering the market, utilities may need to abandon adherence to this principle to create appropriate pricing signals which are cost reflective (Faruqui, 2015). However, if utilities hastily shift tariffs, it may further erode low levels of trust and this will create additional barriers to engaging customers in a positive way.

5   CONCLUSIONS

This chapter has explored the role of the utility and grid-connected electricity pricing in the transition to greater penetration of distributed generation and battery storage. The analysis demonstrates that coexistence, rather than complete grid substitution, is likely to be the reality facing utilities. While there is strong interest in homes and communities going "off the grid," most people anticipate still being able to access the grid for backup (Newgate Research, 2015).

Market analysts are not conclusively forecasting whether or not the strategies adopted by Australian utilities are likely to create value for shareholders. In mid-2015, Stanley (2015) stated that the emergence of batteries and continued uptake of solar PV could reduce the earnings of AGL and Origin (the two publicly listed customer-facing "gentailers") by $30 million to $A40 million, in 2017, and by up to $A100 million in 2020. However, the same analysis indicates that these companies may "claw back" $A20 million per year, and could "surprise further with a compelling competitive response."

This chapter has outlined how utilities within Australia are responding to the emergence of new technologies by offering new products and services. The integrated utilities AGL, Origin, and Energy Australia have all released innovative metering products, embedded generation, storage, and innovative financing offerings, such as PPAs and on-the-bill finance. At the same time, tariffs are being shifted away from "average cost" throughput pricing to more cost-reflective demand based tariffs. Importantly, utilities in different jurisdictions are adopting alternative pricing structures, depending upon their business strategy. Over the coming years, the Australian market will provide evidence as to the success, or failure, of these pricing strategies in embracing and profiting successfully from emerging technologies, and new business models.

Governments have indicated support for these developments. The Clean Energy Finance Corporation recently announced funding of $120 million to expand the range of financing options in Australia (including leasing and PPAs) through Sun Edison, Tindo Solar, and Kudos Energy. The Victorian State Energy Minister recently endorsed publicly the potential of energy storage to change the way households produce and use energy, stating that such a development will "mean affordable clean energy and the jobs that will come out of that, and the technologies, are going to be significant" (Arup, 2015, p.3).

An unanswered question for utilities concerns the impact of the disruption on the ability of the sector to attract capital. During the World Economic Forum Annual Meeting 2014 in Davos, chief executives from leading utilities and energy technology companies identified the risk that some parts of the industry value chain may not attract the necessary investment (Bain & Company, 2014). This is partly due to disruption resulting in higher risk premiums for the sector (Simshauser and Nelson, 2012), the lack of integrated energy and climate change policy, and market designs being inconsistent with new technologies, and a decarbonization agenda (Nelson et al., 2015). This may be an important area for further research as the utility of the future will be influenced by both technology developments and broader energy and climate change policy developments.

REFERENCES

Abbott, M., 2002. Completing the introduction of competition into the Australian electricity industry. Econ. Pap. 21, 1–13.

Adelaide Advertiser, 2015. Solar Raises House Value. Adelaide Advertiser, May 11, 2015, p. 27.

AGL Energy, 2014. AGL Embraces Disruptive Technologies to Meet Changing Consumer Needs. Accessed online at: http://www.agl.com.au/about-agl/media-centre/article-list/2014/november/agl-embraces-disruptive-technologies-to-meet-changing-consumer-needs

AGL Energy, 2015a. AGL Greenhouse Gas Policy. Accessed online at: http://www.agl.com.au/~/media/AGL/About%20AGL/Documents/Media%20Center/Corporate%20Governance%20Policies%20Charter/1704015_GHG_Policy_Final.pdf

AGL Energy, 2015b. AGL is First Major Retailer to Launch Battery Storage. Accessed online at: http://www.agl.com.au/about-agl/media-centre/article-list/2015/may/agl-is-first-major-retailer-to-launch-battery-storage

AMR, 2015. RepTrak Pulse Report. The World's Most Reputable Companies: An Online Study of Australian Consumers, March 2015.

Arup, T., 2015. Government Powers up Battery Focus in Energy Plans. The Age, May 11, 2015, p. 3.

Australian Bureau of Statistics (ABS), 2013. Australian Social Trends. Available at: http://www.abs.gov.au/AUSSTATS/abs@.nsf/Lookup/4102.0Main + Features30April + 2013#back7

Australian Energy Market Operator (AEMO), 2014. 2014 Statement of Opportunities. AEMO Publication, Sydney.

Bain and Company, 2014. The Future of Electricity: Attracting Investment to Build Tomorrow's Electricity Sector. Report to the World Economic Forum.

Bonbright, J., 1961. Principles of Public Utility Rates. Columbia University Press, New York.

Clean Energy Council, 2014. Clean Energy Australia: 2014. Clean Energy Council Publication, Melbourne.

CSIRO, 2013. Residential Electricity Use in Australia. CSIRO Publishing, Clayton.

Energex, 2010. 2010 Queensland Household Energy Survey. Brisbane.

Faruqui, A., 2015. The global movement toward cost-reflective tariffs. Presentation to Energy Transformed Conference, Sydney, May 7, 2015.

Francis, H., 2015. Why Tesla's New Battery is Kind of a Big Deal. The Sydney Morning Herald. Accessed online at: http://www.smh.com.au/digital-life/digital-life-news/why-teslas-new -battery-is-kind-of-a-big-deal-20150509-ggxagc.html

Giarious, C., 2015. Reorienting for the future: delivering energy or service. Panel Session at Energy Transformed Conference, Sydney, May 7, 2015.

Grattan Institute, 2014. Fair Pricing for Power. Grattan Institute Publication, Melbourne.

Graham, P., Dunstall, S., Ward, J., Reedman, L., Elgindy, T., Gilmore, J., Cutler, N., James, G., 2013. Modelling the Future Grid Forum Scenarios. CSIRO, Clayton South, Victoria.

Keane, A., 2015. Solar Still Shines for Household Savings. Adelaide Advertiser, May 11, 2015, p. 48.

King, G., 2015. Origin energy: delivering on priorities. Presentation to the Macquarie Australia Conference, May 6, 2015.

McIntosh, B., 2014. Distributed solar with storage … and disconnection?. Presentation to the 2014 Asia-Pacific Solar Research Conference.

Stanley, M., 2015. Australia Utilities Asia Insight: Household Solar & Batteries. Morgan Stanley Research Note, May 2015.

Nelson, T., Reid, C., McNeill, J., 2015. Energy-only markets and renewable energy targets: complementary policy or policy collision? Econ. Anal. Pol. 46, 25–42.

Nelson, T., McNeill, J., Simshauser, P., 2014. From throughput to access fees: the future of network and retail tariffs. In: Sioshansi, F. (Ed.), Distributed Generation and its Implications for the Utility Industry. Elsevier, Amsterdam.

Nelson, T., Simshauser, P., Nelson, J., 2012. Queensland solar feed-in tariffs and the merit-order effect: economic benefit, or regressive taxation and wealth transfers? Econ. Anal. Pol. 42 (3), 277–301.

Newgate Research, 2015. Community attitudes to energy issues. Briefing for AGL, January 2015.

Riesz, J., Hindsberger, M., Gilmore, J., Riedy, C., 2014. Perfect storm or perfect opportunity? future scenarios for the electricity sector. Distributed Generation and its Implications for the Utility IndustryElsevier, Amsterdam.

Saddler, H., 2013. Power Down: Why is Electricity Consumption Decreasing. Australia Institute Paper, No. 14.

Simshauser, P., 2014. From first place to last: Australia's policy-induced energy market death spiral. Aust. Econ. Rev. 47 (4), 540–562.

Simshauser, P., 2010. Vertical integration, credit ratings and retail price settings in energy-only markets: navigating the resource adequacy problem. Energy Pol. 38 (11), 7427–7441.

Simshauser, P., Nelson, T., 2013. The Outlook for Residential Electricity Prices in Australia's National Electricity Market in 2020. Electricity J. 26 (4), 66–83.

Simshauser, P., Nelson, T., 2012. Carbon taxes, toxic debt and second-round effects of zero compensation: the power generation meltdown scenario. J. Financ. Econ. Pol. 4 (2), 104–127.

Vergetis Lundin, B., 2015. Utilities "Clamoring" to Get Into Home Energy Market. SmartGridNews. Accessed online at: http://www.smartgridnews.com/story/utilities-clamoring-get-home -energy-market/2015-03-24

Wood, T., Blowers, D., Chisholm, C., 2015. Sundown, Sunrise: How Australia can Finally get Solar Power Right. Grattan Institute, Melbourne.

Chapter 7

Intermittency: It's the Short-Term That Matters

Daniel Rowe, Saad Sayeef, Glenn Platt
CSIRO Energy, Mayfield West, NSW, Australia

1 INTRODUCTION

Solar photovoltaic (PV) energy systems are one of the key drivers behind the changing grid; however, there is great concern as to how solar *intermittency* may affect the rollout and realization of solar energy benefits. Already, today, concerns regarding intermittency have resulted in solar energy installation being restricted in certain electricity systems, and questions are being asked as to the viability of solar energy in providing a reliable electricity supply.

In future, utilities could play a major role in managing increasing intermittency, and transacting energy as grid-connected microgrids find their place (see the chapter by Marnay), and as distributed generation grows. A range of countermeasures for intermittency exist, and the importance of their characterization, consideration and application is growing, as increasing renewable penetration, energy storage, electric vehicles, and changing consumer behavior compound. These changes bring both the potential for cost and revenue for utilities, depending on the emergent role of utilities (see chapter by Nelson and McNeill), and the distribution business models adopted (see chapter by Boscán and Poudineh), in the context of a new power system with increased flexibility.

Whilst much is said about intermittency and its effect on electricity networks, the information shared and views expressed are often anecdotal, difficult to verify, and limited to a particular technical, geographical, or social context. Concerns about solar intermittency frequently lack supporting evidence, and do not provide suitable detail for investigation, generalization, and mitigation. Further, detail on the timescales of intermittency, the observed impact, and comparison with other types of intermittency (including load intermittency) is often omitted—increasing the challenge for utilities seeking to adapt and respond to the challenges of intermittency. In fact, there is surprisingly very little real-world data on how intermittency, particularly solar

Future of Utilities - Utilities of the Future. http://dx.doi.org/10.1016/B978-0-12-804249-6.00007-5

intermittency, affects electricity networks. However, solar intermittency *is* regarded as one of the biggest technical challenges facing major players in the changing grid, as utility equipment is exposed to new operating conditions, and the popularity of solar power has resulted in rapid and widespread adoption. As solar penetration grows, so does the potential for different power flow paths (reverse power flow), larger and more rapid changes in power levels due to the combined effect of changes in distributed generation *and* load (ramp rates), and the traditional expectation of voltage "sag" along a line may no longer hold.

Surprisingly, though, significant disagreement exists amongst solar stakeholders regarding the actual impact of intermittency, the timescales of concern, and measures to mitigate or address these challenges. Further, generation intermittency is frequently presented as a significant concern, and a risk to system stability, while load intermittency is accepted as a manageable known that is mitigated by diversity, to be managed in aggregate. This divergence is also apparent in the presentation of intermittency—with evidence of generation intermittency often presented on a per- (or small) system basis, while load intermittency is depicted by aggregated load curves that exhibit diversity and significant averaging. Despite this, intermittency (both load and generation) *is* an important issue to be managed in order to ensure consumer amenity, optimal asset utilization, and efficient equipment operation. The term for the ability of a network's aggregated generation fleet to respond to these changes is referred to as "flexibility."

This chapter explores the issue of solar intermittency, presents stakeholder concerns, and describes what evidence exists (or in fact, what *doesn't* exist) regarding the negative impacts of solar intermittency on grid operations. This chapter also provides detail on: the timescales relevant to solar intermittency and its countermeasures; the importance of high resolution data collection; the need for a fair comparison framework when evaluating solar intermittency against other sources of intermittency; and the importance of an evidence-based linking of solar intermittency with electricity network effects. Detail is also provided on the areas where intermittency has been demonstrated to cause significant problems, and the potential solutions to these issues being investigated around the world.

This chapter is organized as follows: Section 2 identifies the types of intermittency present on electricity networks today; Section 3 defines intermittent generation, and the scales at which grid impacts can occur; Section 4 discusses intermittency concerns, and areas where intermittency has been demonstrated to cause significant problems; Section 5 examines evidence of negative solar intermittency impacts on grid operations; Section 6 outlines intermittency considerations for stakeholders; Section 7 outlines operations impacts of intermittency; Section 8 explores intermittency countermeasures followed by the chapter's conclusions.

2 INTERMITTENCY AND UTILITIES

Intermittency is often associated with solar energy; however, a number of sources of intermittency may be present on electrical networks, including wind and load intermittency. The timescales of this intermittency also vary—including seconds, diurnal (day/night), and seasonal intermittency. The source, magnitude, and frequency of intermittency, and the significance of network impacts, is important to the selection of countermeasures, such as energy storage, load control, and generation control.

One major source of intermittency present on electricity networks, and already managed by utilities, is load intermittency. Fig. 7.1 depicts load intermittency for a residential house in Australia, sampled every second, over a 24-h period, with significant and rapid changes in energy consumption clearly visible. Fig. 7.2, meanwhile, presents a longer-term and aggregated view of load intermittency, with highly averaged changes in energy consumption visible over a duration of days to weeks, characterized by a notable increase in community energy consumption during a holiday period, for a number of years.

Similarly, Fig. 7.3 depicts solar intermittency for a residential building in Australia, over a 24-h period, with significant and rapid changes in solar energy generation clearly visible. This occurrence is not unusual with popular residential rooftop solar PV capacities, often in the 1–5 kW_p range, installed

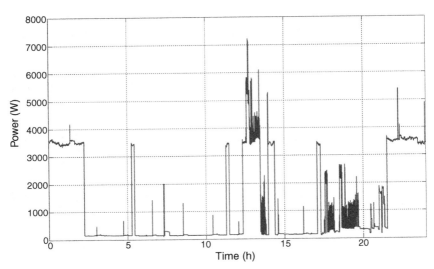

FIGURE 7.1 **A residential house load profile over 24 h showing significant load intermittency.**

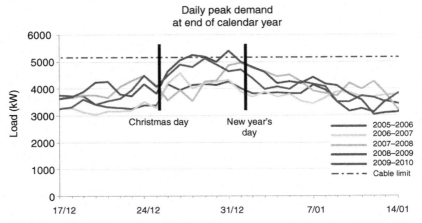

FIGURE 7.2 **A load profile for a holiday location in Australia showing significant load intermittency, with load peaking over the holiday period.** *(Source: Townsville Queensland Solar City (2012).)*

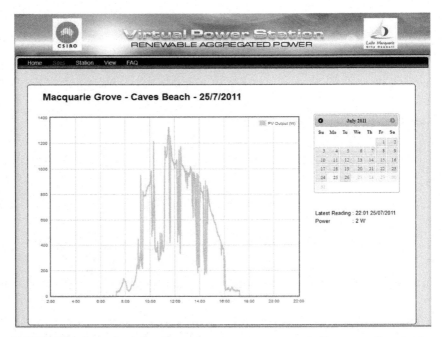

FIGURE 7.3 **A residential rooftop solar generation power profile over a day showing significant solar intermittency.**

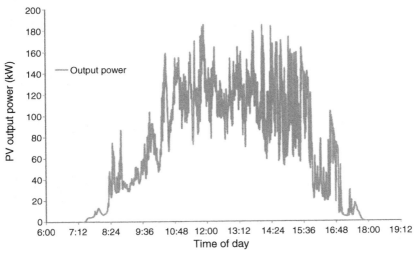

FIGURE 7.4 Solar PV output profile on a partly cloudy day at the DKASC in Alice Springs, Australia.

on rooftops across utility distribution networks. In Australia, there are nearly 1.4 million of these systems, totaling 4 GW$_p$. Meanwhile, Fig. 7.4 illustrates the power output profile of a larger 196 kW solar PV system, sampled every ten seconds. This system is located at the Desert Knowledge Australia Solar Centre (DKASC) in Alice Springs, Australia. Large changes in the magnitude of solar power can be seen to occur due to cloud cover and/or movement, but the percentage change of power, relative to total capacity, is smaller than that of the residential rooftop system with a larger geographical area covered by the system at DKASC. As solar penetration of both large and small systems increase, utilities are tasked with the management of an increasingly dynamic network, requiring a shift in the countermeasures deployed to tackle intermittency. However, solar systems with larger collector areas, such as utility scale solar farms and more distributed solar power, may reduce this burden due to intermittency decorrelation achieved through spatial diversity.

To this point, Fig. 7.5 shows an assessment of how aggregating sites can reduce solar intermittency. Fig. 7.5 shows the variability at a 10 second level for three cases, namely (a) a single node; (b) twenty nodes (referred to as 'combined' in the figure); and (c) ten times the size of a 'combined' system representing 200 nodes. It can be seen that with 20 nodes, there is typically less than 10% variability over a 10 second interval, dropping to around 2% for 200 nodes. In both the solar and load intermittency examples, both the individual-site intermittency, and the aggregate intermittency are currently managed by the electricity grid through diversity, as well as generation (energy), and ancillary services (network stability) mechanisms. However, the utility of the future is likely to deal with significantly more dynamism, as distributed generation,

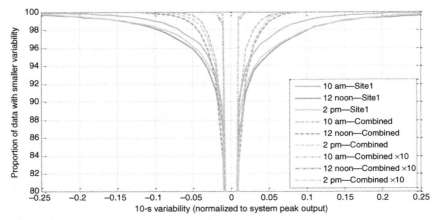

FIGURE 7.5 Assessment of variability of virtual power station output over a 10 second interval for different number of nodes. *(Source: Ward et al. (2012).)*

energy storage, load control and new devices, such as electric vehicles, impact the grid. This dynamism will require increased grid flexibility to accommodate the impacts of new and disruptive technologies, and pricing, on electricity consumption, as addressed by Grozev et al.

3 FRAMING INTERMITTENCY AND POTENTIAL POWER SYSTEM IMPACTS

Comparing intermittency, and identifying the severity or significance of its electrical impacts on utility networks is difficult without characterizing intermittency and its impacts. In order to understand, compare and respond to intermittency suitably, a definition is required that enables discussion about the characteristics of intermittency, and the areas in which both the electrical effects and the countermeasures apply. Understanding the nature of the impact and the suitability and performance of different countermeasures, in different network areas, is important to understanding the potential costs and revenues associated with the management of intermittency by utilities.

In seeking to define high penetration intermittent generation (HPIG), the International Energy Agency (IEA) Task 14 group noted that, "Although up to now, no common definition of 'high penetration PV scenarios' exists, there is consensus amongst the parties developing this Task that a high penetration situation exists if additional efforts will be necessary to integrate the dispersed generators in an optimum manner" (Australian PV Institute, 2013).

CSIRO's 2012 Solar Intermittency: Australia's Clean Energy Challenge report (Sayeef et al., 2012) built on this philosophy, by suggesting that a high penetration, intermittent generation (HPIG) scenario exists, where it is the variability of the intermittent generation, rather than the loads, within a network segment that is the dominant factor in determining the need for substation, network, or control upgrades. This definition can then be expressed mathematically, based on a comparison of the largest net variability, as seen in both loads and generation (Fig. 7.6). Further, the scales at which PV is connected, and the potential power system impacts of intermittency over various intermittency timescales were defined (Figs. 7.7 and 7.8).

These definitions provide a common reference for the characteristics and effects of intermittency to be discussed, and references the relevant Australian network levels where impacts and countermeasures apply. These levels can easily be translated for other regions. Having defined high penetration intermittency, scales and connection points, and potential power system impacts, the availability of evidence regarding the negative impacts of solar intermittency on grid

$$HPIG_\tau \text{ exists if } | Pg_i - Pg_{(i-1)} | \geq | Pl_i - Pl_{(i-1)} |$$
$$\text{for } i \text{ such that } | P_i - P_{(i-1)} | \text{ is maximized}$$

where

Pg_i is intermittent power generation (kW) at time i

Pl_i is load power (kW) at time i

$P_i = Pl_i - Pg_i$ is the net load (kW) at time i

τ is the time interval between time i and $i-1$

Note that the variability of loads and generation needs to be assessed over a timeframe appropriate for the network characteristics under consideration—as is the case when assessing other network performance characteristics such as voltage and frequency fluctuations. This is for the purpose of assessing the time interval τ.

FIGURE 7.6   CSIRO's definition of HPIG expressed mathematically. *(Source: Sayeef et al. (2012).)*

Identifier	Connection point
Small-scale	230 V/400 V (or 240 V/415 V) Low voltage distribution network
Utility-scale	> 230 V/400 V (240 V/415 V) Distribution network
Large-scale	≥ 66 kV Transmission and sub-transmission network

FIGURE 7.7   Scales at which PV is connected in Australia. *(Source: Sayeef et al. (2012).)*

Timescale of intermittency	Potential power system impact
Seconds	Power quality (eg, voltage flicker)
Minutes	Regulation reserves
Minutes to hours	Load following
Hours to days	Unit commitment

FIGURE 7.8   Potential power system impacts of intermittency over various intermittency timescales. *(Source: Sayeef et al. (2012).)*

operations can be examined in order to identify the degree to which intermittency impacts may be experienced by utilities, and the opportunity to respond.

4 WHY DOES RENEWABLE ENERGY INTERMITTENCY MATTER?

The need for generation to occur at the time of consumption has been a longstanding feature of electricity networks, in addition to the paradigms of voltage drop along a distribution line, and one-way power flow. However, this is changing, and increase of distributed generation coupled with storage and load control increases the prospect of a dynamic and complex new mode of grid operation, making conventional network planning and power regulation an increasingly challenging task. In particular, historically, distribution networks have been regarded as a passive termination of the transmission network, and point of connection to customers, but the increase in distributed solar penetration, grid-interactive electric vehicles and storage will make this a particularly active and central part of the network (see chapter by Gimon). A rise in the requirement for power regulation is also likely to increase, due to mismatches between supply and demand induced by renewable generation variability, and changing consumer behavior.

CSIRO has also identified areas where intermittency has been demonstrated to cause significant problems—primarily in "fringe of grid" and off-grid systems, often in rural areas—where the electrical network exhibits high impedance and little interconnection in the presence of high penetration solar power. Data sampled at 10-second intervals for solar power output from a 196 kW solar PV system and voltage on the feeder to which the system is connected to at Alice Springs, Australia, demonstrate the effects of solar intermittency in such an environment, as shown in Fig. 7.9.

Using this data, CSIRO built a model, and modified the penetration of solar (10 and 40% of load) and a lumped grid impedance (of 0.01 p.u for a strong grid, 0.2 p.u. for a weak grid) to investigate effects in four scenarios:

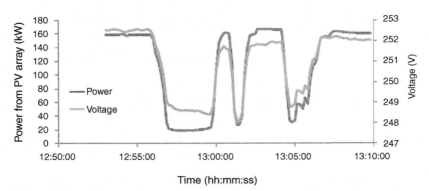

FIGURE 7.9 Actual data from a 196 kW PV system in a remote area showing power and voltage fluctuations. *(Source: Sayeef et al. (2012).)*

- a low penetration of solar in a strong grid;
- a low penetration of solar in a weak grid;
- a high penetration of solar in a strong grid;
- a high penetration of solar in a weak grid.

The resulting impact on system voltage is shown in Figs. 7.10–7.13 showing that when the penetration is low, and it is attached to a strong grid, large rapid swings in power output are not an issue; however, when attached to a rural feeder, where the grid is not strong, an increase in penetration caused an increase in the voltage swings observed. If the penetration were increased on this type of feeder, the voltage swings would potentially begin to impact adversely the operation of this part of the network. It might result in PV inverters tripping off with these high voltage swings, causing a larger power fluctuation and, therefore, worsening the voltage swings. The occurrence of significant voltage swings due to solar intermittency in high solar penetration and weak grid problem areas constitutes a significant concern, given utility power quality requirements including the management of grid voltages within specified limits. Further, the potential

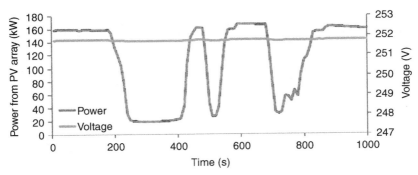

FIGURE 7.10 Voltage at the PV array for low penetration and a strong grid. *(Source: Sayeef et al. (2012).)*

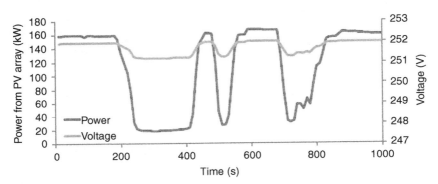

FIGURE 7.11 Voltage at the PV array for low penetration and a weak grid. *(Source: Sayeef et al. (2012).)*

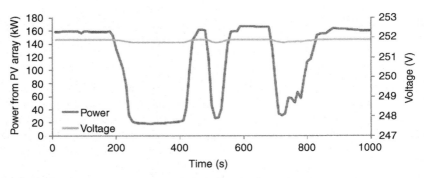

FIGURE 7.12   Voltage at the PV array for high penetration and a strong grid. *(Source: Sayeef et al. (2012).)*

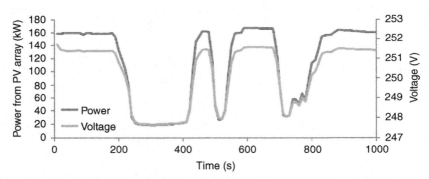

FIGURE 7.13   Voltage at the PV array for high penetration and a weak grid. *(Source: Sayeef et al. (2012).)*

for fluctuating voltages to affect other solar production, worsening these effects, is a risk requiring consideration and management through countermeasures.

CSIRO's Solar Intermittency: Australia's Clean Energy Challenge (Sayeef et al., 2012) outlined specific issues identified by electricity industry stakeholders in Australia, including:

- voltage effects;
- stability effects;
- ramp rate effects;
- solar capacity factor;
- a lack of high resolution solar data.

These issues were found to be representative of industry concerns around the world regarding solar intermittency, albeit that some effects were exacerbated due to the idiosyncrasies of the world's longest interconnected power system, such as the limited level of interconnection.

The potential impact of these common concerns to the operation of electricity grids and the realization of solar benefits necessitates the collection of evidence and data surrounding intermittency, electricity network impacts, and countermeasure efficacy. The need to understand and characterize solar intermittency from the perspective of system planners and operators is critical to the transformation of utility business models, and the formulation and application of an efficient technical and commercial response. Exploration of the issues, opportunities, and solutions surrounding solar intermittency, and its mitigation, may inform and impact the emergent role of utilities of the future.

This stakeholder discussion regarding technical, business, and other changes to the utility operating environment was continued under the subsequent Future Grid Forum, Australia's first extensive whole-of-system evaluation, encompassing the entire energy chain, from generation through to consumption. The public report, Change and choice: The Future Grid Forum's analysis of Australia's potential electricity pathways to 2050 (CSIRO Energy, December 2013), outlined four scenarios that have far-reaching implications for the current and future electricity supply chain, including:

- Set and Forget – where consumers rely on utilities;
- Rise of the Prosumer – where consumers actively design or customize solutions;
- Leaving the Grid – where consumers disconnect from the grid; and
- Renewables Thrive – where storage play a large part in entire electricity system.

This analysis and modeling conducted for this report saw intermittency emphasized, particularly in the renewables thrive scenario, and managed through the deployment of utility-scale and "behind the meter" storage. Updating and building upon the Future Grid Forum, CSIRO and the Energy Networks Association recently launched the Network Transformation Roadmap (NTR) that aims to focus on the 2015–25 decade, and to "identify the new services and technologies that future residential, commercial and industrial customers will value," as well as to "identify the options for regulation, business models and electricity pricing that are best able to support delivery of the future services which customers want, while ensuring an efficient competitive and economically-robust value chain" (Electricity Network Transformation Roadmap Overview, ENA, 2015). The Network Transformation Roadmap links the technical issues and sectoral changes identified by the Solar Intermittency: Australia's Clean Energy Challenge and Future Grid Forum reports, in order to inform utility revenue and business models.

5   EVIDENCE OF NEGATIVE SOLAR INTERMITTENCY IMPACTS ON GRID OPERATIONS

Evidence of solar intermittency's impact on networks is limited. However, a 2008 survey of papers on the likely factors limiting PV penetration was conducted by Sandia National Laboratories (Whitaker et al., 2008), and is summarized

TABLE 7.1 Expected Factors Limiting PV Penetration

References (authors)	Maximum PV penetration	Cause of upper limit
Chalmers, S., et al.	5%	Ramp rates of mainline generations. PV in central-station mode.
Jewell, W., et al.	15%	Reverse power swings during cloud transients. PV in distributed mode.
Cyganski, D., et al.	No limit found	Harmonics.
EPRI report	>37%	No problems caused by clouds, harmonics, or unacceptable responses to fast transients were found at 37% penetration. Experimental + theoretical study.
Baker, P., et al.	Varied from 1.3 to 36%	Unacceptable unscheduled tie-line flows. The variation is caused by the geographical extent of the PV (1.3% for central-station PV). Results particular to the studied utility because of the specific mix of thermal generation technologies in use.
Imece et al.	10%	Frequency control versus break-even costs.
Asano, H., et al.	Equal to minimum load on feeder	Voltage rise. Assumes of LTCs in the MV/LV transformer banks.
Povlsen, A., et al. and Kroposki, B., et al.	<40%	Primarily voltage regulation, especially unacceptably low voltages during false trips, and malfunctions of SVRs.
Union for the Coordination of Transmission of Electricity	5%	This is the level at which minimum distribution-system losses occurred. This level could be nearly doubled if inverters were equipped with voltage regulation capability.
Dispower	33 or ≥ 50%	Voltage rise. The lower penetration limit of 33% is imposed by a very strict reading of the voltage limits in the applicable standard, but the excursion beyond that voltage limit at 50% penetration was extremely small.

Source: Whitaker et al. (2008).

in Table 7.1. The different papers surveyed looked at different types of grid and penetration contexts, with some looking at distribution system penetration, and others looking at central generation penetration.

The maximum PV penetration level accommodated can be seen to be highly situation-dependent, and varying widely, with limits reported between 5 and 50% and, in one case, with no limit. Broad reasons for restricting high PV penetration levels were reported, including:

- ramp rates of conventional generation;
- reverse power swings;
- frequency control;
- voltage regulation.

Other studies are no more certain, with conflicting outcomes from different studies regarding savings and costs associated with solar power presented in Table 7.2.

A good example of output variations that can be expected from a large-scale solar PV system can be seen from the output of a 4.6 MW PV system located in Springerville, Arizona, United States. The extent of intermittency exhibited by such a large system can be seen in Fig. 7.14, where the solar PV output data was sampled every 10 s. Large, abrupt power output drops, from about 4000−500 kW, can be seen to occur over extremely short timeframes.

High ramp rates, associated with solar power output changes due to cloud cover, such as that shown above, have the potential to cause adverse impacts on an electricity network at a high solar power penetration levels. One of the main challenges to the power system is related to the instantaneous penetration of intermittent renewable generation, that is, the fraction of total system load

TABLE 7.2 Conflicting Outcomes in Existing Literature

Study outcomes	Corresponding conflicting outcomes
A study by the New York Independent System Operator (NYISO) reported that significant cost savings could be achieved by integrating intermittent generation (wind in this case) due to the displacement of fuel, primarily natural gas, by wind and by having accurate forecasts.	The POVRY study for Europe reported that the volatility introduced by the highly variable wind and solar output will directly impact on cost. Prices are expected to peak more and become less predictable, representative of the nature of weather systems.
Analysis performed in a Western US study found that savings can be achieved from the introduction of solar power through the displacement of gas and coal fired power generation and a price on carbon. Large forecast errors would, however, cause expensive generation to be brought online.	Simulations performed in the Texas, US (ERCOT) grid where different mixes of wind and solar power were modeled reported that high penetration of intermittent generation will increase system costs due to the upgrade of conventional generation equipment required to achieve increased system flexibility.
Analysis of results in a study carried out by the California Energy Commission on the California Independent System Operator (CAISO) system showed wind power to be more variable than solar power when both were considered in aggregate at 1-h intervals.	It was reported in a Swedish study that the smoothing effect due to aggregation is greater for wind than for solar at the hourly timescale, which conflicts results seen in the CAISO study.

Source: Sayeef et al. (2012).

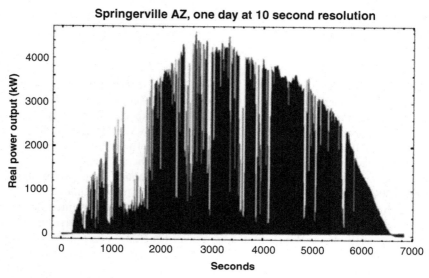

FIGURE 7.14   Power output of a 4.6 MW PV system on a partly cloudy day in Arizona. *(Source: Apt, J. et al.)*

that is being provided by the renewable generation source at a given instant in time (Sayeef et al., 2012). A few studies in the literature have looked into net load with varying penetration levels of intermittent embedded generation, and investigated net load variability due to solar generation. Solar generation is viewed as negative load, and when this is combined with the system load, it yields a *net load*. This corresponds to the power that must be supplied by other sources in the system. When load demand and solar power are both increasing or decreasing coincidentally, the need for other generation sources to vary their output will decrease. However, when load and solar power move in opposite directions at the same time (eg, when load is decreasing while solar power output is increasing), the rate of change increases, and a large variation in net load may result.

The existing electricity system already incorporates significant variability in load demand. This variability and the present variations in net load are managed through generator dispatch, ancillary service mechanisms, and existing utility response mechanisms. As the penetration levels of intermittent renewable generation increase, however, further measures, such as additional ancillary services and utility countermeasure deployment, may be required. A study on the net load variability in the Californian grid was reported in Mills et al. (2009). Hourly change in total load (delta) for load-only and net load (meaning load minus renewable generation) was analyzed for a model of the Californian grid, assuming 33% penetration of renewable energy, wind, and solar in this

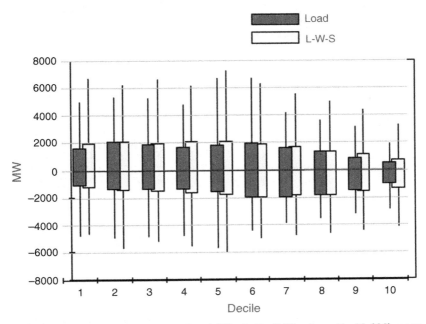

FIGURE 7.15 **Predicted hourly net load variability in the Californian grid with 33% renew-able penetration (note: L-W-S stands for load minus wind minus solar).** *(Source: California Energy Commission (2007).)*

case. Fig. 7.15 illustrates the hourly change for different loading conditions. The data was split into deciles. The first decile is a measurement of delta when load is between 90 and 100% of peak load (top 10% of peak load hours), with the tenth decile being a measurement of delta for loads up to 10% of peak load. The thin lines above and below the bars represent the standard deviation of the positive and negative deltas, respectively. Incorporation of renewables results in an increase in hourly net load variability across the majority of deciles. For instance, the standard deviation of net load variability in the tenth decile is seen to increase by 47%, with the integration of 33% renewable energy (wind and solar).

This analysis was taken further to examine the net load variability for each hour of the day, the distribution of which is shown in Fig. 7.16. At 6 am, the maximum and minimum deltas are similar, but the average hourly variation is seen to increase from 2000 to 2500 MW, when incorporating a 33% penetration of renewables. During the afternoon peak hours, the averages are similar, but the maximum and minimum deltas are significantly larger for net load with solar and wind energy integrated. The standard deviation is also noticeably larger for net load, increasing from approximately 1300−1600 MW, that is, an increase of about 23%.

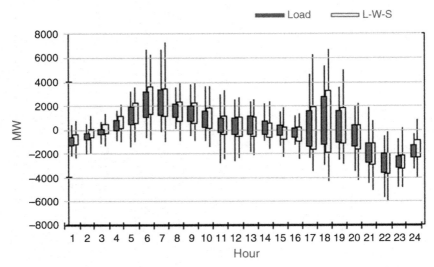

FIGURE 7.16 **Hourly net load variability by hour of day.** *(Source: California Energy Commission (2007).)*

6 INTERMITTENCY CONSIDERATIONS FOR STAKEHOLDERS

Due to the uncertainty of the actual impacts of high penetration intermittent renewable generation on electricity networks and industry concerns, CSIRO sought to provide a resource for stakeholders to reference and understand intermittency considerations and countermeasures.

CSIRO's report, Solar Intermittency: Australia's Clean Energy Challenge (Sayeef et al., 2012), identified the key considerations of:

- a need for high resolution solar data;
- a need to consider spatial diversity as a mitigating factor for solar power, where appropriate, just as with load intermittency; and
- a need for higher system flexibility to accommodate and support higher penetration of renewable generation intermittency.

Although identified in this report, these are global considerations. In particular, sampling rate awareness and a lack of high-resolution solar irradiance and electrical network data was identified as an issue in observing and understanding intermittency. Without high-resolution data, intermittency effects are masked, and characterization of the cause and effect, and efficacy of mitigation measures, is problematic. With limited ability to develop a causal relationship between solar intermittency and electricity network impacts, the selection and implementation of the most efficient and effective countermeasures is unclear. Increased understanding of intermittency effects enables consideration and comparison of a range of countermeasure options, some already used by utilities in the current management of electrical load.

TABLE 7.3 Timescales for Loss of Irradiance Ramps Correlation Versus Distance Between Plants

Distance between plants (km)	Required time scale (min) for loss of correlation
20 < d < 50	>15
50 < d < 150	>30
>150	>60

Source: Mills et al. (2009).

Lawrence Berkley National Labs' (Mills et al., 2009) investigation of spatial diversity identified that irradiance ramps, over timescales of 30 min, were uncorrelated for sites around 50 km apart, with ramps over timescales of 60 min uncorrelated for sites in the order of 150 km apart, and ramps over timescales of 15 min and shorter uncorrelated for all distances between sites, down to the minimum spatial resolution of 20 km between sites. This is presented in Table 7.3, and implies an optimal spacing of 20–50 km between plants, for loss of correlation between solar plants to achieve decorrelation. As with load intermittency, spatial diversity and aggregate management is a mitigating factor for solar intermittency. Consideration of this factor may reduce flexibility requirements.

However, increasing system flexibility, or decreasing the flexibility requirements of the system, is another important determinant for increasing levels of intermittent generation. Due to increased variability of net load due to high penetration of intermittent renewable generation, the number of generator start/stops increases, and sustained load ramps (up and down) steepen. The load following capability will also need to increase above what is required for variation in load alone, due to intermittent renewables (California Energy Commission, 2007). Related decentralized reliability options for utilities are addressed by Woodhouse, and business models and parameters by Boscán and Poudineh, in this volume.

7   OPERATIONS IMPACTS

Historically, conventional generation sources have been controllable, dispatchable, and large-scale, with a dynamic wholesale market driving the trade and supply of reliable electricity. Each generation unit is committed and loaded according to its heat rate, fuel cost and availability, associated transmission losses, and output ramp rate, with the goal of satisfying the electricity demand reliably, at the lowest possible cost (Wan and Parsons, 1993).

The availability and quantity of electricity produced by conventional generation sources, such as coal or gas-fired power stations, can be controlled by system operators, together with the generators' governors and electricity system-wide Automatic Generator Control (AGC) system, or similar. This stands in sharp contrast to solar and wind outputs, as weather variations may lead to highly intermittent generation. The inclusion of such technologies means that conventional generators must follow not only the usual demand variations, but also account for the output variations caused by intermittent generation. Five normal functions of the generation operations that could be impacted are:

- *Load-frequency control*: the amount of load and generation at any given time needs to be matched in order to maintain the system frequency at the desired level (meaning 50Hz in Australia). When load exceeds generation, for instance, the system frequency will drop. When a change in the system frequency is detected, power system operators and AGC/governor interactions will increase or decrease the output of conventional generators, in order to match the load. Intermittent renewable generation technologies generally cannot participate in these system frequency regulations, because their output is independent of system frequency, and output control is typically limited only to curtailment. Thus, as the penetration of intermittent renewable generation systems increase, the capacity to efficiently manage system frequency decreases.

  Distribution utility impact: utilities of the future may play an increasing role in generation, regulation, and load and storage control. In Australia, direct load control of participating consumer air conditioners is already being used for network peak demand management.

- *Load following*: if an increase in solar or wind power is not coincident with the system load increase, other generating units in the system will have to be offloaded, in order to absorb all the solar or wind power. When there is a reduction in solar or wind power production, the output from other units will have to increase, in order to take up the generation slack. Utilities normally use intermediate plants to follow the load. The integration of high penetration intermittent renewable generation may result in increased load-following duties for the conventional generators assigned for system regulation.

  Distribution utility impact: utilities of the future may play an increasing role in generation and load management (ramp rate control), and the operation of storage, resulting in enhanced capacity for load following, shifting, and shaping.

- *Ramping rate*: ramping rate represents the generator's ability to change its output with respect to time. The ramping rate of online generators may need to be increased to follow the sharper system load changes brought about by increased intermittent generation. This may result in some conventional generators not being able to be utilized, if they are not capable of operating at higher ramp rates and, for those that are capable, the increased ramp rate requirement may result in adverse impacts, such as reduced equipment life, and decreased operation efficiency. The increased ramp rate requirement may also result in the need for more peaking plants to be online, with the potential to raise wholesale electricity prices.

Distribution utility impact: utilities of the future may play an increasing role in generation management (ramp rate control), and load and storage control, resulting in enhanced capacity for net load ramp rate control.

- *Unloadable generation*: the down-ramping rate of a generator may be different from its up-ramping rate, both of which are important to meet the normal system load-following requirement. The amount of generation that can be offloaded (down ramping) is called unloadable generation. In order to accommodate the maximum output from intermittent generating technologies, system operators have to make certain that online conventional generators can be backed down quickly enough, particularly when facing a simultaneous sudden increase of intermittent generation output, and a system load decrease. Such an accommodation to absorb energy from intermittent generation cannot be made by tripping off a unit, because the unit may be needed again shortly after being taken offline.

  Distribution utility impact: the emergence of microgrid architectures (see chapter by Marnay) may result in distribution utility operation of localized generation assets. As in the preceding function, the management of generation, load, and storage devices may be executed by the utility to ensure down-ramping capability.

- *Operating reserve*: the impact on the electric system operating reserve is also related to the intermittency of solar and wind generation technologies. Utilities carry operating reserve to guard against sudden loss of generation, and unexpected load fluctuations. Any load and generation variations that cannot be forecast have to be considered when determining the amount of operating reserve. Carrying operating reserves is expensive. If utilities cannot predict the short-term fluctuations of intermittent renewable generation sources, more operating reserves will have to be scheduled in order to regulate the system adequately. This requirement will increase the cost of integrating intermittent embedded generation sources, such as solar and wind systems (Sayeef et al., 2012).

  Distribution utility impact: utility provision of a service incorporating operating reserve may eventuate depending on the emergence of microgrid architectures (as presented by Marnay in this volume), decentralized reliability options (see chapter by Woodhouse), and the business models for power system flexibility (see the chapter by Boscán and Poudineh). In any case, the operating reserve requirements are mitigated by the flexibility increase and net load variability reductions, achievable using the techniques suggested by literature in the preceding section.

Depending on the role of the distribution utility in the provision of these countermeasures, the five impacts mentioned above imply that, in the current paradigm, more units may need to be brought online, or put on regulating duty, fact that may result in increased system operating cost.

8 INTERMITTENCY COUNTERMEASURES

Solar intermittency can be managed through countermeasures, and its effects mitigated by increased grid flexibility and net load variability reduction. It is claimed in literature (Bebic, 2008) that system flexibility can be increased through:

- balancing the generation portfolio;
- introduction of more flexible conventional generation;
- redesign of the power system to enable it to handle reverse power flow from distributed solar PV.

Reducing net load variability reduces the required flexibility of the system, and measures suggested (Bebic, 2008) include:

- energy storage;
- load control;
- increased control and communication;
- ability to curtail intermittent renewable generation; and
- spatial diversity of the intermittent source.

To counter the effects of high penetration renewable generation intermittency on electricity networks, a number of potential solutions are being investigated around the world, including:

- Forecasting—improved renewable generation and load forecasting;
- System flexibility—increasing system flexibility or decreasing the flexibility requirements of the system;
- Curtailment—curtailment of renewable generation if the generation portfolio is not sufficiently flexible to manage increased fluctuations in net load;
- Interconnection—increase in load balancing area that results from increased interconnection;
- Increased control and communication—communication between central control and distributed renewable generation sources for better management of power quality issues;
- Planning—planning and assessment of adequate resources to meet expected demand needs to take into account the requirement for flexibility;
- Voltage regulation—large voltage fluctuations caused by cloud transients would need to be smoothed out to maintain system stability;
- Research and data analysis—further research is necessary to assist in the transition to high penetration intermittent renewable generation, by investigating intermittency issues and countermeasure effectiveness; and
- Energy storage and load response—investigation into energy management systems and assessment of how distributed energy resources, such as residential air-conditioning, could assist in the management and optimization of spinning reserve.

While the choice of measures will affect the economics of solar generation, and the effectiveness of these mechanisms (both individually and cooperatively) requires further investigation, these countermeasures are being developed, trialled, and deployed with increasing speed by an increasing range of market players. In particular, the consumer electronics industry has become increasingly involved in the smart grid, and the technology acceleration delivered to that sector is being transferred rapidly to the grid space. With the uptake of distributed storage, electric vehicles, smart inverters, and energy managers, the opportunity to support the increasing deployment of intermittent generation, such as solar, exists, but a technically and economically optimal solution will only be unlocked if these devices are operated effectively, and cooperation is possible. If controlled effectively, the future network of interactive devices will produce a grid far more flexible and dynamic, and the role of the utility in delivering this flexibility could be significant.

9   CONCLUSIONS

Solar intermittency and its effect on electricity networks is a significant issue that can be managed. However, the issue and relevant mitigation options are poorly understood. Solar energy installation is being restricted in certain electricity systems, and questions are being asked as to the viability of solar energy in providing a reliable electricity supply, due to this lack of evidence. Solar intermittency is able to be addressed, and the business and revenue opportunity for utilities quantified, if the characteristics, impacts, and countermeasures are properly understood.

However, solar intermittency *is* one of the biggest technical challenges facing major players in the changing grid, as utility equipment is exposed to new operating conditions, and the popularity of solar power has resulted in rapid and widespread adoption, resulting in utility operating environment changes. To counter intermittency most effectively and cost efficiently, it is likely that a mixture of gridside and loadside technologies will be employed in future. Advanced information systems, such as forecasting and information computing technology, will assist the grid and utilities of the future in matching supply and demand in real time, and managing the distribution network.

A significant opportunity exists for utilities to identify and respond to solar intermittency and, in the future, utilities could play a major role in managing increasing intermittency and transacting energy as grid-connected microgrids find their place (as noted by Marnay in this volume), and as distributed generation grows. A range of countermeasures for intermittency exist, and the importance of their characterization, consideration, and application is growing, as increasing renewable penetration, energy storage, electric vehicles, and changing consumer behavior compound. These changes bring both the potential for cost and revenue for utilities depending on the emergent role of utilities (see chapter by Nelson and McNeill), and the distribution business models adopted (see chapter by Boscán and Poudineh), in the context of a new power system with increased flexibility.

REFERENCES

Apt, J., Curtright, A. The Spectrum of Power from Utility-Scale Wind Farms and Solar Photovoltaic Arrays. Carnegie Mellon Electricity Industry Center Working Paper CEIC-08-04.

Australian PV Institute, 2013. Task 14 – High-Penetration of PV Systems in Electricity Grids. Available online at: http://apvi.org.au/international-energy-agency-pv-power-systems-programme/task-14-high-penetration-of-pv-systems-in-electricity-grids

Bebic. J., 2008. Power System Planning: Emerging Practices Suitable for Evaluating the Impact of High-Penetration Photovoltaics, NREL Technical Report (NREL/SR-581-42297), February 2008.

California Energy Commission, 2007. Intermittency Analysis Project: Appendix B Impact of Intermittent Generation on Operation of California Power Grid, prepared by GE Energy Consulting (CEC-500-2007-081-APB), July 2007.

Electricity Network Transformation Roadmap Overview, ENA, 2015. Available online at: http://www.ena.asn.au/sites/default/files/electricity_network_transformation_roadmap_overview.pdf

Mills, A., Ahlstrom, M., Brower, M., Ellis, A., George, R., Hoff, T., Kroposki, B., Lenox, C., Miller, N., Stein, J., Wan, Y., 2009. Understanding Variability and Uncertainty of Photovoltaics for Integration with the Electric Power System, (LBNL-2855E), December 2009.

Sayeef, S., Heslop, S., Cornforth, D., Moore, T., Percy, S., Ward, J., Berry, A., Rowe, D., 2012. Solar Intermittency: Australia's Clean Energy Challenge – Characterising the Effect of High Penetration Solar Intermittency on Australia Electricity Networks, CSIRO Report, June 2012.

Townsville Queensland Solar City, 2012. Townsville Queensland Solar City Annual Report. Available online at: http://www.townsvillesolarcity.com.au/PublicationsResources/SolarCityAnnualReport2012/tabid/163/Default.aspx

Ward, J. K., Moore, T., Lindsay, S., 2012. The Virtual Power Station – achieving dispatchable generation from small scale solar, Proceedings of the 50th Annual Conference, Australian Solar Energy Society (Australian Solar Council) Melbourne, December 2012. ISBN: 978-0-646-90071-1.

Wan, Y., Parsons, B.K., 1993. Factors Relevant to Utility Integration of Intermittent Renewable Technologies, August 1993, NREL/TP-463-4953.

Whitaker, C., Newmiller, J., Ropp, M., Norris, B., 2008. Renewable Systems Interconnection Study: Distributed Photovoltaic Systems Design and Technology Requirements. Sandia Report (SAND2008-0946 P), February 2008.

Part II

Competition, Innovation, Regulation, and Pricing

Chapter 8

Retail Competition, Advanced Metering Investments, and Product Differentiation: Evidence From Texas

Varun Rai*, Jay Zarnikau†

*The University of Texas at Austin, Austin, TX, United States of America; †The University of Texas at Austin and Frontier Associates, Austin, TX, United States of America

1 INTRODUCTION

It is argued throughout this book that the utility business model must change in response to competitive pressures, the growth of distributed self-generation, declining load growth, and the introduction of new technologies providing consumers or their suppliers with greater information and control over energy consumption at the individual household level. New, innovative strategies and operations will be required for financial success, and to meet consumer needs and expectations in tomorrow's market environment.

A review of the evolving marketing and pricing strategies among retail electric providers (REPs) in the areas of Texas opened to customer choice may provide a glimpse into future business models. These areas of Texas exhibit perhaps the greatest levels of retail competition in North America, and a large investment in AMI infrastructure facilitates many advance technologies for the monitoring and control of energy use. A wide range of competitive strategies have been adopted by these retailers to attract and retain customers, which in turn, impact consumer expectations, environmental goals, load patterns, and demand response capabilities.

This chapter is organized as follows. Section 2 briefly reviews the development of the competitive retail market in Texas. Section 3 discusses pricing. Section 4 provides an overview of product differentiation as a competitive strategy, while Section 5 examines the products and services now offered in areas of Texas opened to competition and draws comparisons to offerings in areas of the state that continue to be served by monopoly providers. Observations are provided in Section 6, followed by the chapter's conclusions.

Future of Utilities - Utilities of the Future. http://dx.doi.org/10.1016/B978-0-12-804249-6.00008-7

2 BACKGROUND ON RETAIL CUSTOMER CHOICE IN TEXAS

A key impetus for the introduction of retail customer choice in Texas was the belief that competition would provide greater pricing and service options to energy consumers and foster the introduction of new, beneficial technologies.[1] This section discusses the evolution of the competitive market structure, market concentration among retailers, and key regulatory requirements impacting the market.

2.1 Establishment of a Market Structure

Efforts to introduce competition into the retail sector of the state's electricity market were launched in Jun. 1999, with the passage of Senate Bill 7 (SB 7) by the Texas Legislature. This act allowed retail competition in the service areas of the investor-owned electric utilities within the Electric Reliability Council of Texas (ERCOT) power region, as identified in Fig. 8.1,[2] on a commercial basis beginning Jan. 1, 2002. New entrants were permitted to enter the retail market, and compete with retail arms of five former vertically integrated monopoly utility providers. Efforts to introduce competition into Texas' retail electricity markets are detailed in Zarnikau (2005), Adib and Zarnikau (2006), and Wood and Gülen (2009).

The introduction of new pricing plans and service options got off to a very slow start. In the early 2000s, consumers were presented with choices among retailers, but much of the competition focused on price and the reputation of the retailer. An exception was green power with higher renewable energy content, as offered by a few retailers.

The slow introduction to innovative service offerings may be traced to early problems with ERCOT's implementation of systems to track and disseminate information pertaining to the assignment of account numbers to new premises, customer switching, billing data, and disconnection and reconnection of service. The absence of AMI also posed an impediment to the introduction of certain services. It took some time before retail market rules and systems could stabilize, further slowing innovation. Also, rapid increases in prices in the areas opened to competition from 2005 to 2008 led to uncertainties over the fate of the competitive market structure. Thus during the early years of the retail market, retailers and the market in general, were focused on providing basic services to consumers.

After a very difficult transition period, placing ERCOT into the role of Central Registration Agent succeeded in establishing trust in the retail market and

---

1. As noted in an untitled report by the Texas Senate: "...the successful deregulation of other industries has provided customers with different levels of choice and increased services they previously did not have. Proponents of electric restructuring would like an opportunity to see what innovations, such as different rates for different demand times, might occur in a deregulated market." See Page 28 at: http://www.senate.state.tx.us/75r/Senate/commit/archive/IC/IC10REP.PDF

2. In addition to the investor-owned transmission and distribution utility service areas identified in Fig. 8.1, retail choice has also been introduced into the service area of Nueces Electric Cooperative. Sharyland Utilities is transitioning toward offering retail choice throughout its service area, as well.

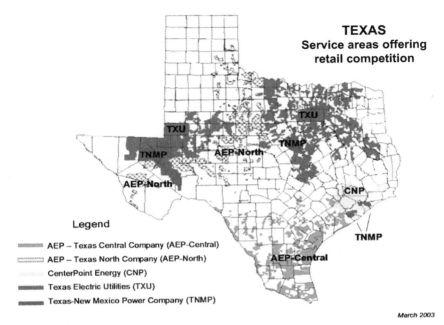

TEXAS
Service areas offering
retail competition

Legend

▬▬▬ AEP – Texas Central Company (AEP-Central)
▭▭▭ AEP – Texas North Company (AEP-North)
        CenterPoint Energy (CNP)
▬▬▬ Texas Electric Utilities (TXU)
▬▬▬ Texas-New Mexico Power Company (TNMP)

March 2003

FIGURE 8.1   Areas initially opened to retail competition. *(Source: Public Utility Commission of Texas (PUCT) at: https://www.puc.texas.gov/agency/about/commissioners/nelson/pp/ Legislative_Briefing_Retail_Competition_020615.pdf)*

reducing potential barriers to entry for REPs interested in competing in the ERCOT market. On Jan. 1, 2007, price-to-beat (PTB) price constraints[3] upon the incumbent utility retail providers expired fully, removing any regulatory oversight over the prices offered by the REPs affiliated with the traditional utility providers, and leading to a reduction in average prices (Kang and Zarnikau, 2009; Swadley and Yucel, 2011). An increase in the production of natural gas through hydraulic fracturing (also known as "fracking") led to lower prices for the ERCOT market's marginal fuel source. This led to lower wholesale electricity prices and a perception that the retail market was fulfilling its promise. In recent years, Texas has been ranked as the most successful restructured electricity market in North America (DEFG, 2015).

2.2   Market Share Among Retailers

In 2015 over 100 REPs competed to serve 5,955,761 residential customers, 1,034,600 commercial customers, and 3,848 industrial energy consumers

---

3. Initially, constraints were placed upon the retail prices that could be charged by retail electric providers associated with the traditional or incumbent utility provider within a particular transmission and distribution utility service area.

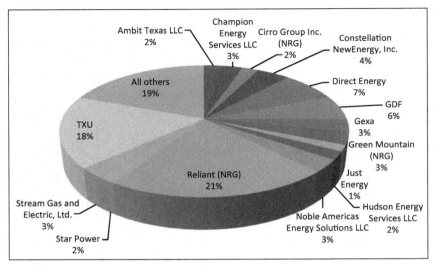

FIGURE 8.2 Approximate residential and commercial market share of REPs based on renewable energy credit requirements. *(Source: Calculated from PUCT Project No. 27706: ERCOT's Annual Report on the Texas Renewable Energy Credit Trading Program.)*

(PUCT, 2015). The market share of the largest REPs to residential and commercial energy consumers is reported in Fig. 8.2.[4] Reliant Energy (the successor to Houston Lighting and Power Company, and now a subsidiary of NRG) and TXU (the successor to TU Electric) together account for nearly 40% of the market share of sales to residential and commercial energy consumers. NRG's market share is about 26%, if its sales through Reliant Energy, Green Mountain, and Cirro are combined. Direct Energy's 7% share of the nonindustrial market is partly the result of its purchase of AEP-Texas' retail operations.[5]

2.3 Other Regulatory and Market Factors Impacting the Retail Market

Other actions shaping the retail market include the establishment of a renewable energy portfolio standard and credit trading program, establishment of the nation's first energy efficiency portfolio standard, and investment in AMI systems.

The State's renewable energy policies affect REPs by setting a minimum "greenness" or renewable energy content to the electricity sold. The restructuring

4. Fig. 8.2 is based on sales data used to allocate renewable energy requirements to each REP. Renewable Energy Credit (REC) requirements are proportionate to each REP's sales. Adjustments were made to remove non-ERCOT utilities from the calculation of percentage shares. Large industrial energy consumers are allowed to opt out of the REC program, which affects this calculation.
5. The vertically integrated Central Power and Light and West Texas Utilities affiliates of AEP opted to sell their retail operations but maintain transmission and distribution operations, following restructuring.

legislation, SB 7, set initial targets. SB 20, in the 2005 legislative session increased Texas' goal for renewable energy to 5,880 MW in 2015 and set a "voluntary" target of 10,000 MW of wind power for 2025, which Texas has already met. Under the program, each load-serving entity must have Renewable Energy Credits (RECs) equal to its share of the renewable energy goals in effect during a given year (Zarnikau, 2011). ERCOT calculates each load-serving entity's share of the statewide goal based on each entity's energy sales. Industrial energy consumers may opt out of the program and have their load excluded from the calculations used to determine REC requirements. As a result of this mandate, all electricity is presumed to possess some minimum renewable energy content of about 5%.[6]

SB 7 also required each investor-owned electric utility [which, within the areas opened to competition, was a transmission and distribution utility (TDU)] to administer programs designed to reduce the energy demand of the ultimate retail energy consumers in their distribution service areas. Minimum energy efficiency goals have been gradually increased over time (Zarnikau et al., 2015). REPs or energy services companies (ESCOs) provide energy efficiency programs and services to consumers, with funds administered by the TDUs.[7] The programs have been extremely successful in meeting the state's policy goals for energy efficiency, although Texas' goals are modest, in comparison with those adopted in some other US states.

Finally, advanced metering systems have been implemented by each of the investor-owned TDUs. These systems can provide a variety of benefits to REPs and consumers in the competitive retail market. ERCOT's financial settlement system now uses actual hourly consumption data, rather than statistical profiles. Thus the impact on utility or system costs of actions taken by consumers to shape their load patterns, or respond to price signals or curtailment requests, can be recognized in ERCOT's settlement system. Hourly consumption data can be used to design new programs and services for the mutual benefit of consumers, REPs, and the market in general.

3 PRICING

There are striking differences between pricing in competitive retail markets versus regulated monopoly markets, with respect to the frequency of price changes, the factors affecting prices, and the variety of pricing plans available to consumers.

Utilities subject to regulation or oversight by a city council change prices infrequently. It is not unusual for the same prices to be in effect for many months

---

6. This rough estimate reflects the RECs required (12,621,749) where each REC equals 1 MWh divided by the load in the competitive sector (252,914,906 MWh) in 2013, as reported by ERCOT in PUCT Project No. 27706: ERCOT's Annual Report on the Texas Renewable Energy Credit Trading Program. Actual renewable generation has been three times higher than the mandated minimum in recent years, allowing for "premium green" products.
7. Funding for these energy efficiency programs is collected by TDUs from REPs (and, ultimately, from retail consumers) through an Energy Efficiency Cost Recovery Factor (EECRF).

or years. In contrast, retailers in Texas' competitive market often change the prices quoted to larger consumers, on a daily basis, and the prices quoted to residential consumers change on a weekly or monthly basis. While components of the price associated with the recovery of regulated transmission and distribution costs and regulatory fees may change infrequently, the generation component is always changing in response to changes in natural gas prices.

Under traditional regulation, the generation charge or fuel factor reflects changes in the utilities' average cost of a broad mix of fuels in order to ensure that the utility has a reasonable opportunity to recover its reasonable and necessary prudently incurred fuel costs. In contrast, retail prices in the restructured ERCOT market follow marginal wholesale power costs closely. These, in turn, tend to follow natural gas prices (Woo and Zarnikau, 2009).

As discussed further in this chapter, the prices quoted in the competitive market may be fixed over some contract duration or may change in concert with market conditions. Dynamic prices tend to enhance economic efficiency by matching prices better to operating costs on a short-term—sometimes, hourly—basis. The indexed prices available in the competitive market adjust prices for changing market conditions—typically on a monthly basis.

The 1- or 2-year fixed price contracts commonly used to serve residential and small commercial consumers are priced to reflect market conditions at the time the retailer quotes the price. The retailer may hedge a portion of the contracted load via electricity or natural gas futures markets. While lacking any short-run price signals to consumers, the resulting prices may nonetheless yield some improvement in economic efficiency over the fixed-price tariffs designed in ratemaking processes by regulatory agencies, with their regulatory lag and use of average embedded costs.

Arguably, pricing in a competitive market leads to efficiency gains by better matching prices to the short-run marginal cost of generating electricity. Regulatory true-ups are avoided, at least for the generation or fuel component of the final price.

A consumers' ability to select from a larger menu of prices also represents an efficiency gain. While any individual retailer may offer a limited set of pricing plans, a consumer's ability to select from a large set of retailers, all offering different prices and pricing plans, provides the consumer with a large set of choices.

4 PRODUCT DIFFERENTIATION AS A COMPETITIVE STRATEGY

It is commonly argued that, in a competitive market, product differentiation or value-added services are needed to attract/retain customers, when price competition is limited. For example, see the broader discussion in Makadok and Ross (2013) about how strategic interactions manifested via product differentiation impact industry structure. Furthermore, dynamic pricing or technologies could make the consumers' load shape more attractive to the retailer, thereby reducing peak generation costs to the mutual benefit of the retailer and consumer.

Previous studies have explored competitive strategies among retailers in electricity markets. Based on survey results from Ontario, Walsh and Sanderson (2008) conclude that a hybrid strategy involving cost, service quality, enhanced communications, and unbundled services, should be adopted when marketing to commercial customers. Stanton et al. (2001) interviewed retailers in Australia and found key elements of their marketing programs to include competitive pricing, price flexibility, price monitoring of competitors, price matching, price leadership, customer service, understanding customer requirements, sales expertise, marketing expertise, and technical assistance. Giulietti et al. (2014) conclude that high search costs (the cost of acquiring information pertaining to offers from competing retailers) inhibit competition. Slim margins and fierce competition among retailers dominate some segments in Norway's market, while monopolistic behavior can be found in other segments (Fehr and Hansen, 2010).

Goett et al. (2000) report the retail service attributes of most value to commercial energy consumers and found that consumers favored local providers and free energy audits, distrusted sign-up bonuses, preferred retailers to make charitable contributions, generally preferred time-of-use rates to real-time pricing, would rather deal with a human customer service person than a voice message system, and had a diversity of preferences regarding generation mix. Woo et al. (2014) describe common rate and service offerings, with a particular focus on exploring the meaning and implications of product differentiation, in the context of electricity supply. They conclude that "applying product differentiation to electricity can greatly induce end-users to more effectively and efficiently satisfy their demand, and to do so in an environmentally friendly way" (Woo et al., 2014).

Many prior studies have focused on the marketing of green energy in restructured markets. Rundle-Thiele et al. (2008) provide an extensive literature survey of strategies involving the marketing of renewable energy. Using focus groups and interviews, Paladino and Pandit (2012) explore the challenges in marketing renewable energy in Australia's competitive market. The role of consumer preferences in the German market, with its high FiT, was explored by Menges (2003). Fuchs and Arentsen (2002) discuss options for breaking consumer trajectories favoring nonrenewable energy. Kim (2013) finds that incumbent utilities exposed to retail competition are less likely to compete effectively on a "green" dimension.

In the ERCOT market, many of same competitive strategies adopted in other competitive electricity markets are emerging, including product differentiation, the introduction of broad pricing options, and a focus on value-added services. Slim margins in a competitive retail market lead to a need for product differentiation, in order to capture higher margins associated with a differentiated and more unique product. There are limits to price competition because of the small margins and many costs (eg, transmission and distribution delivery charges and, to some degree, market-based generation costs) are outside the control of the REP. Such costs are likely to be common across all REPs, so the differentiation based on products or value-added services, discussed in the following section, starts past this "common base."

5   PRODUCT AND SERVICE DIFFERENTIATION IN THE COMPETITIVE ERCOT RETAIL MARKET

It has taken the competitive retailers in the Texas market a while to find their role and develop their marketing strategies. But, product differentiation seems to be picking up, particularly following the completion of AMI deployments in the competitive areas. The first wave of product differentiation through green pricing programs with renewable energy content above that required by the state's goal for renewable energy, may have run its course. The retail sector is now in a second wave of product differentiation, featuring AMI-enabled dynamic pricing, free nights/free weekend time-of-use rate programs, dashboards and other software to provide consumers with information pertaining to their energy use, prepay electricity programs, smart thermostats, assistance with other utilities (eg, telephone service) and trades (eg, HVAC, plumbing, and home security), and links to charities.

5.1   Programs, Plans, and Technologies Offered by REPs

The PUCT's Power to Choose website[8] provides the pricing plans offered in competitive areas, which can be compared with the tariffs of the state's regulated utilities. Inspection of the Power to Choose website for a zip code in Houston (within the CenterPoint service area) suggests availability of the following service options:

- Overall, 322 rate options with a purported range of prices from 5.6 to 14.5 cents/kWh, at a usage level of 1000 kWh/month.
- 52 REPs are offering residential service (although many of the REPs have affiliate relationships with other REPs).
- 23 Prepay plans from 6 REPs (Direct Energy, Frontier Utilities, Penstar Power, Zip Energy, Hino Electric, and Breeze Energy)
- Time-of-Use plans: seven from four REPs (Direct Energy, Champion Energy Services, TruSmart Energy, Clearview Energy). These do not apparently include the "free nights" or "free weekend" plans.
- Fixed rate plans: 268 plans. Available from nearly all of the REPs.
- Variable rate plans: 47 plans from 27 REPs.
- Indexed (market rate) plans: seven plans from three REPs.
- 100% renewable plans: 81 offerings by 28 REPs.

8. Based on a review of residential offerings in May 2015 on www.powertochoose.org

TABLE 8.1 Residential Rate Offerings by Vertically Integrated
Investor-Owned Utilities Not Exposed to Retail Competition

Utility	Rates offered	Structure and notes
Entergy, Texas	Residential Service (RS)	
	Residential Service Time-of-Day (RS-TOD)	Time of day price differences
El Paso Electric Company, Texas	Schedule No. 1 Residential Service	Seasonally differentiated energy charge
	Alternative Time-of-Use rate	Time of day price differences
	Low Income Rider	Customer charge is removed from Schedule No. 1 rate
	Qualified Water Conservation Air Cooling Rider	Not available to new customers
	Off-Peak Water Heating Rider	Not available to new customers
Xcel, Texas (Southwestern Public Service Company)	Residential Service	Seasonal price differences
Southwestern Electric Power Company (AEP)	Residential Service (RS)	Seasonal price differences

Source: Compiled by the authors.

How many plans are offered by the largest REPs? TXU offers 5, Reliant offers 10, and Direct Energy lists 11 plans.

The number and range of plans offered in the competitive market clearly exceeds what is available in areas not exposed to retail competition. Table 8.1 suggests that the residential consumers in the Xcel and Southwestern Electric Power Company service areas within Texas are offered no rate choices. Rate options from the other two investor-owned utilities serving parts of Texas are limited. As suggested in Table 8.2, more rate choices are offered by the state's two largest municipal systems, yet the numbers and diversity of choices continue to pale, in comparison to those offered in the competitive market.

ERCOT's annual survey of load-serving entities may additionally be used to provide insights into the popularity of various types of programs and technologies, in both the competitive and noncompetitive areas served by ERCOT. Between Jun. 2013 and Sep. 2014, there was a dramatic increase in the number of consumers reported to be enrolled in peak rebate programs (ie, programs offering a rebate for demand reduction during high price periods announced by the REP). Yet, due to a dearth of price spikes in 2014, it is unclear how many consumers will actively agree to curtail during a spike in wholesale prices.

TABLE 8.2 Residential Rate and Service Offerings by Vertically Integrated Municipal Utilities Not Exposed to Retail Competition

Utility	Rates/services offered	Structure and notes
Austin Energy	Residential Inside City Limits	Five inclining blocks
	Residential Outside City Limits	Three inclining blocks
	Time-of-Use Option	Three time-of-day periods, two seasons
	Green Choice Option	
	Residential Solar	Bill credit for solar energy generation
	Payment Assistance	
	Energy Efficiency Rebates	Extensive set of programs
	Green Building	Promotes resource conservation in new construction
	Plug-In Electric Vehicles	Provides incentives for charging stations
CPS Energy, San Antonio	Residential Service Tariff	Two-tier inclining block rate
	Energy Efficiency Rebates	Extensive set of programs
	Energy Portal	
	Weatherization Assistance	
	Affordability Discounts	
	Payment Options	
	Residential Demand Response Programs	

Source: Compiled by the authors.

Participation in time-of-use rate programs is similarly strong. Perhaps surprisingly, over 1000 residential consumers are on real-time pricing, a pricing structure more commonly applied to large industrial energy consumers (Table 8.3).

5.2 Value-Added Services Offered by REPs

The value-added services offered by the largest REPs were inspected by visiting each REP's website. As suggested by Table 8.4, a large variety of services are offered, including home security, frequent flier miles, energy information and feedback, carbon offset credits, and opportunities to make charitable donations. Nonetheless, some of the larger REPs appear to be focused on basic electricity service, that is, a no-frills commodity price.

TABLE 8.3 Summary of Results From ERCOT's Demand Response Surveys

Product (code)	ESI IDs 06/15/2013		ESI IDs 09/30/2014	
	Residential	C&I	Residential	C&I
Peak Rebate (PR)	2,410	58	410,675	30,236
Time of Use (TOU)	135,320	328	290,308	3,007
Other Load Control (OLC)	13,606	14	19,232	64
Real-Time Pricing (RTP)	288	4,358	1,001	9,700
Block & Index (BI)	—	23,928	—	6,796
Other Voluntary DR Product (OTH)	169	1,554	57	155
Four Coincident peak advisory (4CP)	—	35	—	247
Total	151,793	30,275	721,273	50,205

Source: ERCOT. See item 5 at: http://www.ercot.com/calendar/2015/2/3/31977-DSWG

TABLE 8.4 Value-Added Services Offered by Largest REPs in Texas

REP	Product/service	Description from website
TXU Energy	MyEnergy Dashboard	Graphs and tools show how and when energy is used to facilitate lifestyle changes and lower costs
	iThermostat	Allows management of a home's energy usage anywhere, anytime
	Personal Energy Advisor	Creates a personalized checklist of tools, tips, projects, and videos to help a customer save energy and money
	Energy Management Alerts	Sends timely email alerts to help manage electricity usage and costs
	TXU Complete Connect^{SM}	Assistance in finding cable, Internet, and phone service providers
	MyHome Protect^{SM}	Warranty service covering common home repairs and replacements
	Surge Protect^{SM}	Two surge protection products with guaranteed warranty service for repairs or replacements due to lightning strikes or power surges
Direct Energy	Benjamin Franklin Plumbers	Plumbing services
	Home Warranty of America	Whole home warranty protection

(Continued)

TABLE 8.4 Value-Added Services Offered by Largest REPs in Texas (*cont.*)

REP	Product/service	Description from website
	One Hour Air Conditioning and Heating	
	Mister Sparky	Electricians
	Nest thermostats	Get a Nest Learning Thermostat™ at no additional charge (a $249 value) with a Comfort & Control plan
Gexa Energy	Frequent flier miles	American Airlines AAdvantage members get a low fixed electricity charge, as well as 15,000 bonus miles for signing up; 2 miles for every dollar spent on the energy charge portion of electricity bill
	MD Anderson Children's Art Project	The MD Anderson Children's Art Project receives $25 when customer selects a Gexa Energy electricity plan
Green Mountain Energy	SolarSPARC™ (Smart People Accelerating Renewable Change)	Green Mountain makes a monthly contribution for each customer to build new solar projects. Plus, customer gets an annual solar credit that grows over time
	Nest thermostats	
	Pollution Free™ Electric Vehicle	Special rate for EV drivers
	Renewable Rewards buy-back program	Receive a credit for any excess energy a customer's distributed energy system exports to the grid
	Carbon Offsets	Calculates carbon footprint, and facilitates purchase of carbon credits
Cirro Energy		All that is mentioned is smart money-saving plans, online and mobile account management, convenient payment options, local customer service, plus some of the lowest fixed rates available
Champion Energy	Smart Track™	Detailed usage information, giving consumer the power to track and manage electricity usage
	Get Connected	Connect electricity, phone, cable, Internet, and security packages all with one phone call
	Home security	Partnership with Alliance Security
	Discount on LED light bulbs	

TABLE 8.4 Value-Added Services Offered by Largest REPs in Texas (*cont.*)

REP	Product/service	Description from website
Constellation NewEnergy, Inc (StarTexPower, a Constellation Company serves Texas Area)	myAccount Dashboard	Manage electricity account by viewing current balance and bill history from a computer or smart phone
	Community grant programs	Investing energy dollars back into the communities StarTex serve through charitable donations and volunteerism
	Rewarding loyalty programs	
Ambit Texas LLC	Ambit Home Services	Two forms of coverage: (1) Ambit AC/Heat Shield for heating and cooling systems; (2) Ambit Surge Protection for lightning or power surge protection
	New customer advantage	Travel voucher gift for new customers
	Power Payback™	Ambit gives advanced notice of an approaching period of extreme demand by email or phone. If customers reduce usage during designated time, Ambit offers a bill credit of $1.00 for every kilowatt-hour saved
	Free Energy	Opportunity to earn Free Energy if one refers 15 customers to Ambit Energy
	Travel Rewards	Earn one travel point for every kilowatt-hour used, or 10 points for every therm/ccf used. Referring friends can earn up to 45,000 points. Points can be redeemed for travel packages listed in the website
Just Energy	SmartStat	A high tech thermostat has Wi-Fi, mobile apps, easy web portal, live weather, and color screen. And SmartStat's Home IQ™ lets you see your monthly estimated energy savings, monthly heating and cooling summaries, details on performance influencing factors such as the weather and average set point, HVAC runtime reports, and more
Hudson Energy Services LLC		All that is mentioned are fixed or variable or mixed rates

(Continued)

TABLE 8.4 Value-Added Services Offered by Largest REPs in Texas (*cont.*)

REP	Product/service	Description from website
Noble Americas Energy Solutions	Nothing noteworthy	All that is mentioned is different pricing options, including different fixed price and index price options. Green energy is also available
Reliant (an NRG company)	Home Solutions®	Services include home generators installation, AC/Heat Protect, AC Tune-Up, Surge Protect, plumbing service, electric line with surge protect, and air filter delivery services
	BabyPower 12 plan	$100 donation to March of Dimes, $100 bill credit, a low price for 12 months, and customer service and support
	eVgo	NRG eVgo offers level 2 in-home charging stations and away-from-home stations
	Solar Sell Back plan	Credit for any excess electricity generation
	Airline Plans	15,000 airline miles for signing up and 500 bonus miles each month for the next 24 months; a low electricity price, and innovative tools to manage electricity usage
	Heart Power plan	A $100 donation to the American Heart Association. Plus, a $100 bill credit, 12 month contract, and customer service and support
	The Reliant Texans plan	A J.J. Watt jersey, a $25 donation split between the Justin J. Watt Foundation and the Houston Texans Foundation, and an invitation to a Houston Texans autograph session
	Reliant Rockets Secure Advantage 12	An autographed Patrick Beverley jersey; invitation to a Rockets autograph session; electricity plan for 12 months; and customer service and support
	The Reliant Cowboys Secure Advantage plan	A term electricity plan, plus: a football autographed by Jason Witten; an invitation to a Cowboys autograph session and customer service and support
	Reliant Rangers Secure Advantage 12 plan	Provides an autographed Texas Rangers baseball

TABLE 8.4 Value-Added Services Offered by Largest REPs in Texas (*cont.*)

REP	Product/service	Description from website
	Reliant Learn & Conserve 24 plan	Free Nest Learning Thermostat™
	Reliant Sweet Deal plan	13 months of electricity for the price of 12
Stream Gas and Electric Ltd	Free Energy Program	Enroll 15 friends and earn free energy credits
GDF (ThinkEnergy)	Energy Education	ThinkEnergy provides energy saving education with blogs and tips online

Source: Compiled by the authors.

A review of the websites of the four investor-owned utilities serving parts of Texas not opened to competition suggest limited availability of value-added services, other than energy information programs. Regulation by the PUCT may make it difficult for these utilities to offer "creative" services.

Residential consumers served by CPS Energy and Austin Energy—the state's two largest municipal utility systems—enjoy a fairly wide range of value-added services to foster their sustainability goals. CPS Energy offers Nest thermostats and an energy portal. Austin Energy established the nation's first green building program. Both utilities offer extensive energy efficiency programs. However, the range of choices and the flexibility of differentially targeted attributes built in those choices is still quite limited in these monopoly municipal utility service areas, compared to the competitive retail parts in the ERCOT market served by multiple competitive retailers.

The net impact of all these pricing strategies, marketing programs, and technologies promoted by the REPs upon the ERCOT market is unclear. As noted in chapter by Levin and Cavanagh and in the Introduction by Sioshansi, many of the contracts offered by REPs have minimum-usage charges which might discourage consumers from undertaking energy efficiency measures (Levin, 2015). With minimum usage charges, consumers cannot achieve cost savings by undertaking measures, which would reduce their consumption below the minimum usage threshold. It would be interesting for future research to explore the implications of such pricing strategies for environmental impacts and for investments in distributed renewable energy systems. On the other hand, the prepay electricity programs offered in the competitive market can have a very large conservation impact (Zarnikau, 2014). Many REPs offer technologies designed to enable consumers to save energy. Programs facilitating consumer response to price signals in the competitive areas have led to a significant reduction in peak demand (Frontier Associates LLC, 2014).

6 COMPETITIVE PRESSURES AND THE RESPONSE BY THE RETAIL SECTOR

Even though it has been over a decade since the introduction of retail choice in ERCOT, the transformation of the ERCOT retail market is still in its early phases. The marketplace is vigorous, with many old and new suppliers and customers switching between them. But the market structure has yet to gel in any sort of equilibrium. New firms are entering the market routinely, even as existing players merge and consolidate. New products are being introduced rapidly, even though positioning (of firms) on existing products has not been established. It is not clear how long this level of flux can be sustained before the market structure crystallizes somewhat.

Two competing factors push in different directions: (1) demand growth and changing demographics in the ERCOT market provide opportunities for retailers to build position continuously, but (2) a relatively high rate of customer switching creates costs and uncertainties for retailers, affording little time to establish position. Technological change, in particular the advent of distributed generation and home energy management systems, further magnifies the challenges faced by retailers in establishing a market position, in the face of a growing market and competition.

Amid this apparent tumult, though, some intended benefits from retail competition in the ERCOT market seem to have been served already clearly. These include: (1) more product choice for customers and (2) higher competition, at least as indicated by the simplistic metric of the number of firms operating, in comparison to the regulated parts of the ERCOT market.

The picture is not clear with respect to competitive pricing. When competition in commodity markets heats up, focus on market segmentation and differentiation starts to take over, leading to product and service diversity and muddling of the price waters. This is a rich area for empirical research, requiring careful control for not only resource factors (such as fuel price or congestion) affecting prices, but also for strategic factors, such as how the *features* of the offered products and services impact prices. The diversity in the offerings takes away the pure commodity nature of the offering, potentially allowing for long term firm-level strategies geared toward commanding higher margins. While empirical proof is weak, it appears as though price competitiveness has improved in the parts of ERCOT opened to retail choice. Perhaps this is because the vigorous nature of firm-level competition that has put customers at the center of value creation in the retail business.

There are early signs of both proactive and reactive innovation by retailers in the ERCOT market. This is encouraging, as innovation is central to value creation, for both suppliers and customers. Incentives for proactive innovation are driven by shifting (and growing) the customer base and by technological change, inducing retailers to introduce new products and services, in anticipation of demand.

Reactive innovation—or to put bluntly, imitation—involves observing the behavior of competitors and that of customers, in order to mold product offerings and firm strategy. It is widely recognized in the innovation literature that imitation is the quintessential mechanism for reaping and spreading the benefits of innovation.

Transparency in the overall market is critical in fueling both proactive and reactive innovation. In ERCOT's retail market, www.powertochoose.org has been the central lynchpin for creating and enhancing transparency.

A related but less appreciated aspect outside the REP circles is the role that advanced data analytics is beginning to play in helping firms chart their product strategy. Such analytics range from tracking customer switching behavior, to analyzing product offerings of competitors, to predicting the impact of weather on load to manage wholesale supply.

These stylistic observations about REPs' strategic behavior are in line with the remark of Steil et al. (2002) on Schumpeter's deep insights around the interlinks between competition and innovation: "...competition under capitalism is...competition through innovation and then trial by actual experience."

The deepest impacts of Texas's retail market transformation are yet to be seen, although some recent events offer some insights into what may be coming. In particular, the conditions ripe for the introduction of *disruptive* market constructs, products, and services in the ERCOT market may be forming. This aspect adds an entirely new dimension to the more common but narrow focus around price competitiveness, when analyzing the benefits retail competition.

Two examples illustrate this point. First, there have been discussions within ERCOT about creating market incentives for aggregating distributed energy resources (DERs). Under the initial concepts being discussed, DERs would be allowed to earn a broad, (locationally) averaged wholesale price (GTM 2015). This could be a significant incentive for DERs, especially in locations with high wholesale prices. This development is interesting, given that, unlike many other markets in the United States, Texas has not had a strong policy support for distributed generation. Examined from that backdrop, it is impressive that largely market-driven forces are increasingly engendering broader structural changes within the market.[9]

This is especially noteworthy, given that low natural gas prices have put downward pressure on incentives to build more capacity in Texas, even though the peak supply situation is expected to be quite difficult over the next decade or so (ERCOT, 2014). Here, Texas has benefited by being on the sidelines, as the solar story has played out over the last decade elsewhere in the United States (especially, California, New Jersey, and Arizona) and the world (Germany, Japan, China, among others). Price and quality for solar products have reached very

---

9. As an aside, this development might be of interest to students of *complex systems*, wherein endogenous factors precipitate internal adaptation, change, and system evolution in often hard-to-predict fashion.

compelling points—these are precisely the most critical ingredients needed for rapid take up in market-driven retail areas like Texas.

Second, there are indications about the entry of *disruptive* retail products. An example of that is the recent product launch by a partnership between MP2 Energy and SolarCity. This product, initially targeted toward solar homeowners in the Dallas-Fort Worth area, *effectively* offers net metering to solar customers (SolarCity 2015). It is too early to comment on how good the product will be for the customers, or the companies offering them, but this product essentially solves a hitherto major barrier for solar customers in Texas, namely the lack of statewide net metering, and it does so based on purely market incentives (as perceived by MP2 Energy and SolarCity). Nearly all other US solar markets have required some form of *ongoing* legislative and/or regulatory intervention to resolve this issue, often engendering acrimonious mudslinging among utilities, solar companies, and consumer groups. That seems to have been avoided in the ERCOT market, at least for the moment, thanks to the underlying incentives in the fiercely competitive retail market in ERCOT.

These two examples suggest that the competitive market design offers tremendous incentives for rapid deployment of disruptive products and services, once certain technical and price points are met.

7 CONCLUSIONS

If electric utilities are subjected to greater competitive pressures in the future, retail business of utilities might look more like the REPs operating in the areas of Texas opened to competition. Energy charges would reflect better the cost of the marginal generation fuel and would change as market conditions changed. Peak rebate programs and other forms of dynamic pricing would become more widespread. Electricity retailers would expand price and service options to meet diverse consumer needs better.

While competition leads retailers to introduce new products and services to attract and retain customers, it is instructive to note that the market structure and environment in Texas has been shaped by government regulation, to a large degree. A Central Registration function was assigned to the system operator. Rules and protocols were developed by the PUCT and ERCOT. Goals for renewable energy and energy efficiency were established by the state government, which also ordered the deployment of AMI and competitive renewable energy zones (CREZ) infrastructure.

Some concluding thoughts on the overall desirability of retail choice in electricity markets, in particular in the presence of externalities (such as harmful emissions), are in order. Specifically, should it not be expected that in retail choice markets, over time, both suppliers and consumers will focus largely on lowest cost products (in dollars per kilowatt-hour), thereby precluding the opportunity to deploy environmentally friendly generation and consumption of electricity? Indeed, there are analysts who, in anticipation of such "race to the bottom" outcomes in retail choice environments, suggest that a better option is to retain the conventional regulated rate-of-return (ROR) model with a strong interventionist regulator that can mandate the level of "greenness" in the mix of electricity sold. Such arguments are tenuous for the following reasons:

- First, as is now well established, the conventional ROR model creates a number of perverse incentives for the regulated utilities that might oppose and scuttle socially desirable efforts to change the electricity generation–consumption system, if such changes would hurt the utilities' business interests. Strong regulators may well be able to deal with such utilities. However, from time to time, regulators themselves are prone to regulatory capture. Maintaining a strong independent regulatory regime that constantly works only in the public interest requires constant public participation, vigilance, and due diligence.
- Second, as argued earlier, there is no reason to expect that retail-choice electricity markets will *necessarily* degenerate into markets where the sole focus is on lowest prices.[10] Several examples of how REPs in Texas are trying to position and differentiate themselves from the rest were provided earlier. Fundamental forces will continue to drive such brand positioning, refinement, and differentiation. The suite of customer-oriented technologies (solar PV, electric vehicles, smart thermostats, home energy management systems, etc.) that are deepening market reach will only help accelerate that process of differentiation. Price competitiveness will still be important, but the main point here is that, in addition to just focus on prices, the market will afford a number of variations in how electricity is sold to customers by catering to their *attribute-specific preferences* on items such as reliability, customer service, simplicity, innovation, trustworthiness, etc. *If* environmental protection is one of those salient attributes, *then* undoubtedly some products will arise to serve those preferences (100% renewable electricity plans are partly an example of this.) However, there is no reason to believe that retail-choice will automatically serve the purpose of improving our environment, unless environmental protection is a clear preference of a segment of the customers (and it is), or is built into the market as an underlying condition of generation and consumption.
- Third, this position presumes that the lowest cost generating units are more polluting, which is increasingly untrue.
- Fourth, and most importantly, regulating at the retail level is not always necessary (or even desirable, for example, due to heightened complexity) to internalize externalities in the electricity supply chain. For example, emissions could be regulated at the generation level, and then all REPs in a given market have to procure from the generation mix that emerges to meet such regulations. In this sense, the retail portion of the market can simply be a taker of whatever mix is provided for by the generation side. An example

---

10. As noted in the following chapter, there is little evidence that retail competition has led to lower prices to consumers for the reasons discussed in Woo and Zarnikau (2009).

is the recently finalized US EPA Clean Power Plan (CPP). The CPP has set state-wise reductions targets for the power sector as a whole, with the most direct and significant impacts on generation technologies. In markets with retail choice, REPs will "work with" the resulting generating mix to develop and sell their electricity products. Thus while they are clearly impacted by the CPP, there is no direct regulation of REPs involved here.

Nonetheless, the effects of minimum-usage charges, as discussed in the following chapter and in the Introduction by Sioshansi, could certainly have an offsetting impact on progress toward environmental goals for some customer segments.

Retail choice can provide an efficient mechanism to leverage competitive dynamics to serve customer demand. In this sense retail choice is best seen at the *mediating mechanism* between supply and demand. *Given* the supply mix and customer preferences, the retail choice mechanism can coordinate an efficient outcome. The nature of the supply and demand themselves, however, may change, owing to changes in regulation, technology, or social norms. Retail choice may allow those changes to be reflected in an efficient manner.

ERCOT's unique retail market structure might not be appropriate for markets of smaller size, inadequate competition at the wholesale level, and lesser faith in market forces. Yet, even in regions where this degree of retail competition is not introduced, dealing with distributed generation and new metering and control technologies will require a customer-centric approach, as discussed in chapter by Gimon.

REFERENCES

Adib, P., Zarnikau, J., 2006. Texas: the most robust competitive market in North America. In: Sioshansi, F., Pfaffenberger, W. (Eds.), Electricity Market Reform: An International Perspective. Elsevier, Amsterdam.

Distributed Energy Financial Group (DEFG), 2015. The Annual Baseline Assessment of Choice in Canada and the United States (ABACCUS).

ERCOT, 2014. Report on the Capacity, Demand, and Reserves in the ERCOT Region, ERCOT. Available at: http://www.ercot.com/content/gridinfo/resource/2014/adequacy/cdr/CapacityDemandandReserveReport-February2014.pdf

Fehr, N.-H., Hansen, P.V., 2010. Electricity retailing in Norway. Energy J. 31 (1), 25–45.

Frontier Associates LLC, 2014. 2013–2014 Retail Demand Response and Dynamic Pricing Project: Final Report. Prepared for ERCOT. http://www.ercot.com/content/services/programs/load/2013-2014_DR_and_PriceResponse_Survey_AnalysisFinalReport.pdf

Fuchs, D., Arentsen, M., 2002. Green electricity in the market place: the policy challenge. Energy Policy 30, 525–538.

Giulietti, M., Waterson, M., Wildenbeest, M., 2014. J. Indus. Econ. LXII (4), 555–590.

Goett, A., Hudson, K., Train, K., 2000. Customers' choice among retail energy suppliers: the willingness-to-pay for service attributes. Energy J. 21 (4), 1–28.

GTM, 2015. Texas Mulls New Grid Markets for Aggregated Distributed Energy Resources. http://www.greentechmedia.com/articles/read/texas-looks-to-distributed-energy-resources-as-market-players?utm_source=Solar&utm_medium=Picture&utm_campaign=GTMDaily

Kang, L., Zarnikau, J., 2009. Did the expiration of retail price caps affect prices in the restructured Texas electricity market? Energy Policy 37 (5), 1713–1717.

Kim, E.-H., 2013. Deregulation and differentiation: incumbent investment in green technologies. Strategic Manag. J. 34, 1162–1185.

Levin, A., 2015. Customer incentives and potential energy savings in retail electric markets: a Texas case study. Electricity J. 28, 51–64.

Makadok, R., Ross, D.G., 2013. Taking industry structuring seriously: a strategic perspective on product differentiation. Strategic Manag. J. 34 (5), 509–532.

Menges, R., 2003. Supporting renewable energy on liberalised markets: green electricity between additionality and consumer sovereignty. Energy Policy 31, 583–596.

Paladino, A., Pandit, A., 2012. Competing on service and branding in the renewable electricity sector. Energy Policy 45, 378–388.

Public Utility Commission of Texas (PUCT), 2015. Scope of Competition in Electric Markets in Texas. Report to the 84th Texas Legislature. January.

Rundle-Thiele, S., Paladino, A., Apostol, S.A., 2008. Lessons learned from renewable electricity marketing attempts: a case study. Bus. Horizons 51, 181–190.

SolarCity, 2015. SolarCity, MP2 Energy Offer Solar to Texas Homeowners for Less than Utility Power Without Local Incentives. http://www.solarcity.com/newsroom/press/solarcity-mp2-energy-offer-solar-texas-homeowners-less-utility-power-without-local

Stanton, P.J., Summings, S., Molesworth, J., Sewell, T., 2001. Marketing strategies of Australian electricity distributors in an opening market. J. Bus. Indus. Mark. 16 (2), 81–93.

Steil, B., Victor, D.G., Nelson, R., 2002. Introduction and Overview. Technological Innovation and Economic Performance Princeton University Press, Princeton, NJ (Chapter 1).

Swadley, A., Yucel, M., 2011. Did residential electricity rates fall after retail competition? A dynamic panel analysis. Energy Policy 39, 7702–7711.

Walsh, P., Sanderson, S., 2008. Hybrid strategic thinking in deregulated retail energy markets. Int. J. Energy Sector Manag. 2 (2), 218–230.

Woo, C.K., Zarnikau, Jay, 2009. Will electricity market reform likely reduce retail rates? Electricity J. 22 (2), 40–45.

Woo, C.K., Sreedharan, P., Hargreaves, J., Kahrl, F., Wang, J., Horowitz, I., 2014. A review of electricity product differentiation. Appl. Energy 114, 262–272.

Wood, P., Gülen, G., 2009. Laying the groundwork for power competition in Texas. In: Kiesling, L., Kleit, A. (Eds.), Electricity Restructuring: The Texas Story. American Enterprise Institute, Washington, DC.

Zarnikau, J., 2005. A review of efforts to restructure Texas' electricity market. Energy Policy 33, 15–25.

Zarnikau, J., 2011. Successful renewable energy development in a competitive electricity market: a Texas case study. Energy Policy 39 (7), 3906–3913.

Zarnikau, J., 2014. How Do Prepay Electricity Programs Impact Consumer Behavior? Distributed Energy Financial Group Prepay Energy Working Group. http://defgllc.com/publication/how-do-prepay-electricity-programs-impact-consumer-behavior/

Zarnikau, J., Isser, S., Martin, A., 2015. Energy efficiency programs in a restructured market: the Texas framework. Electricity J. 28 (2), 1–15.

Chapter 9

Rehabilitating Retail Electricity Markets: Pitfalls and Opportunities

Ralph Cavanagh*, Amanda Levin[†]
*NRDC, San Francisco, CA, United States of America; [†]NRDC Washington, DC,
United States of America

1 INTRODUCTION

Among the most important questions shaping the future energy economy is the role of regulated utilities. Are they about to pass into irrelevance, bypassed by an explosion of innovation that will soon make every building self-sufficient in energy services, through a combination of onsite generation and storage? Will electric utilities survive, but only as passive wires companies connecting customers to a host of independently provided services? Will commodity markets supplant the judgment of regulators, or are there scenarios that could allow for the best of both?

Although these questions now attract what passes for fevered speculation in historically placid energy policy circles, they are hardly new. Commentators have been predicting for decades that technological innovation would disrupt power grids by prompting mass defections, with dire consequences for hometown utilities and their shareholders. When those defections failed to materialize,[1] policymakers sometimes offered a substitute, creating opportunities for new suppliers to take customers away from incumbent utilities without severing those customers' connections to the utilities' distribution grids. The first efforts to supplant regulated electricity service with "retail competition" came in the early 1990s.

This chapter is in part a review of what worked and what didn't over a quarter century of retail competition, with particular emphasis on the handful of

---

1. As one electricity expert memorably put it:

"[C]ount me among the people who get no special thrill from making our own shoes, roasting our own coffee, or generating our own electricity. I don't think my house should be energy independent any more than it should be food independent or clothing independent". (Borenstein, 2015).

Future of Utilities - Utilities of the Future. http://dx.doi.org/10.1016/B978-0-12-804249-6.00009-9

states that were the most ambitious. It aims to provide insight into how emerging energy policies and customer desires affected the new competitive markets. It details strategies for making such markets more consistent with society's reliability and environmental objectives, and reflects on whether "fixing" retail competition is worth the effort, in light of alternative ways to achieve the same ends.

The authors believe that centrally managed grids are here to stay and, in fact, are essential to a clean energy transition that is already well underway. The resilience of the traditional utility model reflects durable realities of grid management, reliability assurance, and resource procurement and integration. Grid bypass remains unappealing to almost everyone with the option to connect or stay connected, for reasons largely irrelevant to every advance in generation and storage; with reliable electric service increasingly essential to every household and business, most are unwilling to delegate ultimate responsibility to any but trusted long-standing partners. Moreover, technological progress in the electricity sector has been and remains much more about opportunities for grid enhancement, than grid displacement. What look like the most tempting invitations to bypass grids also can make them work better at lower cost, from SunPower solar cells to Tesla batteries.

For those committed to a clean energy future, utilities remain the most important investors and integrators. Whether and how to keep retail competition in the mix are questions that occupy the succeeding sections. Regulators exploring ways to redefine "the utility of the future" should be cautious about claims that commodity markets and nonutility actors can replace utilities as long-term system planners, investors, and resource integrators.

This chapter consists of four sections, in addition to the Introduction. Section 2 delves into the results of restructuring—and subsequent regulatory adjustments—in key states. Section 3 covers whether retail choice is a desirable and viable future business model for the clean energy future. Section 4 presents an alternative business model, which embraces utilities and gives them a central role in the clean energy future followed by the chapter's conclusions.

2 WHAT HAS RETAIL ELECTRICITY COMPETITION REVEALED?

2.1 History of Retail Competition in the United States

Retail electric utility service was a firmly established regulated monopoly until the early 1990s, when several states began investigating whether to create the equivalent of competitive commodity markets for both residential and business customers.[2] Retail competition has been controversial from its inception. While its advocates have claimed "explosive growth" in recent years, most recently finding that retail access has doubled between 2010 and 2013, retail competition

---

2. See Cavanagh (1994, pp. 22–24), describing early campaigns in New Mexico, Rhode Island, Texas, Nevada, and Michigan, none of which had reached fruition as of March 1994.

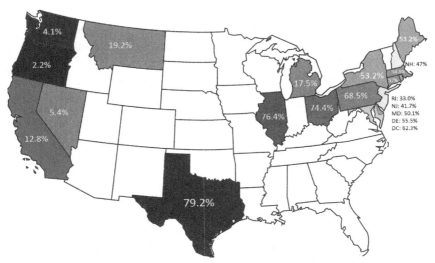

FIGURE 9.1 **Percentage of total load served by an alternative supplier.** *(Source: Data from DEFG, 2015.)*

remains quite limited across the nation (Tweed, 2015). In 2013 it covered about 21% of annual total electricity sales, and only 13% of annual *residential* sales (EIA, 2015a).

As of 2015, 17 states and the District of Columbia have some level of retail electricity competition (EIA, 2012). Of those states, 13 and DC offer uncapped electric retail competition, while 4 states—California, Michigan, Montana, and Oregon—permit, but constrain, retail access. All of these states began restructuring their electric markets in the first wave of restructuring interest, in the 1990s. No states have adopted restructuring since California's energy crisis in 2001 (Borenstein and Bushnell, 2014; Palmer and Burtraw, 2005). In addition, many of the restructured states have implemented additional regulatory mechanisms and consumer protections—such as extending retail price caps, and default service rates (Fig. 9.1).

The extent of customer participation in retail markets varies widely by customer class and state. As shown earlier, rates of customer participation are as low as 2% and as high as 61%. In all states, large commercial and industrial users have been the most active in retail markets. In some states, large customers are the only customer class with retail access. Even when choice is available for all classes, large customers are typically the only *active* consumers in the market. There has been substantial growth in residential switching in the last 2 years, but, in all but one case, the gains have been largely driven by municipal and community choice aggregation. For example, 61% of Illinois residents now receive electricity from an alternate supplier. However, over 70% of those customers switched *passively* through *opt-out* aggregation (ICC, 2015). Significant *active* residential customer participation has only developed in Texas, where 90% of all residential customers have selected their own retail suppliers.[3]

Whether retail markets have improved customer welfare is hotly contested. Procompetition groups, such as Alliance for Retail Choice and the COMPETE coalition, release annual reports claiming significant electricity price savings from retail competition.[4] Groups seeking reregulation or market reforms counter that retail markets have cost consumer billions. At the same time COMPETE was touting Texas's retail record, the Texas Coalition for Affordable Power identified over $22 billion in lost savings for market participants (TCAP, 2014). After a retrospective study of the last 20 years of restructuring efforts and results, the Lawrence Berkley National Laboratory concluded that there was no conclusive evidence that restructuring had any measurable impact on state electricity prices, either for better or worse (Borenstein and Bushnell, 2014). In every state, fluctuations in natural gas prices were the overwhelming driver in price changes.

Far less contentious has been the elimination of utility monopolies over electric generation ownership, and the promotion of wholesale competition, led by the Federal Energy Regulatory Commission. This chapter does not discuss wholesale markets, a development, which the authors fully support. Over 40% of electric generation capacity is now owned independently of utilities, and utility access to increasingly vibrant wholesale markets for short- and long-term purchases is now essentially universal across the nation (EIA, 2015b). Independent power producers, along with commercial and industrial facilities, also are the predominant sources of renewable energy, supplying 88% of renewable energy capacity and generation in 2013 (EIA, 2015b,c). Even skeptics of retail competition have embraced the opening of the wholesale electricity marketplace.[5]

2.2 California Experience

In Apr. 1994 the California Public Utilities Commission (CPUC) launched a tragically flawed electricity restructuring initiative with the best intentions and, ultimately, the worst imaginable results (CPUC, 1994). The CPUC wanted to "replace economic regulation with the discipline of market forces"

---

3. Note: This includes customers who stay with the incumbent-affiliate provider, but chose a non-default plan from the incumbent. Residential customers with a non-affiliate provider varies by region, from 60% to 83%. (See *Ercot Market Share Report*). For context, 97% of large users and 92% of small non-residential users have utilized their retail choice.
4. For example, see Annual Baseline Assessment of Choice in Canada and the United States (ABACCUS).
5. See, for example, Cavanagh (1994 at pp. 25-28) (joint declaration from March 1994 of consumer groups, independent energy service companies, independent power producers, and public interest groups, endorsing trends toward more competitive wholesale markets while uniting in opposition to retail competition proposals).

(CPUC, 1994, p. 37). For decisions about new resources of every kind, the CPUC's mantra was "let the market decide," independent of utility planners and investors (CPUC, 1994, p. 50). The CPUC framed its goal as follows: "Both entry into and exit from such a business, as well as the terms of such contracts, are left to the genius of the marketplace and the will of market participants."[6]

In Aug. 1996, by a rare unanimous vote, this vision was enacted largely intact by the California legislature. Utilities ceded responsibility for electricity resource development. But no effective substitutes emerged. Electricity supplies tightened, as "the genius of the marketplace" failed to respond as anticipated, and a memorable electricity reliability and price disaster unfolded suddenly in 2000, and continued into 2001.

Wholesale electricity prices that previously had ranged between 2–3 cents/ kWh soared to at least 15 cents, on average, from Jun. through Aug. 2000. That average price then doubled again in Dec. 2000 and Jan. 2001.[7] Two of the three major distribution companies were on the brink of insolvency by early 2001. PG&E ultimately filed for bankruptcy, and Southern California Edison teetered on the brink for months. At the same time, notices of supply emergencies became routine throughout the state, as operating reserves dropped below 5% for weeks on end. An all-out emergency statewide energy conservation campaign was required to spare the state between 50 and 160 h of rolling blackouts (NRDC and Silicon Valley Manufacturing Group, 2003, p. 8).

Not only did the marketplace fail to respond to support investments in new generation capacity, energy efficiency, renewable energy sources, and other resources in the public interest also languished. In the early years of restructuring, energy efficiency funding nationwide decreased dramatically for energy efficiency programs. In nominal dollars, it decreased from about $1.8 billion in 1993 to $900 million in 1998 (ACEEE, 2005).

In California the CPUC responded to the 2000–2001 crisis with comprehensive, aggressive energy policies, which restored utilities' resource procurement responsibilities. State energy efficiency spending and savings have increased greatly since the restructuring debacle, for both investor-owned and publicly owned utilities. And renewable energy development has flourished, thanks to evolving statutory targets and utilities' revived ability to offer long-term purchase contracts.[8]

---

6. CPUC (1995) as modified by CPUC (1996) at p. 8.
7. See Smith (2000). ("[t]he average cost of power, per megawatt hour, was $185 in August, $117 in July and $167 in June.") A price of $1.50 per kWh cleared the California Power Exchange's day ahead market for deliveries at 6 am on Dec. 13, 2000, according to the PX website (www.calpx. com). The weighted average cost of system power purchased through the Power Exchange from Nov. 20 through Dec. 20, 2000 was 28 cents/ kWh; for the period Dec. 20 through Jan. 22, 2001, it rose to 29.4 cents/kWh. These weighted averages are reported by Green Mountain Energy, in the form of retail electricity bills received by the author for those months.
8. In CPUC (2006), at p. 23, noting the importance of long-term contracts for new renewable energy development across the nation, and "not simply the result of circumstances unique to California."

2.3   Pennsylvania Experience

Pennsylvania was also one of the first states to undergo restructuring, with a gradual rollout of customer choice between Jan. 1999 and 2001 (EIA, 2010). As part of the transition, electricity rates were frozen and incumbent utilities were allowed to gradually recover "stranded costs" associated with their generation assets.

Though rate caps were set to expire in 2005, only two of the utilities actually lifted their caps. One of them—Pike County Power and Light—saw rates climb 50% overnight (Kleit et al., 2011). The commission required the other utilities to extend their freezes until the end of 2010, effectively delaying competition until 2011 (Kleit et al., 2011). Customer participation has increased in recent years. As of Jul. 2014, 36.5% of residential customers and 37.6% of all customers were with an alternative supplier.

Despite the promise of innovative pricing, true time-of-use (ToU) or real time pricing (RTP) has floundered. Of the nine competitive areas, alternative suppliers only offer ToU pricing for residential and nonresidential customers in two areas (PA PUC, 2014, Section 3). In comparison, four regulated utilities offer ToU. Nonresidential customers have more opportunities, though real-time pricing or hourly pricing options, available through alternative suppliers in seven of the competitive areas, or with the utilities in eight. But customer engagement has been extremely low and dropping. In total, only 1% of customers elect ToU (PA PUC, 2014, Section 3).

The other "innovative" pricing option, variable rate plans, have created a host of issues for customers and the state. A significant number of residential and small nonresidential customers were lured by low, promotional rates into such plans. Others were enrolled automatically after their fixed price plans expired. A lack of regulations and customer education led to widespread and dire results during the polar vortex of 2014. Rates and bills more than tripled for many customers. Suppliers charged anywhere from 20 to 44 cents/kWh, when the standard offer price was 8 cents/kWh (McCloskey, 2014). Many bills—including those of low-income residents—skyrocketed above $1000, leading some to have to default on other obligations in order to pay for heating. Agencies were flooded with more than 60,000 calls, complaints, and inquiries (McCloskey, 2014).

This incident exposed flaws with variable pricing and within the market (McCloskey, 2014). Disclosure statements were often vague. There were no regulatory limits on variable rate plans, such as how many times or how much the energy charge could change within the billing period. Customers also didn't have the right to access the price they would be charged before receiving their bill. Under a deluge of customer calls and complaints, providers' customer service faltered, with many customers being unable to reach a representative. State law required customers to contact suppliers before sending a complaint to the PUC, leaving many unable to file such complaints. In addition, the crisis highlighted Pennsylvania's barriers to switching suppliers, which kept many unhappy customers captive for more than a month after they tried to opt out (McCloskey, 2014).

Government response has been swift, with rulemakings clarifying disclosure requirements and requiring accelerated switching (now 3 days). Five suppliers have been hit with formal complaints from the state government. The Commission's Bureau of Investigation has brought additional complaints against two suppliers (McCloskey, 2014). In addition, legislators introduced five separate measures to regulate suppliers and variable rate plans (McCloskey, 2014).

2.4 Illinois Experience

Illinois has undergone a lengthy process to implement restructuring, but a truly competitive retail electricity market is still elusive. Limited commercial and industrial opportunities began in Oct. 1999, and the residential sector followed in May 2002 (EIA, 2009). By the end of the rollout for commercial and industrial customers, only 12% had switched and there were very few alternative suppliers in the market (EIA, 2009). The competitive market was not developing nearly as quickly as envisioned. In response, rate caps for residential and small customers were extended until 2006 to allow for the creation of a functioning market (ICC, 2002).[9]

When the extended rate freeze ended in Jan. 2007, prices skyrocketed and calls for reregulation mounted. Residential rates in ComEd territory increased by 26%, while those in Ameren territory increased by 55% (Maryland Public Service Commission, 2008). Yet customers failed to migrate. By Sep. 2007, only one residential customer in ComEd territory had found an alternative supplier (Maryland Public Service Commission, 2008). Even small commercial and industrial customers had lackluster engagement, with less than 10% on alternative suppliers in Ameren territory. In response to public outcry, the state began a process of reregulation, creating the Illinois Power Agency (IPA).

The IPA now negotiates power purchases on behalf of customers receiving power from the distribution utility (ie, not with an alternative supplier), and is the only investor in long-term renewable energy contracts to meet the state's Renewable Portfolio Standard. But due to municipal aggregation and the exodus of C&I customers from incumbent utilities, IPA's role in renewable energy development has declined significantly. In fact, the IPA indicated in 2012 that its renewable energy procurement would have to pause until 2018, despite the availability of more than $130 million in dedicated contributions by retail electricity providers (Roberts, 2012).

While customer participation in retail markets has increased greatly in recent years, this has been primarily driven by municipal aggregation. Alternative

---

9. The Report is available on the Illinois Commerce Commission's website in a section containing reports, etc., by Chairman Mathias.

suppliers accounted for 79% of all electricity use, including 60% of all residential customers in Jun. 2015 (ICC, 2015). However, 70% of these residential customers simply failed to opt out of municipal aggregation programs. The high proportion of municipal-driven switching has led to large market concentration, with winning aggregation bids almost all going to the few dominant power companies in the state (ICC, 2015).

Since Jul. 2014, the growth in municipal aggregation has reversed. Of the 740 communities that opted initially for aggregation, 121 abandoned their programs since mid-2014 (ICC, 2015). Most notably, the city of Chicago—the nation's largest municipal aggregator—has announced that it will not renew its contract. Despite initial savings in the early months of Chicago's program, many customers saw higher bills for most of the contract period—an average of $80 more on annual household energy bills (Daniels, 2014). In fact, the Illinois Office of Retail Market Development estimated that customers with ComEd saved $73.4 million between Jul. 2014 and 2015, compared to those on alternative suppliers (ICC, 2015). According to Mayor Rahm Emanuel, "continuing the program would not have resulted in substantial savings for consumers while returning residents to ComEd will reinvigorate funding for investment in the state's renewable energy portfolio" (Daniels, 2015).

2.5 Ohio Experience

Ohio's competitive wholesale and retail market opened Jan. 2001 (PUCO, 2015). During a 5-year market development period, government aggregation was the only significant competitive force (PUCO, 2015). In response, as the development period ended, the state stepped in and worked to establish rate stabilization plans with incumbent utilities, with the aim of moving gradually to market-based prices.[10]

Facing an acute shortage of competitors, legislators amended the restructuring mandate in 2008. As of 2015, the transition to competition remains troubled. Distribution utilities are still allowed to own and operate generation, and all fought in 2015 to reregulate or rate-base aging coal plants that have struggled in the competitive market (Gearino, 2015). These utilities also have imposed "financial integrity" or "stability" charges on their distribution and transmission services, which are used to prop up their generation arms (Weston, 2013).

As Ohio's distribution utilities invest in grid modernization and smart meters, they have maintained sole rights to interval data. Electric marketers are unable to offer innovative pricing without this data, leaving the incumbent companies with sole ability to offer these services. The continued relationship between the distribution and generation arms of the utility has also impaired

---

10. PUCO (2015); Market-based pricing would only be allowed given a demonstration that competition was effective.

customer participation. Customers often believe that the utility has a preference for their affiliate, and worry that service reliability will decline if they switch (Weston, 2013).

An Ohio PUC investigation in 2014 uncovered large discrepancies and inconsistencies between service territories, posing large market barriers to customer or alternative supplier engagement (PUCO, 2014). Bill language was not standardized; the price-to-compare (or standard offer rate) was calculated differently by each utility; and utility-supplier terms concerning purchase of receivables and shared data varied by territory. Lack of consistent market rules and oversight created an unfair, confusing environment, and large barriers to any significant participation or meaningful benefits for the state's electricity users.

2.6 Texas Experience

Texas has the nation's most active retail electric market. Much of the state—about 6.7 million homes and businesses—has had full customer choice since 2002. According to the Public Utilities Commission (PUCT), the state now has more than 190 providers in Texas, with at least 50 in any given area, offering over 300 residential plans.[11] In reality, however, many of these providers are owned by a small number of large holding companies. The top six holding companies serve 89% of both residential customers and residential sales.[12] For customers as a whole, these top six provide power to 88.5% of the total and account for 63.6% of total sales. While the market is highly concentrated, it is active. Over 90% of residential customers have participated in the market, and in 2014, there were over 3.2 million switches (ERCOT, 2015a,b).

Unlike other competitive electricity markets in the country, Texas has no regulatory mechanisms to ensure resource adequacy, such as minimum reserve requirements or a forward capacity market. New generation investment decisions are based solely on wholesale prices in the spot market. The lack of a minimum reserve requirement or capacity market has created issues for The Electric Reliability Council of Texas (ERCOT). While total electricity consumption and demand in other states have flattened, or even decreased, in the last decade, demand and consumption in Texas is still growing rather quickly. The state's statutorily constrained energy efficiency programs—with an aggregate impact in 2013 of less than 0.19% of retail sales—have not been able to significantly slow Texas's growing demand and new generation has struggled to keep pace.

Original forecasts for 2015 predicted a reserve margin below 10% (Newell et al., 2012). Afterward, the spot market price ceiling was raised from $3000 to $9000 MWh^{-1} to attract new generation investments. This measure delivered

---

11. Phone call with Scott Hudson and Sydney Sieger, TXU. December 18, 2014.
12. NRDC Analysis of EIA Data.

significant short-term results (Newell et al., 2012). Previously mothballed generation was put back online and 2000 MW of new planned generation was announced. The revised report now predicts a reserve margin of 15.7% for 2015. However, the increased cap will not ensure resource adequacy in the longer term. The reserve margin is predicted to dip below the planning target by 2019, and below 10% by 2022, without additional generation or a reduction in peak demand (Newell et al., 2012).

The growing electric load and demand, paired with unusual weather in recent years, have increasingly pushed ERCOT to its limit, requiring emergency load shedding and demand control. However, many retail providers (REPs) effectively encourage wasteful and ill-timed consumption.

REPs have developed novel and diverse offerings to attract and retain customers. These have allowed engaged customers to benefit greatly from competition. In the aggregate, however, the current offerings are detrimental to the long-term financial health of customers and the environment. While the offerings are unique, they are often not innovative in the way imagined by retail competition's proponents.

To attract customers, many REPs have designed offerings that pander to shortsighted desires for cheap energy rates, rather than on service or customer options that reduce energy bills. Many customers see energy as a pure commodity. It is not—ever more true as new technology and home services are developed. Energy is a service. Clean energy, energy efficiency assistance, customer care, digital and mobile tools, and convenience are all important elements that should be considered—but are not—when choosing energy supply.

The perception that energy is a pure commodity is only furthered by the PUCT's PowerToChoose website. This allows customers to compare all available plans and restrict offerings based on selected characteristics. However, the default search sorts results by the average cost per kilowatt-hour (at consumption of 1000 kWh). This average cost per kilowatt-hour includes both fixed and variable charges. The cost shown is simply the total expected bill at 1000 kWh divided by 1000 kWh. Providers and customers effectively treat fixed charges as variable costs.

REPs need to have low average costs to appear at the top of the search. This puts an organization emphasis on cost-cutting measures, rather than on investing in ways to offer a convenient, pleasant, and holistic electrical service customer experience. Some REPs have also developed workarounds, using miscellaneous or additional fees not included in the average cost calculation to artificially lower the average kilowatt-hour cost the website shows. These include minimum consumption fees, service fees, payment processing fees, and high consumption credits. Due to these types of fees and the confutation of fixed and variable charges, the average cost per kilowatt-hour for the same plan can vary as much as 40% between monthly consumption of 500 and 1500 kWh, creating customer confusion and distrust, when bills don't match the advertised cost (Schurnman, 2014).

In addition, many REPs offer plans that tout free energy or reward customers for consuming more to try to entice customers. Many use promotions throughout the year to garner customer interest, such as Reliant's "Sweet Deal." Like a "Baker's Dozen," after 12 months of service on this plan, a customer gets the 13th month free. A number of REPs—Reliant, Direct Energy, CFL, WTU, TXU, TruSmart, and Clearview—also offer "free power" plans. These plans have no energy charge, and often no T&D charge, during certain times of the week. Current offerings include free nights, free Saturdays, free Sundays, and free weekends. TXU's Free Nights plan and Reliant's Free Weekends Plan, both charging a higher rate during the day, are two of the few designated ToU plans in Texas. In fact, of the 117,570 residential customers on a ToU plan in 2014, over 100,000 of them were on TXU's Free Nights Plan (Frontier Associates, LLC, 2014). The traditional ToU pricing structure, where there are high peak prices, is rare in Texas due to low customer interest (Delurey, 2013).

Reliant also has developed a novel offering known as the "Predictable 12" Plan. Under this plan, customers pay a fixed amount—determined beforehand, based on historical consumption—regardless of a customer's monthly electricity consumption. Reliant designed this plan to give ultimate bill security to customers, but this new plan has quickly been dubbed the "all you can eat plan" (Texas Electricity Ratings, 2014). There is no incentive for customers to invest in energy efficiency and no penalty for keeping the AC on at 60°F all summer—even if not at home. During peak summer hours, this plan provides an almost perfectly perverse price signal.

A few REPs offer material rewards for electricity consumed. Reward deals like Fuel Rewards—where every $50 a person spends on a utility bill earns them $.05 off each gallon at a Shell station, and travel rewards—where the customer earns points toward cruises and resort stays for kilowatt-hours consumed—are pervasive.[13] A few offer more unique benefits. Southwest Power & Light and YEP Energy both offer charity plans, where every 1000 kWh consumed result in either one tree being planted by the Arbor Day Foundation, or a $3 donation being made to the USO's Warrior and Family Care Program. In addition, both Ambit and Stream offer Refer-A-Friend programs that give a "free energy" credit, calculated by multiplying the average daily energy charge for any referred customers by the number of days in the billing cycle, on every monthly bill.[14] Customers can receive a total credit up to their entire energy charge portion of the bill every month, giving them an incentive to encourage *their referred friends* to consume more.

There are also a few common practices that reward customers that consume more or penalize those who consume less. Most REPs—more than 75% in the Houston and Dallas areas—include minimum consumption fees, or bill credits,

---

13. For example, see http://ww2.ambitenergy.com/rates-and-plans/ambit-advantages/travel-rewards
14. For more information, see Ambit Energy's "Earn Free Energy" page, http://ww2.ambitenergy.com/rates-and-plans/ambit-advantages/free-energy

for some or all of their plans for residential, single family customers.[15] Minimum consumption fees, which can range from $4.95 to $19.95, are tacked on to customer bills—in addition to paying for energy consumed—when they do not consume a threshold electricity amount, usually set at 999 kWh or less a month (Bledrzyckl, 2013). They are a penalty for being energy efficient. To the same effect, some REPs instead waive a "base" or "service" charge if consumption is above 999 kWh, or offer a bill credit for "high consumption." For example, Gexa Energy's Choice Plan and Spark Energy's Smart Saver Plan provide customers with a $60 bill credit if consumption is above 999 kWh.[16] These fees and credits push customers to consume enough, even if unnecessary, to reach the threshold in order to save money. They penalize customers for using energy smarter.

The offerings developed due to customer's focus on a low average cost per kWh, and not the actual bill, dull incentives for consumers to invest in energy efficiency technologies or shift behavior to reduce consumption. However, not all offerings run counter to cleaner and smarter energy use.

Some of the larger REPs have invested significant resources into customer tools, new platforms, and energy-efficient innovative offerings. They want to become "trusted energy advisors," creating loyal customers by offering unparalleled service and helping customers help themselves. By providing additional services, the relationship goes beyond the bill and beyond the energy charge. These customer-centric services and offerings highlight the potential for the market to embrace DERs and profit.

The most common energy saving offerings in the retail market are plans that include free smart thermostats. Some of these smart thermostat plans have a demand response (DR) component or automated energy efficient mode included, but not all. These REPs have begun to develop new tools to use the wealth of new data. For example, Direct Energy has just released a new billing format, which uses smart meter data to breakdown the bill by household appliance power use.[17] Customers see the monetary cost of heating and cooling, lighting, washer, refrigerator, etc. The larger REPS also offer "budget watch" services—weekly emails that provide an overview of daily and weekly usage, including an estimated monthly bill amount, a comparison to customer's historical consumption, and tips on how to conserve energy easily (Reliant Energy, 2010). REPs have also developed online energy tools to help customers control power use within the home. By analyzing past household consumption data and customer home appliance information, these tools perform a "virtual audit" that identifies

15. NRDC Analysis, conducted on January 15, 2015 on the *PowertoChoose* website.

16. Gexa EFL: http://www.gexaenergy.com/UI/Handlers/DynamicRatePdf.ashx?pdfid=29022& rkey=8480|1640|1002|15225; Spark EFL: https://www.sparkenergy.com/Document/pdf/EFL/ EFL_21216.pdf

17. See Direct Energy and Bounce Energy's "Direct your Energy" program.

significant energy users. From this, the tool provides customized feedback and advice on how to reduce energy.

Some positive pricing changes have also appeared in recent years. Reliant, Green Mountain Energy, TXU, and TruSmart also offer plans with an inverted tier pricing design.[18] While some of these plans still include minimum consumption fees, it is a move in the right direction toward incentivizing energy efficiency.[19] In addition, TXU has released a ToU plan with high peak pricing from 1 to 6 pm every weekday during summer and fall.[20]

While the competitive market is beginning to offer positive and novel rate offerings and customer service, innovation has mainly been detrimental to energy efficiency. Dynamic pricing has largely failed in the state due to a lack of customer acceptance, and there has been a proliferation of offerings designed to make power appear cheaper. In practice, this has led to a market that rewards energy hogs, and penalizes those who conserve, stifling the deployment of energy efficiency and distributed energy resources.

3   IS RETAIL COMPETITION THE ANSWER?

The debate over retail competition is both unending and repetitive. While the restructuring discussion dissipated after the California energy crisis in 2001, it has recently gained traction. With the rise of new customer-sided technologies, and the increasing affordability of clean energy, some advocates have again taken up the banner of retail choice.[21] They argue that the utility monopoly model is stifling progress and innovation—hampering attempts toward a cleaner and distributed energy future. They believe retail markets would produce cheaper, cleaner energy and a much faster pace of innovation and progress. These arguments are almost identical to the claims made in the last round of restructuring.

Each side can pull facts and stories to support assertions that restructuring has been a complete failure or an emerging success. As in the chapter by Zarnikau and Rai. This finding is supported to some extent by our own review of US retail markets. However, a proliferation of retail products is inconsequential by itself. What matters much more is the content of these options and actual customer engagement. As demonstrated most clearly in Texas, a large majority of the plans appeal to customers' assumed obsession with finding the lowest short-term per kWh costs, rather than helping them use energy smarter or invest in cleaner

---

18. For example, see Reliant's Conservation Plan or Green Mountain Energy's Pollution Free Conserve Plan.

19. For example, TruSmart Energy's Conserve and Save 950 plan has a minimum usage fee of $14.95 if usage below 950 kwh.

20. See TXU's Energy PowerStart Plan.

21. For example, see Dave Robert's piece "19 tweets that explain how we're stifling electricity innovation" and the recent Arizona Commission Inquiry into Retail Electric Competition (Docket E00000W-13-0135).

energy. Many rely on gimmicks, promotions, and sometimes dubious methods to attract customers. So yes, retail choice has opened the way for offerings that allow customers to consume as much as they want with a fixed bill, but is that a good thing for our energy system or broader society?

Evidence has shown that restructuring doesn't result in more renewable generation or customer-sided energy efficiency, as many proponents had argued (Nooij and Baarsma, 2009; Palmer and Burtraw, 2005; Joskow, 2006). It also hasn't led to wide adoption of dynamic, market-responsive pricing. In fact, a competitive retail market seems to be a hindrance on both counts. Customers generally want simplistic, nonvolatile rates; dynamic pricing, while economically desirable and arguably integral for a functioning market, doesn't sell in the retail marketplace. Blunted price signals result in unresponsive energy demand, and suboptimal consumption. If policymakers want increased renewable generation or customer-sided energy efficiency, they need to continue support and expansion of clean energy policies, such as renewable portfolio and energy efficiency standards, building codes, and benchmarking requirements.

Significant research has been done to understand the market dynamics and barriers to clean energy development and adoption (Zuckerman et al., 2014; Brown, 2001). Wholesale price volatility and other uncertainties about cost recovery have scared off investors. Long-term power agreements to hedge market risk are much harder to procure in competitive retail markets, further inhibiting new capital investment. Customers face high information and transaction costs.

Competition can breed retail innovation, but the innovation has too often reflected customers' time and information constraints and undermined their long-term economic, reliability and environmental interests. There has been innovation in customer engagement tools, which regulated utilities, and regulators should incorporate and push in noncompetitive markets. Retail electricity choices can benefit individual customers who are proactive and informed, but on the whole, the retail market produces outcomes that are not socially desirable. Regulators in every state have had to intervene to ensure reliability and promote clean energy, and the greatest progress in energy-saving, innovative customer pricing has originated in regulatory decisions. Abundant experience shows that retail competition is not a promising route to a clean energy future.

4 A BETTER MODEL: UTILITIES AS CLEAN ENERGY PARTNERS

Regulators and legislators should give priority now to finding ways to better motivate and involve utilities in the clean energy transition. Utilities and the grid that they have helped build and improve over the last century are vital resources for everyone engaged in creating a clean energy future. In fact, the centralized grid—and the role of the utility to maintain it—is becoming increasingly important in the 21st century. Creating a robust and resilient grid is necessary to take full advantage of clean energy opportunities and their climate, economic, and resiliency benefits. In addition, utilities have the knowledge, experience, and strong relationships to serve as a bridge between consumers and these opportunities.

Increasing penetration of DER and other energy management technologies will disrupt the current utility business model, and require changes in both utility thinking and grid management. As more customers gain control over their energy use, the importance of the utility as a commodity trader diminishes. However, the utility's role in overseeing the local grid and providing energy services becomes more critical. DER not only improve grid resiliency and function, but the grid confers greater benefits to those who own customer-sided energy technologies. Unlocking the full value of DER requires the nation's centralized grid and partnership with the utilities that oversee and maintain it.

Many utilities are already exploiting their potential as clean energy partners. California utilities have begun partnering with SolarCity, Tesla, and others to provide customers with new opportunities in solar, batteries, and electric vehicles. Arizona utilities are piloting new tools that use personalized information to help customers decide energy matters, such as whether to install rooftop solar, and what rate tariff is the best fit for their consumption patterns. Other utilities have begun to act as middlemen—connecting interested customers to contractors, purchasing bulk orders of solar panels to reduce customer costs, and offering financing and on-bill loan programs to reduce customer barriers to clean energy. Utilities have been, and can be, instrumental in pushing clean energy deployment and innovation. However, the current utility business model often fails to motivate them to, or worse, discourages them for embracing clean energy. Future utility business models must correct this, and motivate utilities to be clean energy partners.

The best way to make utilities effective clean energy partners is by rewarding them for clean energy advancements through performance-based regulation.

Traditional regulation treats utilities as commodity providers—stymying the transition to a clean energy future. Performance-based regulation would treat utilities as *service* providers. No longer would the goal of the utility be to boost kilowatt-hour sales; the goal would be to offer the best customer experience. Regulators' societal objectives and goals can become the heart of utility regulation. The focus would no longer be on whether consumers paid the appropriate amount for energy already used, but on how to pay for the energy services and energy system society wants. Performance-based regulation would incentivize utilities to be innovative while also giving utilities and their investors the financial assurance required to fund all necessary capital and grid investments for a clean energy future.

Utilities can quicken the nation's transition to a clean energy system. Utilities' knowledge and experience put them in the best position to empower customers to take control of his or her energy use and take advantage of new clean energy opportunities. However, the transition to a clean energy partner requires a change in the utility business model. Not through bypassing the utility, but by providing incentives that make it profitable for utilities to be more innovative and forward-looking. Performance-based regulation is a utility model that can achieve this. Under performance-based regulation, the utility's main objective would be to improve customer access and opportunities behind-the-meter (eg, energy efficiency, demand response, solar, and energy storage) and be responsive to customer desires and needs.

5   CONCLUSIONS

More than two decades ago, a wave of enthusiasm for retail competition broke across the United States. But its momentum soon ceased, and has not been revived. As a clean energy transition unfolds, with its host of new challenges and opportunities, we see no new grounds for enthusiasm about a largely failed experiment. Indeed, in important respects, retail competitors can be seen to have pushed in exactly the wrong direction, with products exemplifying a philosophy of "all you can eat," rather than "all cost effective energy efficiency," or "continuous improvement in environmental and reliability performance."

   This is hardly surprising, since the case for retail competition was grounded from the beginning in a commodity model of electricity commerce, which puts paramount emphasis on minimizing unit costs, and maximizing kilowatt-hour sales, to the exclusion of other societal interests. Yet, some are now invoking those very interests to argue for giving new providers more opportunities to displace "monopoly" utilities.[22]

   Such claims have been made before. This chapter has canvassed the results, for the jurisdictions most active on direct access, and the record is clear. In terms of demonstrated progress on energy efficiency, renewable energy, and the grid enhancements that add value to both, direct access has failed to deliver. This does not mean that direct access products cannot be improved, as we have indicated. Nor is it an argument against robustly competitive wholesale electricity markets—which we support. But for those who care about reliable and affordable electricity service, with less pollution, motivated utilities remain essential partners. Encouraging them to persevere in such efforts, with regulatory and business model reforms, remains a paramount objective for all engaged in the evolution of the electricity sector.

---

22. See, for example, SB 286 (Hertzberg) in California, which proposes to expand direct access for nonresidential customers "especially to provide options for acquiring electricity from renewable sources of generation." Section 1 (g) (as amended, Apr. 29, 2015).

REFERENCES

ACEEE (American Council for an Energy-Efficient Economy), 2005. Third National Scorecard on Utility and Public Benefits Energy Efficiency Programs: A National Review and Update of State-Level Activity.

Bledrzyckl, C., 2013. Texas Electricity Consumer, Beware of REP Fees. Texas Ratepayers' Organization to Save Energy. August 12, 2013.

Borenstein, S., 2015. Is the Future of Electricity Generation Really Distributed? May 4, 2015. https://energyathaas.wordpress.com

Borenstein, S., Bushnell, J., 2014. The U.S. Electricity Industry after 20 Years of Restructuring. Energy Institute at HAAS.

Brown, M., 2001. "Market failures and barriers as a basis for clean energy policies". Energy Policy 29, 1197–1207.

Cavanagh, R., 1994. The Great "Retail Wheeling" Illusion—and More Productive Energy Futures. E Source Strategic Issues Paper.

CPUC (California Public Utility Commission), 1994. Order Instituting Rulemaking and Order Instituting Investigation, I. 94-04-032.

CPUC, 1995. Decision 95-12-063.

CPUC, 1996. Decision 96-01-009.

CPUC, 2006. Decision 06-10-019.

Daniels, S., 2014. Emanuel's Power Pact Could zap Chicago Homeowners. Crain's Chicago Business, May 16, 2014.

Daniels, S., 2015. Chicago Sending City Households Back to ComEd. Crain's Chicago Business, April 21, 2015.

DEFG (Distributed Energy Financial Group LLC), 2015. Annual Baseline Assessment of Choice in Canada and the United States (ABACCUS).

Delurey, D., 2013. Case Study Interview: Reliant Energy—Bill Harmon. Association for Demand Response and Smart Grid, Prepared for the National Forum on the National Action Plan on Demand Response: Program Design and Implementation Working Group.

EIA, 2009. Illinois Restructuring Active. http://www.eia.gov/electricity/policies/restructuring/illinois.html

EIA, 2010. Pennsylvania Restructuring Activity. http://www.eia.gov/electricity/policies/restructuring/pennsylvania.html

EIA, 2012. State Electric Retail Choice Programs are Popular With Commercial and Industrial Customers. http://www.eia.gov/todayinenergy/detail.cfm?id=6250

EIA (US Energy Information Administration), 2015a. Sales, by State, by End-Use, by Provider, Annual Back to 1990, (Form EIA-861).

EIA, 2015b. Existing Nameplate and Net Summer Capacity by Energy Source, Producer Type and State, (Form EIA-860).

EIA, 2015c. Net Generation by State by Type of Producer by Energy Source, (Forms EIA-906, EIA-920, and EIA-923).

ERCOT, 2015a. About ERCOT. http://www.ercot.com/about

ERCOT, 2015b. Historical Number of Premises Switched. Data available at: http://www.ercot.com/mktinfo/retail/index

Frontier Associates, LLC, 2014. 2013–2014 Retail Demand Response And Dynamic Pricing Project Final Report prepared for Staff of ERCOT.

Gearino, D. AEP's Plan for Guaranteed Coal-Plant Income Rejected. Columbus Dispatch, February 26, 2015.

ICC (Illinois Commerce Commission), 2002. Report of Chairman's Summer 2002 Roundtable Discussion Re: Implementation of the Electric Service Customer Choice and Rate Relief Law of 1997.

ICC, 2015. Office or Retail Market Development 2015 Annual Report.

Joskow, P., 2006. Competitive Electricity Markets and Investment in New Generating Capacity. Working Paper 06-14. AEI-Brookings Joint Center for Regulatory Studies.

Kleit et al., 2011. Impacts of Electricity Restructuring in Rural Pennsylvania.

Maryland Public Service Commission, 2008. State Analysis And Survey On Restructuring And Reregulation: Final Report. PSC #01-01-08.

McCloskey, T., 2014. Retail Choice in Pennsylvania And the Impacts of the Polar Vortex, Presentation, NARUC Subcommittee on Accounting and Finance.

Newell, S., et al., 2012. The Brattle Group prepared for the Electric Reliability Council of Texas. ERCOT Investment Incentives and Resource Adequacy.

Nooij, M., Baarsma, B., 2009. "Divorce comes at a price: An ex ante welfare analysis of ownership unbundling of the distribution and commercial companies in the Dutch energy sector". Energy Policy 37, 5449–5458.

NRDC and Silicon Valley Manufacturing Group, 2003. Energy Efficiency Leadership in California: Preventing the Next Crisis.

PA PUC (Pennsylvania Public Utility Commission), 2014. Retail Electricity Choice Activity Report 2013.

Palmer, K., Burtraw, D., 2005. The Environmental Impacts of Electricity Restructuring: Looking Back and Looking Forward. Resources for the Future.

PUCO (Public Utilities Commission of Ohio), 2014. Staff Report, Case No. 12-3151-EL-COI. In the Matter of the Commission's Investigation of Ohio's Retail Electric Service Market.

PUCO, 2015. History of Electric Regulation in Ohio, Presentation.

Reliant Energy, 2010. Smart Electricity Grid Reaches Texas. Press Release, January 19, 2010.

Roberts, D., 2012. How to Make Illinois Into a Clean Energy Leader. Grist, October 19, 2012.

Schurnman, M., 2014. Power to Confuse: Sneaky Fees Obscure Costs for Texas Electricity Shoppers. The Dallas Morning News, September 6, 2014.

Smith, R., 2000. Probe of California Power Prices Begins, But New Plants Aren't Seen as Solutions. Wall Street Journal, September 11, 2000.

Texas Coalition for Affordable Power (TCAP), 2014. Deregulated Electricity in Texas: the History of Retail Competition.

Texas Electricity Ratings: The Blog, 2014. Reliant's New Unlimited Electricity Plan: A Closer Look. June 17, 2014.

Tweed, K., 2015. Retail choice has doubled in the U.S.—Does it matter for Electric Industry Innovation? Greentech Media, July 15, 2015. https://www.greentechmedia.com/articles/read/does-it-matter-that-electric-choice-has-doubled

Weston, B., 2013. Comments by the Office of the Ohio Consumers' Counsel, Case No. 12-3151-EL-COI. In the Matter of the Commission's investigation of Ohio's Retail Electric Service Market.

Zuckerman et al., 2014. "Are Recent Forays into Electricity Market Restructuring a Threat to Energy Efficiency?". 2014 ACEEE Summer Study on Energy Efficiency in Buildings.

Chapter 10

Residential Rate Design and Death Spiral for Electric Utilities: Efficiency and Equity Considerations

Rasika Athawale, Frank Felder
Center for Energy, Economic, & Environmental Policy, Edward J. Bloustein School of Planning and Public Policy, Rutgers University, New Brunswick, NJ, United States of America

1 INTRODUCTION

The electric distribution industry in the United States has witnessed two distinct trends since the last recession: flat or declining demand (Fig. 10.1, Introduction to this volume), and capacity addition of new distributed energy resources (DER). DER can be both on the supply side (those that generate electricity, such as rooftop solar, and combined heat and power plants), and on the demand side (those that reduce electricity consumption, such as energy efficiency and demand response solutions).

Increase in penetration of DER is a result of improved economics due to various federal, state, and local policies and support, even though the capital costs of DER are substantially high, leading to less installed capacity per initial dollar invested, as compared to conventional energy sources (Beck and Martinot, 2004; Doris, 2012). Supporting policies notably include: (1) market preparation policies, such as net-metering and interconnection standards, (2) market creation policies, such as mandates or Renewable Portfolio Standards (RPS), and (3) market expansion policies, such as incentives in the form of tax rebates and grants (Doris, 2012).

Investments in decentralization have, in a way, created a partial substitute for the conventional utility grid and supply services. Between Jan. and Dec. 2014, the net-metering eligible customer-generators supplied cumulatively 723,611 MWh to the electric distribution companies in the state of New Jersey.[a]

---

a. EDC Semi-Annual Reports on net-metering and interconnection, NJCEP, http://www.njcleanenergy.com/renewable-energy/programs/net-metering-and-interconnection

Future of Utilities - Utilities of the Future. http://dx.doi.org/10.1016/B978-0-12-804249-6.00010-5

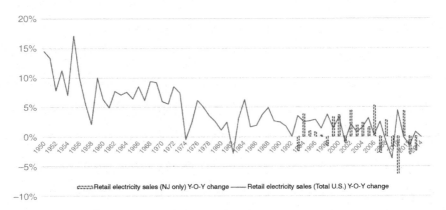

FIGURE 10.1 Year-on-year change in retail electricity sales in the US and New Jersey. *(Source: EIA, 2015, Monthly Energy Review, Mar. 2015; EIA, State Energy Profile.)*

That is approximately 1% of the total retail electricity sales in New Jersey for the year 2014.[b] At the national level, customer-sited photovoltaic (PV) systems generate only 0.2% of the total electricity consumption (Satchwell et al., 2015).

Now faced with some "real competition," the electric utilities have understandably not remained silent (business models under consideration by utilities to challenge this competition are discussed in chapter by Nillesen and Pollitt in this book). Their contention is that, while the benefits from DER are generally not questionable (although the mechanism for quantification could be), the costs must also be captured (Edison Electric Institute, 2013). Households with rooftop solar PV reduce their total consumption with little or no change in their peak demand, but contribute less toward the actual cost of grid services (Fig. 10.2 depicts the average load for a residential consumer in New Jersey on a summer weekday, and hourly generation from a 1 kW solar system). The overarching issues faced by electric utilities have been succinctly summed up in the chapter by Sioshansi in this book.

The current structure of utility tariffs, favoring volumetric rates, coupled with policies such as net-metering, is argued to lead to a revenue death spiral situation (Felder and Athawale, 2014). Needing to promote a mix of resources that can efficiently meet supply requirements, regulators have created a revenue imbalance, which favors the owners of distributed resources, and puts additional burden on the nonparticipants. Redesigning rate structures so that network costs are recovered through fixed charges (eg, customer and demand charges) is one solution, but this must be done with an eye on efficiency and equity considerations.

---

b. EIA, Projections based on growth rate in Annual Energy Outlook, http://www.eia.gov/electricity/state/newjersey/

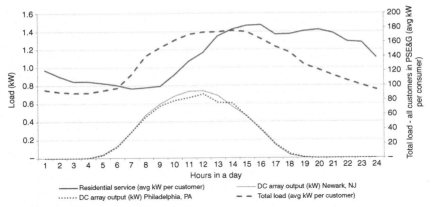

FIGURE 10.2   Load profile of customers in PSE&G service area and DC output from a solar system installed in Newark and Philadelphia. *(Sources: Sample residential and commercial load shape in the PSE&G service area is presented as average kW per customer, https://www.pseg.com/business/energy_choice/third_party/load_profiles.jsp; DC output from a 1 kW solar system installed in Newark and Philadelphia is calculated using NREL PVWatts® Calculator, http://pvwatts.nrel.gov/pvwatts.php. Data for Jul. 15, 2014.)*

The chapter is organized as follows. In Section 2, the major past and present themes governing electric tariff design are discussed, exploring how utilities recover costs through a combination of fixed and variable charges. In Section 3, the actual pricing structure followed by distribution utilities (using the state of New Jersey as an example) is discussed to quantify effects of current net-metering policy on various types of residential consumers, including low-income consumers. This is followed by projections of results for 2028, taking into account the effects of other policies, such as aggressive solar mandates. The insights from these analyses are then used to illustrate the magnitude of gain and loss for winners and losers, under possible scenarios of rate adjustments, followed by chapter's conclusions.

2   COST RECOVERIES VIA FIXED AND VARIABLE COMPONENTS

2.1   A Brief History of Electricity Tariffs

Capital intensive industries, such as electricity, natural gas, and communications. have used nonlinear pricing since the early 1900s (Joskow, 2007). Sherman and Visscher (1982) have shown that rate-of-return regulated firms, such as electric utilities, chose to adopt multipart tariff structures so as to augment their profits, and not because such tariff design is considered the best option for ensuring economic efficiency and equity. A widely accepted method of charging consumers for electricity services has been a two-part tariff, in which a "fixed" access fee recovers the utility's investment cost in setting up the transmission and distribution (T&D) network, and a "variable" usage fee ideally sets marginal price equal to the marginal cost of power generation. Within the two-part structure, however, the industry was initially divided on the methodology for charging fixed costs to consumers. The "Wright System" led to revenue maximization and monopoly-building outcome, and was advocated ferociously by dominating industry trade associations and large central power station owners; in contrast, the "Barstow System" was considered suitable for productive efficiency and short-term profit maximization (Yakubovich et al., 2005). Growth was the primary emphasis under the Wright System, which did not differentiate demand charge based on time of usage. The Barstow System recommended pricing as a measure to flatten the load curve and, in turn, propagated production efficiency. The Wright System later got adopted universally by regulators as the baseline-pricing model, and helped large integrated utilities fight competition for peak load consumers from "isolated generators," such as those owned by urban electric railways and municipalities.

A shift was observed in the 1970s, in the United States. By that time, two factors were evident: (1) benefits from economies of scale and technological progress had been nearly reached, and (2) change in residential consumer demand from increased usage of centralized air conditioning led to a wide difference in peak and base load requirements. Volume discounts or declining block rate pricing[c] was a popular tariff design to incentivize increasing use of electricity, so this concept was phased out to introduce rates differentiated by consumption, season, and time of day. Increasing energy efficiency was the key driver for such changes to the overall tariff design.

Another significant regulatory change during that period was the enactment of the Public Utilities Regulatory Policies Act (PURPA)[d] in 1978, which required utilities to buy power from qualifying facilities—some set up as distributed sources of generation. The Act allowed qualifying facilities to be compensated at full "avoided costs" that the utility would have otherwise incurred in generation or procurement from another source.

2.2 Rate Design: Fixed Versus Variable

As illustrated in Fig. 10.3, the fundamental principles of system sustainability, economic efficiency, and consumer protection guide the tariff for distribution services. Together, these ensure capital attraction, consumer rationing, and fairness to ratepayers, while taking into account the revenue-related attributes, cost-related attributes, and practical-related attributes for a sound rate structure (Bonbright et al., 1988).

---

c. During 1976, the electric tariff in the city of Los Angeles was a fixed charge of $1.50 per month and a declining block rate with 4.6 cents/kwh for the first 150 kWh, 3.5 cents/kWh for the next 250 kWh, 3.2 cents/kWh for the next 600 kWh, and 2.9 cents/kWh for all consumption in excess of 1000 kWh per month (Acton et al., 1978).

d. 16 U.S.C. § 824a-3 (2012).

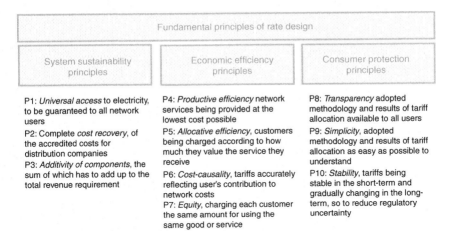

FIGURE 10.3   **Fundamental principles of rate design.** *(Source: Picciariello, A., Reneses, J., Frias, P., Soder, L., 2015. Distributed generation and distribution pricing: why do we need new tariff design methodologies? Electr. Power Syst. Res. 119, 370—376.)*

Some of these principles, which sound perfectly well in theory are often hard to establish and quantify in practice. For instance, the cost-causality principle proposes that network users be charged for their share of contribution to costs. Does that mean that a newly connected load (and consumer) should pay for entire system upgrade costs, which in effect, have been necessitated due to its addition on that part of the network? Welfare policies and political intervention often lead to compromise on other principles, such as equity, by demanding lower cost allocation to economically challenged consumers.

To maintain simplicity and to reflect the real costs as closely as possible, some utilities rely on a three-part tariff structure, which differentiates usage based on consumers' contribution to system peak demand, aggregate consumption, and time of consumption. The three building blocks of the tariff, as shown in Fig. 10.4, are required at a minimum to approximately depict the total costs incurred in meeting power supply, quality, and reliability. In the absence of time-of-day pricing, consumers do not necessarily care about their contribution to peak demand, and only care about the total aggregate usage in a month, irrespective of the hour when electricity is consumed. Such tariff design also introduces a type of cross-subsidy, from consumers who do not contribute toward peak load, to consumers who significantly contribute to the distribution power system's peak demand (Borenstein, 2005). Providing real-time time-of-day pricing signals to achieve operational efficiencies and encourage energy efficiency is one of the basic premises of smart grid investments.

There could be a fourth dimension that captures consumer characteristics and, hence, contribution toward utilities' fixed and variable costs. For instance, network investment required in providing electric supply to a consumer in rural

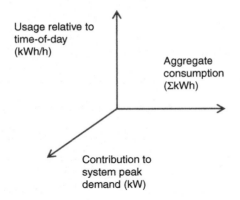

FIGURE 10.4 **Building blocks of tariff.**

area can be different than those incurred in meeting the demand of a similar consumer, but in a dense urban setting.[e] (See the chapter by Grozev et al. in this book for a discussion of residential network tariffs, and its impact on energy consumption patterns.)

In reality, however, the traditional tariff structure is designed in a way where the utilities' fixed costs are recovered primarily over a volume of usage. No doubt, a one-unit replacement of supply from a utility by a one-unit generation by a DER affects the overall revenue recovery of the utility. This type of a structure, which ties utilities' financial health directly to the volume of electricity sales, has long been recognized as one of the key barriers for growth of distributed resources and efficient network planning (Carter, 2001; Anthony, 2002).

Using New Jersey as an example, all four electric distribution utilities[f] charge almost all costs on a per unit consumption basis, to residential consumers. As seen in Fig. 10.5, all four utilities levy a small fixed portion as a monthly service charge ($/month), while the supply, as well as delivery charges, are volumetric ($/kWh). Out of a total average illustrative bill of one hundred dollars paid by a residential consumer, about 60% goes toward recovery of supply charges, while approximately 35% goes toward recovering delivery charges. The fixed

---

e. Hydro One in Canada classifies residential consumers into three categories and charges different fixed delivery rates: (a) Urban High Density consumers (contains 3000 or more customers, with at least 60 customers for every kilometer of power line used to supply energy to the zone) $16.64 per month; (b) Medium Density (contains 100 or more customers, with at least 15 customers for every kilometer of power line used to supply energy to the zone). $24.07 per month; and (c) Low Density (the remaining area not covered by Urban or Medium Density areas) $33.03 per month.

f. Public Service Electric and Gas (PSE&G), Jersey Central Power and Light (JCP&L), Atlantic City Electric (ACE), and Rockland Electric Company (REC).

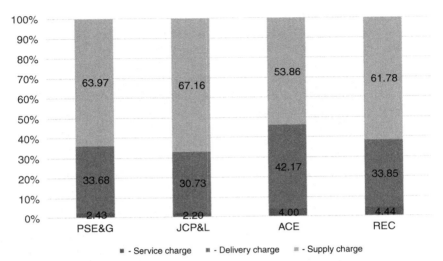

FIGURE 10.5 **Network and supply costs comparison of NJ utilities.** *(Source: Respective utility tariff for electric service.)*

monthly charge for a residential consumer of PSE&G is $2.43 per month, irrespective of total consumption; similarly, for JCP&L it is $2.20 per month. Rockland Electric charges the maximum fixed service charge ($4.44 per month) to a residential consumer in New Jersey.

Not all utilities, though, charge such low fixed rates (Fig. 10.6). For a residential consumer in Hartford, the Connecticut Light and Power charges as high as $19.25 per month as consumer charge, and a residential consumer availing supply from ConEdison in New York City pays $15.76 per month, as fixed charge. Regulators in some states have recently approved moderate

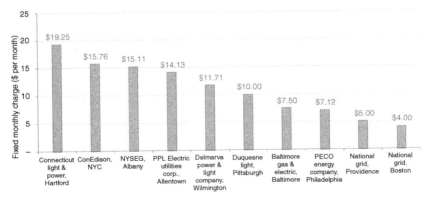

FIGURE 10.6 **Comparison of fixed monthly charges for utilities in the Northeast Region.** *(Source: Respective utility tariff for electric service.)*

to significant increases in the fixed charges of utilities. In late 2014, the Public Service Commission of Wisconsin approved fixed charges for residential consumers of Madison Gas & Electric to go up from $10.50 per month to $19 per month—an increase of 81%. The Public Service Commission of Wisconsin in its Order[g] dated Dec. 23, 2014, noted that the rates should provide a correct signal to a consumer about the costs (fixed and variable) incurred by the utility in providing access to the grid, and quality and reliable power supply. Other utilities, such as the Arizona Public Service, have asked for a 320% jump in fixed charge for DG consumers—that is from the current $5 per month to $21 per month.

Elsewhere,[h] in Canada, the Ontario Energy Board suggested revenue decoupling for low volume customer classes, through fixed rate design. The Board believes this is the most effective rate design for ensuring that rates reflect the cost drivers for the distribution system, and one that best responds to the current environment of increased penetration of distributed generation (Ontario Energy Board, 2014). The Board's draft report suggests that a fixed charge in effect is beneficial to a low volume consumer, as it makes the bill more predictable and stable. It lays out three options for revenue recovery:

- a single monthly charge, which is same for all consumers within a rate class;
- a fixed monthly charge with the size of the charge based on the size of the electrical connection; and
- a fixed monthly charge with the size of the charge based on the usage during peak hours.

---

g. 3270-UR-120; Application of Madison Gas and Electric Company for Authority to Change Electric and Natural Gas Rates http://psc.wi.gov/apps35/ERF_view/viewdoc.aspx?docid=226563

The Order states "…It is reasonable to consider a variety of factors in determining which utility costs should be treated as electric fixed costs, including Commission policies, fairness, and economic efficiency over the short and long term, when setting fixed charge rates for residential and small commercial consumers. …. *While the revenue to be recovered from each class is a separate determination, the increases proposed for the fixed charges are intended to better align the costs to serve individual customers with the revenues received from that customer….* .If the fixed charge is too low, the customer will receive an incorrect price signal that the cost to provide access to the electric system is lower than it actually is to the utility. They will also receive an incorrect signal that the variable cost to provide energy is higher than it actually is to the utility. Setting price signals correctly is important because those signals influence customer behavior, which in turn influences how the utility incurs costs… .*More importantly, the purpose of rate design is not to subsidize the payback of energy efficiency measures or renewable energy…* the Commission has determined that the proposed fixed charge increase for residential consumers will not disproportionately disadvantage lower-income groups to a significant extent."

h. In the city of Mumbai, India, for an average consumption of 250 units per month, a residential consumer pays about 5.12% of the total bill as fixed charge (INR 75 per month). The same for a below poverty line consumer is INR 5 per month, which translates to 9% of the monthly bill for a consumer. Electricity Tariff for Reliance Infrastructure Distribution Consumers (India), http://www.relianceenergy.in/pdf/Reliance_Energy_Tariff.pdf.

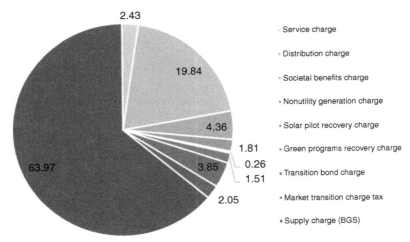

2.43

Service charge

Distribution charge

19.84

Societal benefits charge

Nonutility generation charge

4.36

Solar pilot recovery charge

1.81 Green programs recovery charge

63.97   3.85   0.26

1.51 Transition bond charge

2.05 Market transition charge tax

Supply charge (BGS)

FIGURE 10.7   Components of an electric bill (Utility: PSE&G). *(Source: PSE&G tariff for electric service, as of Mar. 1, 2015.)*

Several public policy program expenditures and historical true-up costs are embedded into the delivery charge, and are thus recovered on a volumetric basis, using the underlying principle that one who consumes more should pay more. Delivery charge in a PSE&G residential bill contains various program expenses, such as Societal Benefits Charge (SBC), Nonutility Generation Charge, Solar Pilot Recovery Charge, Green Programs Recovery Charge, Transition Bond Charge, and Market Transition Charge Tax. The share of these "nondelivery fixed charges," totaling $33.68 out of a $100 electricity bill, is shown in Fig. 10.7.

There may or may not be more economically efficient methods of recovering these other costs. However, considering the principle of simplicity of rate, and the time and effort consumers, especially residential consumers, devote to understand their electricity bills, it is easy to bundle all costs unrelated to supply into the delivery volumetric charge. Some of these charges (such as the SBC, in this case) are merely collected by the utility, and then passed on to the state government's Office of Clean Energy (OCE) that funds various renewable energy and energy efficiency programs. According to the OCE, from 2001 to 2006, about $787 million of SBC funds were collected from electric and natural gas consumers in the state. About 30% of the money the SBC collected supports the Universal Service Fund (USF) that provides assistance to low-income consumers.[i]

---

i. http://www.njcleanenergy.com/files/file/FAQs_pdf_4.pdf.

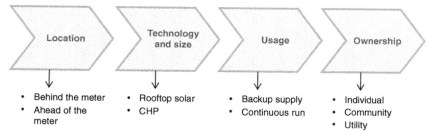

FIGURE 10.8 Characteristics of a DER.

3 WINNERS AND LOSERS: A CASE OF NEW JERSEY

3.1 Supporting Policies and Incentives for DER

Costs of, and benefits to, a DER should theoretically be based on the actual value provided by a particular DER project. State and federal policies and subsidies vary, depending upon several factors, and are preset for different characteristics of a DER, as shown in Fig. 10.8.

New Jersey has been a frontrunner in clean and renewable energy capacity addition. The Solar Energy Industry Association's 2013 report ranks New Jersey first in distributed, nonutility solar installations, and third in cumulative installed capacity nationwide.[j] The state has adopted a renewable portfolio standard (RPS), setting a target of 22.5% of electricity procurement from renewable energy resources, by the year 2021 (State of New Jersey, 2011). An Energy Resilience Bank has been launched to support investments in microgrids and distributed generation projects that can increase reliability and resiliency. New Jersey has successfully transitioned away from solar rebates (offered under the Clean Energy Program) to a market-based incentive mechanism that compensates owners of PV with tradable Solar Renewable Energy Certificates (SRECs).[k]

In response to the oversupply of SRECs in 2011–2012 compliance periods, the state passed the Solar Act of 2012[l] that accelerated the RPS compliance schedule, and changed the SREC requirement from a fixed volume target to a target based on percentage of load served. The Solar Act of 2012 also lowered the solar alternative compliance payment (SACP) schedule, under which the load-serving entities pay to buy retiring SRECs, so as to meet their annual solar RPS compliance obligations. The electric utilities offer solar financing programs that provide long-term SREC price support for participating system owners.

Net metering for residential and small commercial consumers has been prevalent in New Jersey since the introduction of the Electric Discount and

---

j. US Solar Market Insight: 2013 Year-in-Review, GTIM Research/SEIA.
k. 1 SREC = 1 MWh of solar electricity.
l. http://www.njleg.state.nj.us/2012/Bills/S2000/1925_R4.PDF.

Energy Competition Act (EDECA) in 1999.[m] The New Jersey Board of Public Utilities further relaxed the net metering rules in Sep. 2004, by (1) increasing the maximum customer-generator eligible capacity to 2 MW from 100 kW, and (2) by including solar thermal, fuel cells, geothermal, wave or tidal, biomass, and methane gas landfills into the spectrum of eligible technologies, in addition to the existing solar PV and wind. Currently, there is a cap of eligible capacity addition up to 2.5% of the state's peak demand, proposed to be raised up to 7.5% under the New Jersey Bill A-2420.[n] At the end of 2011, there were 12,907 customers in New Jersey who had net-energy meter solar PV systems, with an installed capacity of 507.36 MW.[o]

3.2   Quantifying Change in Tariff for Residential Consumers

On average, a residential consumer in New Jersey uses 750 kWh of electricity per month.[p] This compares to a consumption of around 600 units per month by a low-income consumer,[q] and approximately 1875 kWh per month by a high usage consumer, typically one with a large single-family house.[r] Total retail sales in New Jersey, in 2012, were 75,052,914 MWh, out of which 38% was consumed by the residential sector alone.[s] Roughly one third of the total number of households in New Jersey is eligible for the Low-Income Home Energy Assistance Program[t] (LIHEAP), using the federal maximum LIHEAP income standard for eligibility (greater of 150% of HHS poverty guidelines, or 60% of State median income). As compared to the whole of the United States, New Jersey is home to 3% of LIHEAP eligible households. Low-income consumers are provided several types of incentives, as shown in Table 10.1 Median household income in New Jersey is $71,637,[u] meaning that a monthly income of less than $3,581 for a household makes it eligible for LIHEAP benefits.

The average monthly electric bill for different types of residential consumers is calculated as a first step to quantify the net effects of volumetric rate recovery, and increased penetration of DER within a utility service territory. Residential consumers are categorized based on their usage pattern. For simplicity, it is

---

m. http://www.energycentral.net/article/05/08/net-metering-new-jersey.

n. http://www.utilitydive.com/news/new-jersey-bill-proposes-tripling-the-net-metering-cap-to-75/318830/.

o. http://www.seia.org/research-resources/net-metering-state for a state by state comparison of net-metering policies, please see http://freeingthegrid.org/.

p. EIA Average NJ household electricity consumption = 8,902 kWh/year.

q. Low income does not necessarily mean low consumption (ACEEE, Myth of Low-Income Energy Efficiency Programs: Implications for Outreach), but assumption is 20% lower than average.

r. Opower, top 1% of homes consume 4 times more electricity than average; a proxy large house uses 2.5x more energy than an average consumer.

s. EIA, State Energy Profiles.

t. The Low-Income Home Energy Assistance Program (LIHEAP) is a federally funded program that helps low-income households pay their home heating and cooling bills.

u. 2008–2012 American Community Survey.

TABLE 10.1 New Jersey: Number of Households Eligible and Served Under Various Low-Income Assistance Programs

For New Jersey	No. of HHs
HHs served under Universal Service Fund (2013)[a]	212,898
HHs served under Fresh Start (2013)[a]	14,564
HHs served under Lifeline (2013)[a]	304,534
HHs served under Comfort Partners (2013)[a]	11,760
Total number of LIHEAP eligible HHs (2011)[b] Federal Income Standards	1,044,279
Total number of HHs in New Jersey (2013)[c]	3,578,141

[a]*LIHEAP Clearing House, http://www.liheapch.acf.hhs.gov/dereg/states/njsnapshot.htm*
[b]*LIHEAP Home Energy Notebook, for Fiscal Year 2011, published Jun. 2014, https://www.acf.hhs.gov/sites/default/files/ocs/fy2011_hen_final.pdf*
[c]*US Census Bureau, http://quickfacts.census.gov/qfd/states/34000.html*

assumed that (1) a high usage consumer is a potential candidate to install a DER (such as a rooftop solar panel), and (2) by doing so, her monthly draw from the utility, post net metering, is half of the original. Fig. 10.9 shows the monthly bill for all types of consumers as of today, who all pay similar service charges, though their share of delivery and supply charges vary, based on their consumption. These calculations are done using the tariff rates for PSE&G, the largest utility in New Jersey. A low-income consumer pays a total electric bill of $113 per month, an average consumer pays $141 per month, and a high-usage consumer pays $348 per month. Note that the service charge (fixed dollars per

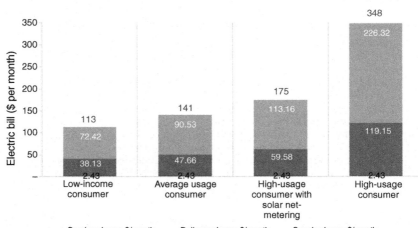

FIGURE 10.9   Current consumer bills (2014) for different residential usage.

month) remains the same for all types of residential consumers, irrespective of their usage. That would mean that a low-income consumer spends 3.15% of income toward electricity bills, while an average income consumer would spend around 2.36% of his income on electricity per month.

In the above example, the high-usage consumer with DER and net-metering benefits pays only 50% of the delivery charges. This means that under a full revenue recovery scenario, delivery charges for the rest of the consumers will have to increase correspondingly, so that the utility has the opportunity to recover 100% of its delivery costs (including embedded "nondelivery fixed charges," such as SBC). This effect of increase in delivery costs for other consumers can be estimated by calculating the total deficiency of delivery charge recovery from all DER-owned consumers in the state.

Reporting and tracking of SRECs generated within New Jersey is managed through the Generation Attribute Tracking System (GATS), administered by PJM Environmental Information Services. Of the 31,337 solar installations, totaling 1418 MW in New Jersey, around 31,233 systems are behind-the-meter installations with a cumulative capacity of 1163 MW. Electricity generated by these solar systems is eligible for net metering, and they can produce together approximately 125,604 MWh per month.[v] For these many units now sold less by the utilities, the respective lower recovery of delivery charge is equivalent to $8.15 million per month. That translates into a 2% increase in delivery charge for all consumers, arrived at by allocating the shortfall over the total retail sales in New Jersey.[w] That means the new delivery charge for all residential consumers would be 0.0648 $/kWh, instead of the previous 0.0635 $/kWh. Fig. 10.10 depicts the change in customers' bills as of today, for the four types of residential consumers, with an increase in delivery charges. All residential consumers would see a 0.7% increase in their monthly electricity bills.

Similar results for the year 2028 yield a 4.3% increase in delivery charge to recover the delivery charges shortfall of 262,994 MWh of solar energy generated per month, as mandated by the New Jersey Solar Act of 2012. Total retail sales per month in the year 2028, based on the average growth rate between 1991 and 2012, is projected to be 76,973,773 MWh.[x] For all residential consumers, in 2028, this translates to around 1.45% increase in their monthly electricity bills, when compared to their bills in 2014. Under a hypothetical aggressive solar mandate of, say, 10% of retail sales in 2028, the increase in monthly bills for all consumers would be around 3.80%, with delivery charge going up from 0.0635 per unit today to 0.0706 per unit.

---

v. Solar generation from behind the meter systems = capacity × solar array's inverter estimated efficiency (80%) × 4.5 (NREL's average hours of sunlight per day for New Jersey) × calendar days for month (NJ OCE Utility net-metering reports http://www.njcleanenergy.com/files/file/Utilities/Net%20Metering%20Reports/PSEG%20Semi-Annual%20Net%20Metering%20Report%20-%20January%202014-December%202014%20%20%20%202172015.pdf).

w. NJ total retail sales (2012, EIA) = 6,254,410 MWh/month.

x. Projections based on EIA assumption for growth rate in Annual Energy Outlook, 2014.

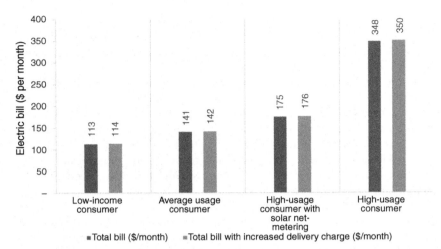

FIGURE 10.10 Changes in consumers bills (2014) for a possible scenario of increase in delivery charges for all consumers.

3.3 Quantifying Tax Effects for New Jersey State Government

Government revenues, in the form of tax collection, also get affected due to higher penetration of DER, especially solar distributed generation systems. At county level, this could be as a result of lower property tax collection, where home owners are allowed to exempt 100% of the value of home solar power systems from property taxes.[y] For the state government, there could be a reduced collection of the absolute amount of tax collected on the sale of electricity (with volumetric utility rates, the state's taxes are based on the total amount of the electric bill; any reduction in sales leads to lower billing and collection and, therefore, lower tax dollars).[z] In addition, the state government could in fact need to increase its subsidy to low-income consumers (assuming (1) low-income consumers are nonparticipants, that is, they do not invest in installing a DER; and (2) the subsidy provided to low-income customers is adjusted so that their expenditure on electricity does not exceed 6% of their income.)[aa] Finally, for the federal government, there could be an impact on the total income tax collection, as residential sized projects are eligible for a 30% Investment Tax Credit under Section 25D, allowing homeowners to apply the credit on their annual income.[bb]

---

y. In New Jersey, local property tax exemption is allowed, if the system is used to meet on-site electricity, heating, cooling, or general energy needs. http://www.seia.org/policy/finance-tax/solar-tax-exemptions

z. Retail sale of electricity is subjected to Sales and Use Tax (SUT), currently at 7%. http://www.state.nj.us/treasury/taxation/enernot.shtml

aa. Subsidy to low-income consumers is provided from the Universal Service Fund. http://www.state.nj.us/bpu/residential/assistance/

bb. The Federal ITC is applicable for systems placed in service on or after Jan. 1, 2006, and on or before Dec. 31, 2016. For more details, please see http://energy.gov/savings/residential-renewable-energy-tax-credit

In Section 3.2, it is assumed that the utility delivery charge shortfall is recovered by adjusting the delivery charges to account for reduced sales to consumers. Thus, for the state government, there is no absolute tax collection loss. However, the same is not true for supply charges, meant to recover the Basic Generation Service (BGS) charges. The utility does not recover BGS for every single unit less sold; this, in turn, also reduces the government tax collection. For New Jersey in 2014, that meant a tax loss of $1.06 million per month, or $12.74 million per year for 1,507,248 MWh of production by customer-generators. Results for 2028 yield that the NJ State Government would lose $26.67 million, assuming no change in supply charges.

The burden of low-income assistance would also go up, as the electric bill for low-income consumers increases with an increase in delivery charge. For instance, a low-income consumer whose monthly bill is $100 receives an assistance of $60, after paying $40; and where $40 does not exceed 6% of the consumer's monthly income. If the electric bill of such a consumer goes up by 1.5% (Section 3.1 for year 2028), and the consumer still pays equivalent to 6% of income, then the government's share actually goes up by 2%. As detailed in Table 10.2, DER capacity addition, along with various prevalent supporting policies and incentives, leads to certain winners and losers, and the extent of gain and or loss depends upon the mechanism followed for utility revenue recovery and rate adjustments. For simplification purposes, it has been assumed that sales remain constant (as projected), even with an increase in tariff for all or certain category of consumers. Note that these calculations do not factor in any change in the SREC prices, which may increase with any increase in tariff for the participating consumers (who may like to increase their revenue by REC sales to provide for higher costs) and which, in turn, shall increase the tariff for all other consumers (as SREC prices are allocated to all electricity sales).

4 CONCLUSIONS

Pricing for electricity has always been nonlinear, unlike many other goods that have a single price. Various alternatives or tariff design options exist within this nonlinear methodology for pricing of electricity supply services. The main concern of this chapter is not to opine on desirable tariff design. Instead, this chapter provides a quantitative comparison of tariff increase for participating and nonparticipating residential consumers, under the assumption that the utility is allowed to recover its revenue by allocating costs in the same fashion as today. Further effects of such rate increases on low-income consumers, and on government's tax collection, are also analyzed, and results for a future period are projected.

There is a possibility that the utility of the future shall advocate for radical changes to rate design, with an intent to adopt and adapt to the growing influence of DER. It may be worthwhile to conduct further research on the quantification of effects of an alternate rate design on various categories of consumers,

TABLE 10.2 Winners and Losers Under Possible Scenarios of Rate Adjustments

Possible scenarios for rate adjustments

Effect on stakeholders	No full revenue recovery for the utility		Revenue recovery through increase in variable charge (all customers pay)		Revenue recovery through increase in fixed charge (only participants pay[a])	
	2014	2028	2014	2028	2014	2028
Consumer						
Bill for high usage consumer with solar and net-metering	No change	No change	+ 0.70%	+ 1.45%	+ 60.80 $ per month	+ 62.12 $ per month
Bill for high usage consumer	No change	No change	+ 0.70%	+ 1.46%	No change	No change
Bill for average usage consumer	No change	No change	+ 0.69%	+ 1.45%	No change	No change
Bill for low-income consumer	No change	No change	+ 0.69%	+ 1.44%	No change	No change
Government						
Loss of SBC	$12.39 million per year	$25.94 million per year	Nil	Nil	Nil	Nil
Loss of SUT	$19.44 million per year	$40.70 million per year	$12.74 million per year	$26.67 million per year	$12.74 million per year	$26.67 million per year
Utility						
Loss of revenue (network related; not generation)	$95.78 million per year	$200.55 million per year	Nil	Nil	Nil	Nil

[a]Assuming 0.6% consumers as participants in 2014 (PSE&G has a consumer base of 2.2 million and 12,504 solar systems installed in its service area). Total number of consumers in NJ is estimated at 4 million, out of which 2.2 million served by PSE&G, 1.1 million served by JCP&L, 545,000 consumers of Atlantic City Electric, and 200,000 consumers of Rockland Electric. Participant share for 2028 is assumed to be 1.2%.

as well as government tax collection. It is essential to project results for a future year, assuming DER penetration goals are met, since the scale and size of DER capacity would have a strong impact on the utility revenues and rates. Various nuances are encountered in actual calculations, such as the methodology adopted for determination of total dollars for the Universal Service Fund and fallback mechanism, in case of under recovery; effect of reduction in sales due to utility-sponsored and ratepayer sponsored energy efficiency programs, along with estimation of increase in delivery charge due to such program expenditure; changes in customer behavior and elasticity of demand, with increase in prices, and increase in certain components of prices, and so on. As a further investigation, the effect of changes in net metering policy (cap on generation, calculation of avoided costs, credit period, etc.) on the gains and losses to stakeholders can be studied.

REFERENCES

Acton, J.P., Mitchell, B.M., Sohlberg, R., 1978. Estimating residential electricity demand under declining-block tariffs: an econometric study using micro-data. The Rand Corporation.

Anthony, A., 2002. The legal impediments to distributed generation. Energ. Law J. 23, 505–524.

Beck, F., Martinot, E., 2004. Renewable energy policies and barriers. In: Cleaveland, Cutler (Ed.), Encyclopedia of Energy. Academic Press/Elsevier Science, pp. 365–383.

Bonbright, J., Danielsen, A., Kamerschen, D., 1988. Principles of Public Utility Rates, second ed. Public Utilities Reports.

Borenstein, S., 2005. The long-run efficiency of real-time electricity pricing. Energ. J. 26 (3), 93–116.

Carter, S., 2001. Breaking the consumption habit: ratemaking for efficient resource decisions. Electric. J. 14 (10), 66–74.

Doris, E., 2012. Policy Building Blocks: Helping Policymakers Determine Policy Staging for the Development of Distributed PV Markets. National Renewable Energy Laboratory, Golden, CO, Prepared for 2012 Renewable Energy Forum, May 13–17, 2012. http://www.nrel.gov/docs/fy12osti/54801.pdf.

Edison Electric Institute, 2013. Since Net-Metered Customers Are Both Buying and Selling Electricity, They Are Relying on the Grid More Than Customers Without Rooftop Solar or Other DG Systems. Edison Electric Institute.

Felder, F., Athawale, R., 2014. The life and death of the utility death spiral. Electr. J. 27 (6), 9–16.

Joskow, P.L., 2007. Regulation of natural monopolies. Handbook of Law and Economicsvol. 2, no. 2Elsevier BV, pp. 1227–1348.

Ontario Energy Board, 2014. Draft Report of the Board, Rate Design for Electricity Distributors, March 31, 2014. http://www.ontarioenergyboard.ca/oeb/_Documents/EB-2012-0410/EB-2012-0410%20Draft%20Report%20of%20the%20Board_Rate%20Design.pdf

Satchwell, A., Mills, A., Barbose, G., 2015. Quantifying the financial impacts of net-metered PV on utilities and ratepayers. Energy Policy 80, 133–144.

Sherman, R., Visscher, M., 1982. Rate-of-return regulation and two-part tariffs. Quarterly J. Econ. 97, 27–42.

State of New Jersey, 2011. NJ Energy Master Plan. http://nj.gov/emp.

Yakubovich, V., Granovetter, M., McGuire, P., 2005. Electric charges: the social construction of rate systems. Theory Soc. 34, 579–612.

Chapter 11

Modeling the Impacts of Disruptive Technologies and Pricing on Electricity Consumption

George Grozev*, Stephen Garner†, Zhengen Ren*, Michelle Taylor†, Andrew Higgins‡, Glenn Walden†

*CSIRO, Clayton, VIC, Australia; †Ergon Energy, Fortitude Valley, QLD, Australia; ‡Dutton Park, QLD, Australia

1 INTRODUCTION

In Australia, until 2009, electricity consumption had increased year by year consistently, for more than a century (Ren et al., 2016). In the five financial years since 2008–09, the net electricity used from the grid in the National Electricity Market that covers all eastern and Southeastern states, in each year, has been less than the year before, with a total reduction of 8%, that is, from 210.5 TWh in the financial year 2008–09, to 193.6 TWh in 2013–14 (AER, 2015a). This decline has been attributed to several factors, including rapid uptake of solar photovoltaic (PV) arrays supported by government policies, energy efficiency programs, and other government policies related to climate change and renewable generation, structural change of the economy away from electricity intensive industries, and the response of residential consumers to higher electricity prices. As noted in chapter: What is the Future of the Electric Power Sector? by Sioshansi, the declining electricity consumption is happening in other developed countries as well—the growth of electricity sales in the United States has stagnated, and in many states has become negative. Chapter 1 provides demand projections for other countries, including China and OECD averages. In most cases, the decline in demand is frequently happening in combination with rising retail tariffs.

In this context, distribution utilities in Australia face declining network utilization (energy transported per unit of asset capacity) which, given a predominantly fixed asset cost base, places upward pressure on network prices and, hence, retail electricity prices. Of specific interest, in relation to energy consumption patterns and distributed energy resource (DER) adoption, are the

Future of Utilities - Utilities of the Future. http://dx.doi.org/10.1016/B978-0-12-804249-6.00011-7

distortional effects of residential electricity prices which are principally volume (energy) based, while the majority of the underlying electricity supply costs are essentially fixed asset costs, that is, not volume based, and designed to meet the maximum demand. The predominantly flat residential tariff structures lead to larger afternoon and evening peak demand across the networks. Understanding the vicious circle of reducing electricity consumption contributing to rising prices, and mechanisms to redress this distortion, is critical to electricity businesses (Chapter 1). For more information about alternative pricing models, and recent tariffs proposals in Australia, see the chapter by Nelson and McNeill, and for rate design and history of the tariffs in USA, that by Athawale and Felder, in this volume.

In April 2015, Tesla announced mass production in the United States of two new, affordable lithium-ion batteries for residential and industrial use—7 and 10 kWh (Tesla, 2015). These batteries are priced at $3000 and $3500, respectively, not including inverter, which created a lot of excitement, and it is considered a step-change in relation to the price drop of home batteries.

Batteries, together with hydro pump storages and thermal storages, are balancing options for a grid with intermittent PV and wind generation, and they compete with demand response aiming to change normal demand patterns, in order to reduce peak electricity demand, and to increase network utilization (Gils, 2014).

It is expected that home batteries will change significantly the way residential customers buy and sell electricity. The Australian Energy Market Operator (AEMO, 2015) estimated in June 2015 that the installed battery capacity in the regions of the NEM will grow to 3.4 GWh in 10 years, even without considering retrofitting of existing solar PV with battery storage.

Nelson and McNeill summarize comprehensively how utilities within Australia are faced by these multiple challenges, due to the emergence of new technologies. In their chapter, the authors also describe how the biggest energy utilities in Australia try to adapt to this new and rapidly changing situation, by offering new products and services, including new, more cost-reflective tariffs, which could be quite different in different jurisdictions.

The utilities in Australia are not only following with a keen interest in the developments in the area of home storage. A number of them are considering different options to utilize home storage, and to enter this new, potentially big market. Ergon Energy (2015a, c) in Queensland has initiated a 12 month trial of battery storage, home energy management systems, and different tariffs in 10 homes in Townsville. The retailers Ergon Energy, ActewAGL in the Australian Capital Territory (ACT), and Red Energy in Tasmania will start selling an 8 kWh lithium-ion home battery in October 2015 (Macdonald-Smith, 2015). Electricity grid battery storage devices called Grid Utility Support System (GUSS) are expected to be deployed by Ergon Energy (2015b) in rural Queensland in the next couple of months. GUSS, based on the lithium-ion technology, will be used on constrained single wire, high voltage distribution lines, known as Single

Wire Earth Return. The batteries will be charged from the grid overnight, and discharged during peak electricity demand, improving the quality and reliability of electricity supply.

This chapter consists of five sections, in addition to the Introduction. Section 2 summarizes the modeling approach undertaken in this study. Section 3 provides a brief description of several electricity price tariffs modeled in this study, and indicates potential customer response in combination with solar PV Feed-in Tariff (FiT). Section 4 covers proposed battery charging and discharging rules that may be applied in different customer circumstances. Section 5 describes some modeling results, and provides examples of annual electricity cost for two house types, with several options of solar PV plus batteries, and two occupancy patterns, followed by chapter's conclusions.

2 MODELING APPROACH

This chapter presents the modeling approach used by CSIRO and Ergon Energy (Fig. 11.1). The study was carried out in 2014, and it tested a hypothesis that adoption of residential "network tariffs" (cost-reflective tariffs) will reduce or eliminate the distortional effects of volume based tariffs on residential energy consumption patterns. It tested the effect of three existing and five proposed new tariff structures on the future uptake of solar PV and battery storage options, in the region of Townsville, to 2025, and the subsequent effect on peak demand and average electricity consumption. The new tariffs involved changes to the energy charge and daily supply charges, and included time-of-use (ToU), capacity or critical peak charge to encourage lower electricity consumption in peak periods.

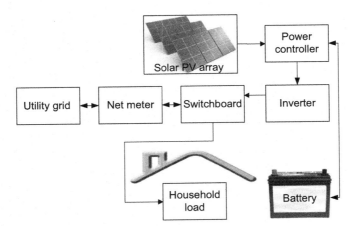

FIGURE 11.1 **The distributed energy resources considered in this study include: solar PV arrays and home batteries in combination with household electricity demand and different electricity price tariffs.**

TABLE 11.1 Modeled Scenarios

Tariff	Solar PV FiT	Solar PV and battery options	Dwelling type	Occupancy type
T11	FiT1 (8 ¢/kWh)		Detached house	All day occupied
T11 + T31		No PV, no battery		
T11 + T33		2 kWh battery		
T12		1.5 kW PV		
1A		1.5 kW PV + 2 kWh battery		
2A	FiT2 (44 ¢/kWh)	3.5 kW PV	Apartment	Evening occupied
2B		3.5 kW PV + 8 kWh battery		
2C		5.5 kW PV		
3A		5.5 kW PV + 16 kWh battery		

Information about all modeled scenarios is summarized in Table 11.1. Scenarios are defined as combinations of options—one from each column. The total number of scenarios is defined by the following multiplication: 9 tariffs × 2 FiT × 8 PV and Battery options × 2 Dwelling types × 2 Occupancy types = 576.

The future uptake to 2025 of solar PV and batteries, for the region of Townsville in Queensland, for the same study carried out by CSIRO and Ergon Energy, is reported by Higgins et al. (2014a). The authors described an application of a choice-diffusion model to estimate the future trends. The uptake of battery storage connected to solar PV ranged between 3 and 5.4% of households at 2025, depending on the price tariff, with the larger PV + battery options being more popular. A sensitivity analysis in this study shows that battery price was a major driver to uptake, while typical financial subsidies to purchase price have a lower effect.

The modeling approach applied in this research project has five main subsystems shown in Fig. 11.2 (Ren et al., 2013; Higgins et al., 2014b; Platt et al., 2014).

A description of the electricity tariffs studied is given in Section 3. Both currently existing and several proposed residential electricity tariffs are modeled. The proposed tariffs are assumed to be more cost-reflective, with a network component structured to recover the substantially high total fixed cost to own and operate the network, and a retail component—to recover the substantially variable energy production and supply costs. However, there is no expectation that one or several of the proposed tariffs will be immediately offered to customers. Rather a scenario-based approach is utilized to understand better the

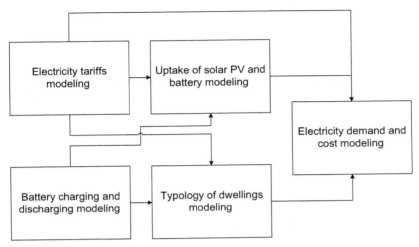

FIGURE 11.2   Main modeling subsystems.

cost, electricity consumption, and the impacts under given assumptions, and uptake of DERs.

The uptake of solar PV in combination with different battery options is based on the application of a choice-diffusion model (Higgins et al., 2014a) with a primary goal to forecast the stock of solar PV and battery options at incremental time intervals of 3 months, through the study period. This uptake model is outside of the scope of this paper. The battery charging and discharging rules are discussed in Section 4. The important point is that the battery charging and discharging rules are also influenced by the value of the solar PV FiT, when the household has also a solar PV.

The typology of dwellings modeling is based on the approach of Ren et al. (2012, 2013), which considers house type, vintage and size, family type, occupancy time, etc. In this research study, the dwelling typologies are extended to consider different electricity tariffs, and different solar PV FiT (8 ¢/kWh and 44 ¢/kWh). Based on the projected household load profiles, and given a specific FiT, the annual electricity cost for several of the modelled dwelling typologies is estimated.

3 TARIFF DESCRIPTION AND ANTICIPATED CUSTOMER RESPONSE

Optimal pricing of distribution networks is a major policy issue in Australia, at the moment, and the debate has focused on "cost-reflective" tariffs for distribution networks (see the chapter by Biggar and Reeves in this volume). Recent rule changes in Australia require distribution networks to charge cost-reflective tariffs that are set according to long-run marginal cost. However,

Biggar and Reeves suggest that the retail wholesale electricity price should reflect the short-run marginal cost of producing and delivering electricity to the customer location. Biggar and Reeves argue that policymakers should simultaneously ensure that retailers have access to hedging instruments to help them convert dynamic wholesales charges into a range of retails tariffs that customer prefer.

In this chapter, the basic conceptual model for residential electricity tariffs is that they would be built from a combination of a network (N) component and a retail/energy (R) component, namely, $(Total) = $(N) + $(R). The network component is structured to recover the substantially high total fixed cost to own and operate the network, while the retail component recovers the substantially variable energy production and supply costs. Additional "signaling information" could be built in to pricing, such as the concept that power (kW) and/or energy (kWh) demand has varying value/cost, depending on a range of factors, including the time and location where the electricity is consumed.

In this study, several existing and several potential future tariffs were modeled to estimate the cost of electricity for different types of customers. The main existing residential electricity tariffs in Queensland are described in Table 11.2 (DEWS, 2015b). The retail electricity prices are regulated by the Queensland Competition Authority; however, customers may pay a different price if they are on a market contract with their retailer. The main current tariff is T11—it is a flat tariff regardless of the ToU and amount of electricity consumed. It is the default and most popular currently used tariff. T12 is a ToU tariff with off-peak, shoulder, and peak periods defined. There are also two control load tariffs—T31 and T33—usually used for hot water heating, pool pumps, or other major loads.

There are also two FiTs that use net metering to measure the export to the grid from residential solar PV panels. Net metering to measure the export to the grid is dominant in Australia, like in the United States (Chapter 1). FiTs in Australia were initially introduced by the States or Territories. FiTs in Queensland were firstly proposed in 2007 by the State Government (Queensland Government, 2007) in the Climate Smart 2050 policy. Initially, the solar PV FiT was very generous, providing 44 ¢/kWh of exported electricity to the grid (DEWS, 2015a). After July 2012, new installations of solar PV can have only 8 ¢/kWh FiT. The 44 ¢/kWh rate is legislated to expire on July 1, 2028 and it is subject to several conditions. These two different FiTs motivate very different customer behavior in terms of load shifting, as explained later in this section.

Due to the unexpectedly high initial uptake of solar PV by early adopters supported by very generous FiTs, network utilities began losing revenues and, at the same time, wealth from customers who didn't have solar PV began to be transferred to these early adopters. It is estimated that the Renewable Energy Target (RET) contributed about 4% to household electricity bills in the financial 2013–14 year (CER, 2015). This is approximately equal to $60 per year for an

TABLE 11.2 Main Existing Residential Electricity Tariffs in Queensland

Tariff	Short description	Long description
T11	Residential, flat	This tariff is the default tariff for residential customers, regardless of their size
T12	Residential, ToU	Depending on the time of day, energy is priced differently. Customers must have a ToU capable meter to access the tariff
T31	Off-peak, super economy	Specified connected appliances are controlled by network equipment, so supply will be permanently available for a minimum period of 8 h at the absolute discretion of the distribution utility, but usually between the hours of 10 pm and 7 am.
T33	Off-peak, economy	Specified connected appliances are controlled by network equipment so supply will be available for a minimum period of 18 h per day during time periods set at the absolute discretion of the distribution utility
FiT1	Solar PV (net with FiT)	This tariff is part of the Queensland Solar Bonus Scheme: FiT rate of 44 ¢/kWh, paid by the distribution utility
FiT2	Solar PV (net with FiT)	This tariff is part of the Queensland Solar Bonus Scheme: FiT rate of 8 ¢/kWh, paid by the distribution utility

average electricity bill. However, the RET includes both small-scale and large-scale systems, and solar PV is one, nevertheless the main component, of the small-scale residential systems.

Queensland is the leading state in Australia in terms of the number of solar PV installations and total solar PV capacity installed, as noted by Nelson and McNeill. In Queensland, there are more than half a million residential customers with installed solar PV (Ergon Energy, 2015c). The number of small-scale solar PV installations, according to the Clean Energy Regulator (CER, 2015) is about 440,000, most likely due to up to one year allowed delay in registration of the corresponding certificates. These installations in the region of Townsville are approximately 27% of all residential customers, or close to 17,000 at the end of April 2015—growing from 13.5% in 2013. In some suburbs of Townsville, this percentage uptake is up to 36% (APVI, 2015). Currently, distribution charges correspond to approximately 42% of the annual electricity bill for Ergon Energy customers (AER, 2015b).

The plausible future tariffs modeled in this study are combinations of retail and network components (Table 11.3). The three retail components are fairly

TABLE 11.3 Modeled Residential Electricity Tariffs

Tariff	Energy Charge		Daily supply charge	Capacity charge	Critical peak charge
	Flat	ToU			
T11	√		√		
T12		√	√		
1A	√		√		
2A	√			√	
2B		√		√	
2C	√		√		√
3A	√		√	√	

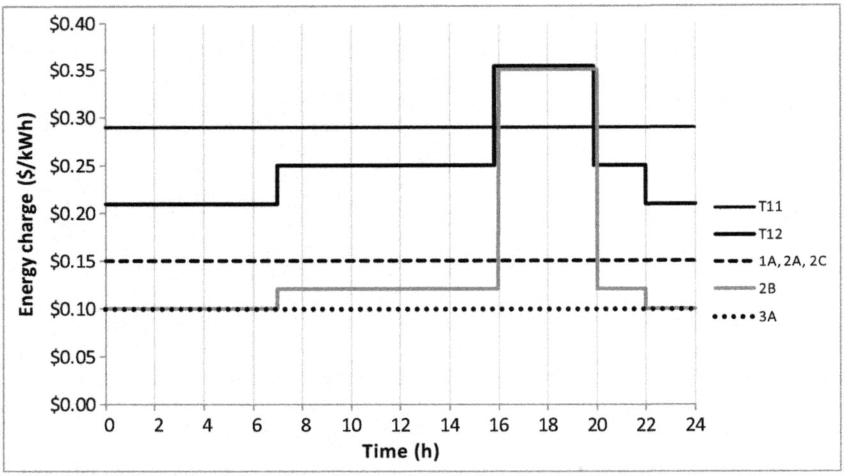

FIGURE 11.3 Energy charge for the considered electricity tariffs.

"standard" concepts related to electricity consumption measured in kilowatt-hour (Fig. 11.3):

● Flat, where the anytime cost per kWh is constant;
● ToU, where the cost per kWh is shaped to reflect the different economic cost of electricity through the day with peak, shoulder, and off-peak periods;
● Critical Peak Pricing (CPP), where the anytime cost per kWh is constant, except for prenotified periods where a higher cost per kWh applies—typically these periods are notified one day in advance, are of one or two hours duration, occur on up to twelve days per annum, and have a "moderate" price differential.

The three network component concepts are related to electricity demand measured in kilowatts.

- Fixed, where the cost to connect is constant—this exists in current tariffs, but at a relatively low rate, compared to the actual underlying costs;
- Capacity, where the anytime price per kW peak demand is constant; and,
- Peak Demand, where the anytime price per kW demand is constant, except for prenominated times of higher peak demand when a higher rate applies, and the charge that applies is the greatest of the calculated charges for the period—conceptually, the peak demand periods would be nominated at the start of an annual regulatory period (June), would be for the duration of the residential peak period (4−5 h), 5−7 days per week, and have "high" price differentials.

Note that the conceptual tariff to be modeled also includes an additional fixed daily charge (Table 11.4).

In Townsville, many households with solar PV installations signed up to the 44 ¢/kWh FiT. This FiT has encouraged households to shift their electricity demand away from the daytime period to maximize income generated from their PVs (Fig. 11.4). This increases the peak demand from the network between 4 and 8 pm. In Jul. 2012, the 44 ¢/kWh FiT was reduced to 8 ¢/kWh for new customers. This new tariff encouraged households to shift electricity demand to the

TABLE 11.4 Network Component Charge of the Two Existing and Five Proposed Tariffs

Tariff	Fixed daily charge $/day[a]	Daily capacity charge ($/kW/day)	Monthly demand charge for maximum peak in the periods 4–9 pm (Dec.–Mar.) ($/kW/month)	Monthly demand charge for maximum peak in the periods other than 4–9 pm (Dec.–Mar.) ($/kW/month)
T11	0.55			
T12	1.25			
1A	4.00			
2A		0.50		
2B		0.50		
2C		0.50		
3A	1.25		30.00	3.00

[a]All dollar values are assumed in Australian dollars (AUD), except explicitly stated otherwise.

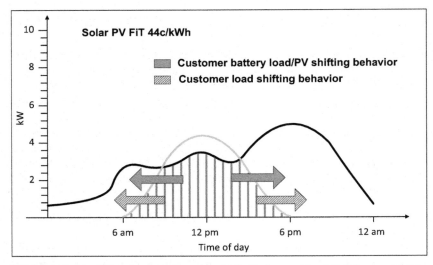

FIGURE 11.4 Anticipated customer response in the Business as Usual (BAU) case with Solar PV and high FiT.

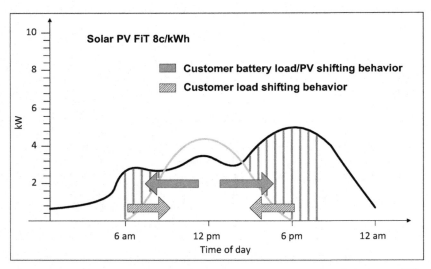

FIGURE 11.5 Anticipated customer response in the BAU case with Solar PV and low FiT.

daytime period, to minimize electricity bought from the grid at a usually much higher price than 8 ¢/kWh (Fig. 11.5). With the introduction of capacity charge and ToU tariffs, the goal is to incentivize households to shift demand from the peak periods (Fig. 11.6 and Fig. 11.7). Households purchasing electric vehicles

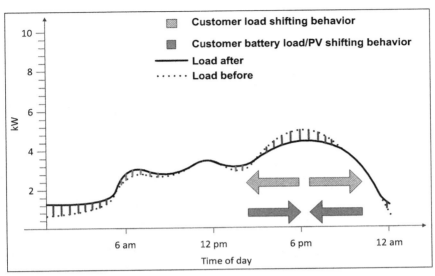

FIGURE 11.6 Anticipated customer response of load smoothing in case of capacity tariffs.

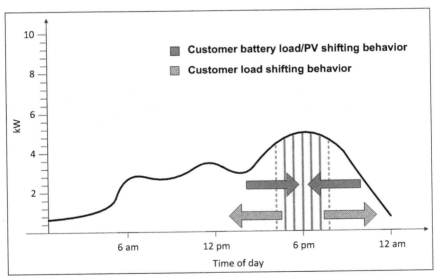

FIGURE 11.7 Anticipated customer response of load shifting in cases of CPP, ToU and peak demand tariffs.

(EV) under tariff T11 are more likely to charge the vehicles in late afternoon or evening, upon arriving home from work. The capacity charge and ToU tariffs are likely to encourage households to charge the vehicles much later in the evening (Fig. 11.8).

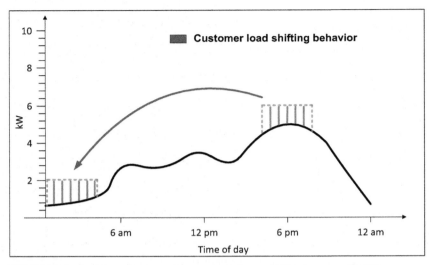

FIGURE 11.8 Anticipated customer response of charge shifting in cases with EV.

4 BATTERY CHARGING RULES

Timing of battery charging and discharging is a key determinant of both load shifting and electricity cost/income to the household. It is assumed that households will use their battery to maximize income generated from electricity exported to the grid, and reduced import from the grid due to self-consumption and load shifting from peak periods. Use of the battery will also heavily depend on which solar FiT the household is on. In general, households on the 44 ¢/kWh tariff will use the battery to maximize electricity exported from their PV, while those on 8 ¢/kWh tariff will do the opposite (minimize the import/use from grid, storing the excess from solar PV generation for later use).

In terms of battery charging and discharging rules, it is necessary to consider the potential customer response (see the previous section). Examples of anticipated customer behavior in the case of the flat T11 tariff are described in Table 11.5 for different combinations of solar PV, battery, and FiT. The presented examples are indicative only, with some typical traits expected. It can be seen that the customer load shifting behavior is very different for customers with solar PV, when using two different FiTs.

In the case when this response is facilitated by use of a residential battery, a corresponding example of battery charging rules for customers with T11 and two solar PV FiT is provided in Table 11.6. For simplicity, the day is considered as a sequence of 24 hourly time periods, and the battery can only charge or discharge for any particular hour. The smart power controller depicted in Fig. 11.1 is outside of the scope of this paper, and it is assumed that the battery charging and discharging rules can be very flexible. For example, the battery

TABLE 11.5 Anticipated Customer Behavior in the BAU Case with a Flat Tariff (T11)

PV and battery option	Customer behavior
Battery only	No action (except to charge the battery to be prepared for a blackout)
PV only (44¢/kWh FiT)	Customer moves discretionary load (eg, clothes washing/drying, pool pump, (possibly) air conditioning, etc.) out of day light hours to minimize internal use and maximize solar export, that is, high FiT received
PV only (8¢/kWh FiT)	Customer moves discretionary load (eg, clothes washing/drying, pool pump, (possibly) air conditioning, etc.) in to daylight hours (possibly limiting some loads to not exceed the solar output, eg, air conditioning) to minimize solar export, ie, total bill paid
PV (44¢/kWh FiT) + battery	As per PV only (44 ¢/kWh FiT) and will discharge the battery to further minimize the load on the network during daylight hours by using the battery to match the load up to the battery's kW rate to limit of the battery's kWh capacity. The battery will be recharged to full kWh capacity overnight
PV (8¢/kWh FiT) + battery	As per PV only (8 ¢/kWh FiT) and will charge the battery to further minimize the export of solar energy to the grid by using excess solar energy to charge the battery up to the battery's kW rate until the battery is fully charged (kWh capacity)–it is not charged from the grid. The battery is used to meet the load at night up to its kW rate for the amount of the stored capacity in the battery (ie, the kWh of stored excess solar energy from the day). The battery will stay at the lowest "safe" state of charge overnight to maximize the opportunity to stores excess solar energy the next day, hence since the battery is charged with excess solar, which is then used to meet the load, the battery will return to minimum "safe" state of charge each night unless the excess solar energy is significantly in excess of the battery capacity

may be charged from the grid with the same power level—far below the maximum charging power level—over several hourly intervals. Other important assumptions are that the battery may be discharged to the house, following the household load profile, or the battery may be charged from the excess of solar PV generation.

As T11 is a flat tariff, there are only two periods with different charging and discharging battery rules—the period with solar radiation, when the solar PV panels can generate electricity, and the rest of the day. For the region of Townsville, the usual hours with daylight when solar PV can generate electricity are between 6 am and 6 pm. For the period with solar PV generation, the charging

TABLE 11.6 Battery Charging and Discharging Rules for a Flat Tariff (T11)

Hour beginning at	Battery + PV (8¢/kWh FiT)	Battery + PV (44¢/kWh FiT)
12 am	Discharge the battery to the house with power following the household demand	Charge the battery from the grid evenly over one or more adjacent time periods
1 am	As for the previous period	As for the previous period
...	...	...
6 am	Charge the battery from the net PV generation (after satisfying the household demand)	Discharge the battery to the house with power following the household demand
7 am	As for the previous period	As for the previous period
...	...	...
6 pm	Discharge the battery to the house with power following the household demand	Charge the battery from the grid evenly over one or more adjacent time periods
7 pm	As for the previous period	As for the previous period
...	...	...
11 pm	As for the previous period	As for the previous period

and discharging rules presented in Table 11.6 are usually different from the periods before (midnight to early morning) or after (evening to midnight). It is suggested that the battery will be used differently also in relation to the two FiTs. In the case of low FiT (8 ¢/kWh), the battery will be charged from the excess PV generation, and later discharged to house to minimize the import from the grid (Fig. 11.5). In the case of high FiT (44 ¢/kWh), the battery will be discharged following the household load during the hours with solar PV generation, to reduce the import from the grid, and to maximize the export to the grid from the solar PV generation (Fig. 11.4).

The battery charging rules for tariff 1A are very similar to the rules for T11, as 1A is also a flat tariff. Tariff 2A is also flat, however it is also a capacity-based tariff, and the network cost component is proportional to the maximum daily demand measured in kW. In the case of tariff 2A, the discharge of the battery will happen during the period of maximum demand. This usually happens between 4 and 9 pm for the region under consideration.

Tariff 3A is a demand tariff, with a high $/kW/month charge proportional to the half-hour maximum demand occurring between 4 and 9 pm during all summer days, from Dec. to Mar. Understandably, the battery will be charged before this time period and discharged during the usual summer peak between 4 and 9 pm. For readability of this chapter, and due to size limitations, other detail descriptions of battery charging and discharging rules are not included.

5 RESULTS

Electricity demand and cost models calculate hourly and annual values of electricity demand, electricity consumption, and cost of electricity to the consumers in Townsville. Results are generated at a fine-grained Statistical Area 1 (SA1) of around 100–200 houses, with electricity consumption at an hourly time-step over 365 days of the year.

The annual cost of electricity for each of the modeled tariffs, for detached house and apartment, for day or evening occupied, with a different FiT and different configurations of solar PV and batteries. are shown in Figs. 11.9–11.13. All of these dwellings are assumed to be located in Townsville, and their heating, ventilation, and cooling energy required are related to the corresponding tropical climate zone. The initial investment cost to purchase solar PV panels and batteries are not included or considered in these figures. For each of these charts on Figs. 11.9–11.13, it is clear that the annual cost of electricity will be reduced if bigger solar PV panels and a bigger battery are used. This is easy to understand, given that a bigger solar PV panel will generate more electricity, and a battery provides additional flexibility to shift load to other time intervals.

If the two different FiTs (8 ¢/kWh and 44 ¢/kWh) are compared for the same type of dwelling and the same type of occupancy (Figs. 11.9 and 11.11), it is clear that the higher FiT will reduce the annual cost of electricity more than

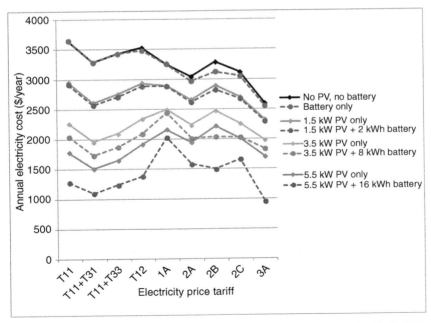

FIGURE 11.9   Annual cost of electricity for detached house, day occupied with 8 ¢/kWh solar PV FiT.

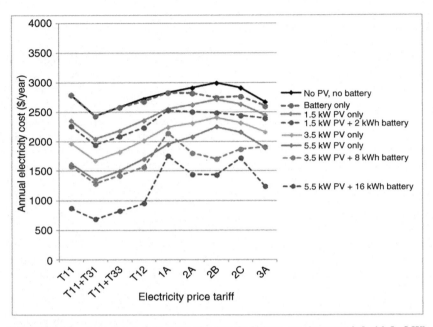

FIGURE 11.10 Annual cost of electricity for detached house, evening occupied with 8 ¢/kWh solar PV FiT.

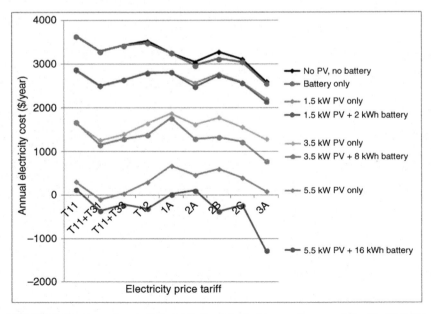

FIGURE 11.11 Annual cost of electricity for detached house, day occupied with 44 ¢/kWh solar PV FiT.

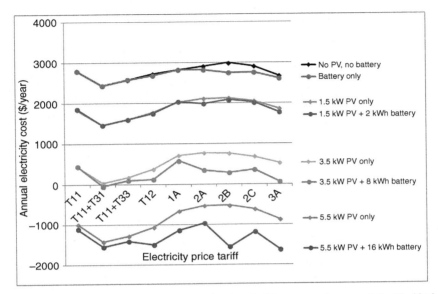

FIGURE 11.12   Annual cost of electricity for detached house, evening occupied with 44 ¢/ kWh solar PV FiT.

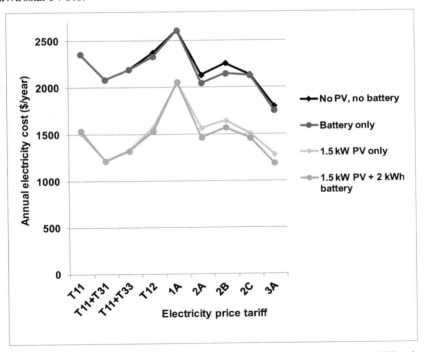

FIGURE 11.13   Annual cost of electricity for apartment, day occupied with 44 ¢/kWh solar PV FiT

the lower FiT. In case of big solar PV panels and high FiT, the annual cost can be negative, meaning that instead of a customer paying electricity bill, the utility will give money back to the customer for electricity exported to the grid (bottom lines in Figs. 11.11 and 11.12).

To observe the impact of the occupancy patterns on the annual cost of electricity it is necessary to compare a couple of dwellings with all other settings the same (Figs. 11.9 and 11.10, or Figs. 11.11 and 11.12). Generally, if the house is occupied during the day, the electricity consumption and the associated cost will be higher than in the case with a similar house and behavior patterns, but occupied only during the evening, night and morning, and not during the normal business hours of the day.

For annual cost of electricity when the dwelling is an apartment, only small solar PV panels and a small battery are considered and modeled—1.5 kW PV and 2 kWh battery (Fig. 11.13). This choice is based on the limitations of space—both roof space and space for the battery.

Perhaps the most important observation is that the proposed new tariffs don't necessarily increase the annual cost of electricity and, in the cases when increase is evident, it is not significant, and in many cases is in the order of 10%. These can be seen on all charts shown in Figs. 11.9–11.13, when comparing the values for existing tariffs T11, T11 + T31, T11 + T33, and T12 with the proposed new, or hypothetical tariffs, 1A, 2A, 2B, 2C, and 3A. This modeling accounts only for very limited load shifting, in the case of the proposed tariffs. It is anticipated that under the cost-reflective tariffs that have some elements of ToU, capacity charge or critical peak charge, the customer will aim to shift some of the less important electricity load to time intervals outside of the peak periods, which will reduce further the charges, specifically if related to demand or consumption during peak periods. The scenario results demonstrate that cost-reflective tariffs, in combination with DER, can improve network utilization, and potentially put downward pressure on retail prices. The analysis also demonstrates that electricity consumption could drop by more than 10%, in the next decade.

CSIRO and many industry participants in the Future Grid Forum, which developed potential electricity pathways for Australia to 2050 (CSIRO, 2013), estimate that onsite generation will grow between 20 and 70%, in the next 10 years—by 2026—according to four different scenarios constructed and analyzed. Electricity consumption supplied by central generation will stay more or less the same, or slightly grow in the next 10 years.

This analysis and the results in this chapter are broadly consistent with the findings of Chapter 1 about average electricity consumption in other countries. For example, the growth of electricity sales in New York for the period 2014–24 is estimated to be only 0.16% per annum, and the projection of electricity demand growth in the United States to asymptotically decline to 0.9% by 2040. These changes for utilities will translate in utility revenue loss in billions in dollars for developed countries, including the United States and Europe, as noted by Sioshansi. So, utilities need to adapt to these new conditions, and offer new products and services to survive.

6 CONCLUSIONS

This chapter outlines a modeling approach for DER that integrates models for household electricity demand, solar PV, home batteries, and a set of existing and potential electricity price tariffs, including solar PV FiT. Based on analysis of typical households, the results demonstrate that the proposed new tariffs don't necessarily increase the annual cost of electricity and, in the cases when increase is evident, it is a small percentage, usually. The modeling results reveal that cost-reflective tariffs in combination with DER can improve network utilization, and potentially put downward pressure on retail prices.

The analytical approach is an example of how integrated spatial and temporal modeling contributes toward the understanding of impacts due to uptake of solar PVs and batteries. Improved understanding, models, and data are required for constructing plausible future grid scenarios that may be useful for multiple purposes, including assessment of alternative price tariffs, changes in electricity consumption and peak demand, as well as network utilization.

As electricity utilities face declining electricity consumption, declining network utilization, and introduction of new disruptive technologies, such as solar PV, batteries, smart meters, and power control devices, proper pricing of the networks becomes more and more critical for their survival. These new DER represent not only threats, but also provide opportunities for utilities to offer new products and services. In this context, it is even more important to model and simulate new scenarios, new products, and new tariffs, before they are introduced into the marketplace. It is particularly important to understand the impact of new electricity price tariffs on different groups and clusters of customers, as some of these groups may be severely affected by proposed changes, due to their electricity demand patterns, low income, or capacity to adapt to new conditions. The dual purpose of the modeling approach is to assess the potential impacts of new products and tariffs on the operation and revenue of the utility itself.

ACKNOWLEDGMENTS

The authors are grateful to Ergon Energy and CSIRO for supporting this study.

REFERENCES

AEMO (Australian Energy Market Operator), 2015. Emerging technologies information paper: National electricity forecasting report.Available online at: http://www.aemo.com.au/News-and-Events/News/News/2015-Emerging-Technologies-Information-Paper

AER (Australian Energy Regulator), 2015a. Electricity supply to regions of the National Electricity Market. Available online at: http://www.aer.gov.au/node/9778

AER (Australian Energy Regulator), 2015b. Final Decision: Ergon Energy determination 2015−16 to 2019−20. Available online at: http://www.aer.gov.au/system/files/AER%20-%20Final%20 decision%20Ergon%20Energy%20distribution%20determination%20-%20Overview%20 -%20October%202015.pdf

APVI (Australian PV Institute), 2015. Mapping Australian photovoltaic installations. Available online at: http://pv-map.apvi.org.au/historical#9/-19.4433/146.8433

CER (Clean Energy Regulator), 2015. Available online at: http://www.cleanenergyregulator.gov.au/ RET/About-the-Renewable-Energy-Target

CSIRO, 2013. Change and Choice: the Future Grid Forum's Analysis of Australia's Potential Electricity Pathways to 2050. Available online at: https://publications.csiro.au/rpr/download?pid=c siro:EP1312486&dsid=DS13

Department of Energy and Water Supply (DEWS), Queensland Government, 2015a. FiT in Queensland. Available online at: https://www.dews.qld.gov.au/energy-water-home/electricity/ solar-bonus-scheme/feed-in-tariffs

Department of Energy and Water Supply (DEWS), Queensland Government, 2015b. Electricity prices. Available online at: https://www.dews.qld.gov.au/energy-water-home/electricity/prices

Ergon Energy, 2015a. Ten homes in a Townsville street are trialing battery storage, home energy management systems and alternative electricity tariffs which could help shape the future of our network. Available online at: https://www.ergon.com.au/about-us/news-hub/talking-energy/ technology/battery-storage-the-future-for-electricity-networks

Ergon Energy, 2015b. Battery storage systems arrive soon. Available online at: https://www.ergon. com.au/about-us/news-hub/talking-energy/technology/battery-storage-systems-arrive-soon

Ergon Energy, 2015c. The Battery Conversation. Available online at: https://www.ergon.com.au/ about-us/news-hub/talking-energy/technology/the-battery-conversation

Gils, H.C., 2014. Assessment of the theoretical demand response potential in Europe. Energy 67 (0), 1–18.

Higgins, A., Grozev, G., Ren, Z., Garner, S., Walden, G., Taylor, M., 2014a. Modelling future uptake of distributed energy resources under alternative tariff structures. Energy 74, 455–463.

Higgins, A., et al., 2014b. Modeling future uptake and impacts of distributed energy resources and tariff offerings in Townsville, CSIRO Report.

Macdonald-Smith, A., 2015. Energy retailers climb on board with Panasonic for battery trials. Available from: http://www.smh.com.au/business/energy-retailers-climb-on-board-with-pana-sonic-for-battery-trials-20150602-ghellb.html

Platt, G., Paevere, P., Higgins, A., Grozev, G., 2014. Electric vehicles: new problem or distributed energy asset? In: Sioshansi, F.P. (Ed.), Distributed Generation and its Implications for the Utility Industry. Elsevier, Oxford, UK, 2014.

Queensland Government, 2007. Climate Smart 2050 Queensland climate change strategy 2007 a low carbon future. Available online at: http://www.enviro-friendly.com/ClimateSmart_2050.pdf

Ren, Z., Paevere, P., McNamara, C., 2012. A local-community-level, physically-based model of end-use energy consumption by Australian housing stock. Energy Policy 49, 586–596.

Ren, Z.G., Paevere, P., Grozev, G., Egan, S., Anticev, J., 2013. Assessment of end-use electricity consumption and peak demand by Townsville's housing stock. Energy Policy 61, 888–893.

Ren, Z., Grozev, G., Higgins, A., 2016. Modeling impact of PV battery systems on energy consumption and bill savings of Australian houses under alternative tariff structures. Renewable Energy, 89, 317–330.

Tesla, 2015. Powerwall – Tesla home battery. Available online at: http://www.teslamotors.com/ powerwall

Chapter 12

Decentralized Reliability Options: Market Based Capacity Arrangements

Stephen Woodhouse
Pöyry Management Consulting, Oxford, UK

1 INTRODUCTION

Reliability of our electricity supply is of vital importance to modern society, but energy markets are being challenged by new uncertainties. In the decades since liberalization, most European markets have relied on market incentives to keep the lights on, although generally starting from a situation with wide reserve margins and—at times—using out-of-market tactics to smooth the path.

As decentralized generation reduces output, revenue, and profitability of conventional generation, the 'energy-only' wholesale market is now under fire in many countries. The growth of national capacity mechanisms is challenging the development of the integrated European energy market, which relies on spot energy price to determine flows between price areas. There are obvious parallels with the discussions in other chapters (eg, chapters by Nelson and McNeil, Athawale and Felder, and Grozev et al.) on how to charge customers for fixed costs. This debate applies principally to liberalized wholesale markets, and is part of the dialogue about how utilities can survive the transition to a world in which customers and decentralized generators provide an increasing share of the electricity consumed. The concepts outlined are intended to deliver a value for flexibility not just capacity, which is an essential step toward inclusion of demand and customers, in the smart electricity market of the future.

Most decentralized generation delivers energy, but not reliable capacity. The balance of cost for conventional generators is shifting away from variable costs, with fixed costs becoming more important. Existing utilities are seeking new ways of covering their fixed costs of generation, as their total sales of electricity fall, and, as a consequence, capacity remuneration mechanisms (CRMs) are being developed in several countries, including France, Spain, Italy and the United Kingdom, and under discussion in several others. Meanwhile, the European

Future of Utilities - Utilities of the Future. http://dx.doi.org/10.1016/B978-0-12-804249-6.00012-9

Commission is seeking to ensure that the integrity of electricity trading between countries, based on spot prices, is not undermined by national CRMs.

This chapter considers some of the underlying of features of existing and proposed CRMs, and offers a new model—*decentralized reliability options (DROs)*—for consideration as a potential basis for a European CRM blueprint. Decentralized ROs would permit a wide degree of freedom to implement designs, which meet national needs, without causing significant distortions between markets, while allowing a transition to a 'smarter' future where customers can determine their chosen level of reliability with reduced administrative input. This could form the basis of the hedging products and customer offers described in the chapter by Biggar and Reeves.

The remainder of this chapter sets out more detail on the new concept of decentralized ROs. Section 2 describes the changes in wholesale market design in Europe, which are being driven by the rise in decentralized generation. Section 3 describes how a decentralized reliability option scheme would work. Section 4 assesses the ideas against alternatives. Section 5 presents next steps, followed by the chapter's conclusions.

2 BACKGROUND

The established cycle of investing in new baseload capacity, and using older low-merit plants for peaking operation and reserve, appeared sustainable, in a world of steadily growing demand. Prices were generally expected to return to long-run marginal costs and reward investment, despite weak supporting evidence. The main commercial risk related to market price (for electricity, fuel, and latterly CO_2). Price risk was hedged through forward contracts spanning a few years or vertical integration against a retail portfolio, while more risk-averse investors sought long-term power purchase agreements. Volume risk was not a major concern.

Hundreds of new power plants have been built around the world under these conditions. What has changed?

The theory behind an energy-only market is straightforward. Capacity is needed until the point where its marginal cost[1] equals its marginal value[2] (ie, of avoiding capacity shortage). Any market intervention to

---

1. In principle, under an "energy-only" market, prices are expected to follow short-run marginal production costs most of the times, but when the capacity margin becomes tight, the prices should also reflect the possibility of scarcity. In the short term, these scarcity prices provide incentives for imports from neighboring areas, and for reduction in price-sensitive demand. Over the long term, the returns should balance the marginal cost of capacity with its marginal value. In this context, the marginal cost of capacity includes amortized investment cost (net of other revenue), converted to a cost per MWh of delivered energy. The key unknown is the number of hours in which the marginal capacity is needed.

2. Ideally, the marginal value would be defined by price-responsive customers, but in practice most demand does not face spot prices, and electricity markets generally use a deemed "value of lost load" and an administrative process for disconnections at times of scarcity.

limit the hours of scarcity or the level of pricing in these hours will lead to "missing money," that is, a situation where, systematically, the market arrangements fail to reward adequate investment to meet the desired security standard.

In Europe, a sustained reduction in demand due to energy efficiency and the prolonged impact of the financial crisis has been coupled with a sharp increase in weather variable renewable generation—principally wind and solar. These new technologies make a greater contribution to energy delivery (MWh) than to peak demand (MW). Any new build generation can no longer expect to achieve baseload operation, and is heavily reliant on capturing peak (scarcity) prices. Meanwhile, the number of peak hours diminishes, as the variable renewable generation covers some (but not all) of the demand peaks. For new plants to recover investment costs from fewer peak hours, the prices in these hours would have to be far higher than previously encountered.

Increased price risk is compounded by *volume risk*. The output of most generation will depend on patterns of wind and solar generation. However, the standard market contracts—for firm patterns of delivered energy—do not give market participants appropriate tools to hedge their risk through forward contracting. Such a combination of price and volume risk is dealt with in other commodity markets, by trading options, but options are not widespread in European energy markets. Irrespective of whether they face "missing money," electricity markets appear to have "missing contracts."

For most European markets, day-ahead markets have become the predominant source of spot pricing and dispatch patterns, with intraday trading and re-dispatch at low levels. As variable renewable generation grows its market share, its contribution to forecast error will grow sharply. For thermal generators, this means that, in addition to lower output levels, the timing of production will become more unpredictable, based on short-term weather patterns. This uncertainty continues even close to delivery time, as the weather forecast changes. Such variations are expected to be met by flexible capacity, able to respond to such fluctuations at short notice, as the forecasts and market prices change within day.

Are national capacity markets the answer? Market participants and the financial community across Europe are now questioning whether reliance on infrequent scarcity pricing is a credible basis for investment. Concerns about brownout risks, and a threat to security of supply are raised, and policy makers are increasingly coming to believe in a need for a separate mechanism to reward capacity. CRMs, once a feature only of the countries at the edges of Europe, are now in process, or under serious debate, in the largest European electricity markets, including Great Britain, France, and Germany, as well as the more established capacity markets in Spain, Italy, and Ireland.

There is widespread discussion on whether "missing money" is inevitable, or due to avoidable market distortions; for example, limits on the formation of scarcity prices, price caps or other interventions that prevent scarcity from occurring.

Such distortions exist in many markets, and are generally intended to protect consumers. Consumer protection is an essential part of a ROs scheme.

Many CRMs under consideration in European markets are national, different in design from neighboring countries, and with no arrangements yet in place[3] for cross-border participation. Yet, this outbreak of national markets takes place in the face of strong agreement by virtually all stakeholders to complete the European Internal Market for Electricity effectively.

Should CRMs reward all capacity or the right type of capacity? Most CRM schemes under construction fail to recognize the additional value that flexible capacity brings to the system, treating all available capacity alike. The focus of CRMs has traditionally been on generation adequacy, interpreted to mean ensuring that sufficient capacity is on the system to meet peak demand.

In order to incentivize the right type of capacity markets in a world with increasing levels of weather variable generation, CRMs will need to be able to cope with emergent system performance requirements, and should consider flexibility, in particular, as one of the parameters of their design.

The EC has designed a Target Model[4] to govern cross-border trading of electricity. It includes a coordinated process to determine price areas,[5] and allocate forward capacity rights for interconnection, with any unused physical capacity being released to the day-ahead market,[6] under "use-it-or-sell-it" rules. All areas are to be coupled in a single pan-European day-ahead market. There will be a coordinated continuous intraday market until close to real time, with cross-border trading permitted as long as unused interconnector capacity remains. All participants should face balancing responsibility, with organized markets for energy balancing, which determine imbalance prices. The heart of

---

3. UK Government's Department for Energy and Climate Change (DECC) has confirmed that in the next capacity auction to be held for Great Britain later in 2015, interconnectors will be eligible for one-year contracts.

4. The European Target Model for Electricity is a combination of (under construction) legally binding European Codes, covering Forward Capacity Allocation of transmission rights (FCA), Capacity Allocation Congestion Management (CACM) and Energy Balancing (EB). In addition to the Target Model, the vision of the Internal Market for Electricity is being built through the creation of a coordination bodies for European regulators (ACER) and a European TSOs (ENTSO-E), which have obligations on cross-border network coordination, planning and pricing. Aside from the Target Mode, ACER and ENTSO-E are creating a series of additional Codes, which will improve coordination in planning and operational timescales. Equivalent processes are happening for European gas markets.

5. Price areas are intended to reflect network congestion (subject to national approval) and the areas may join or subdivide countries. For convenience, this chapter uses the terms "cross-border" and "interconnector" to denote the price areas and the network capacity between them.

6. Physical forward capacity rights give the holder an option to nominate a physical flow between the two price areas. If the right is not exercised, the capacity is allocated to the day-ahead market, and the holder receives payment for the use of the capacity based on the price differential (if any) between the two areas.

the concept is cross-border trading in which market prices alone determine the flow of energy.

Progress toward the Target Model is underway, but key components are already in place. Day-ahead market coupling is now active in most parts of the EU,[7] and intraday and balancing markets are progressing slowly.

However, national CRMs will challenge the effectiveness of the Internal Market for Electricity. The Target Model sets day-ahead electricity prices as the governor of trade flows, but most designs of CRM risk distorting these prices at critical times. Instead of scarcity prices, most CRMs are intended to provide a supplementary revenue stream to reward capacity. To date, cross-border participation has not been implemented in any of the national CRMs under design, and there has not been an attempt to create a regional CRM.

European organizations are aware of the threat, which the uncoordinated development of CRMs brings to the Internal Market for Electricity. ACER and ENTSO-E have each published their own analysis, noting the potential risk of market distortions arising from national CRMs.

The EC has produced a Staff Working Document (European Commission, 2013), and has imposed revised State Aid Guidelines (European Commission, 2014b) to cover capacity adequacy mechanisms. These documents set out a number of criteria intended to limit the potential negative impact that a national CRM would have on the Internal Market for Electricity.

A coordinated approach to CRMs could be highly beneficial. If security of supply is threatened, national political intervention is inevitable, with the threat that existing investments are undermined. Electricity markets need to move to a sustainable model for investment, with less reliance on policy and regulatory decisions. Not all European markets are considered to need a specific CRM, but others are certain to proceed. The EU needs to find a blueprint for a CRM, which permits national governments to take action to protect electricity reliability without counteracting the internal market, or causing distortions to trade at national borders.

3 DECENTRALIZED RELIABILITY OPTIONS—HOW DO THEY WORK?

Capacity schemes take various forms, often categorized into "targeted" (paying specific types of capacity) or "broad" (paying market-wide); and price-based (prices are centrally determined and paid to all) or quantity-based (prices arrived at through competition between providers). A decentralized reliability option scheme is a market-wide, quantity-based scheme.

---

7. At the time of writing (May 2015), day-ahead electricity market coupling is in place across 19 European countries (covering approximately 85% of European power consumption). There is a separate Czech-Slovak-Hungarian-Romanian market coupling process, with plans for continued integration.

At its simplest, a decentralized reliability option scheme introduces a set of contracts between capacity providers and (indirectly) consumers. Retailers are required to buy ROs to meet their demand at critical times. Sellers of ROs commit their availability at critical periods, and forego revenue from price spikes, in return for which they receive a stable revenue stream.

The contracts are a hybrid between a call option (which is essentially commercial) and a physical commitment to make capacity available to the system at key times. The call option introduces a financial settlement (aside from the physical commitment and penalty arrangements), whereby the seller of the option returns the difference between the reference market price and the strike price, if any, to the buyer (European Commission, 2013). Customers benefit from security of supply to an agreed standard, and their exposure to scarcity pricing is reduced in return for an up-front fee.

Although many details of the scheme could be altered to suit local circumstances, an outline design for a decentralized ROs scheme is presented in Fig. 12.1.

The mechanics may be summarized as follows.

The TSO makes available forecasts and information on its view of the capacity balance, from several years ahead until close to delivery, in order to aid transparency and price discovery:

- these are forecasts, and do not define obligations for buyers or sellers of ROs;
- however, the maximum contribution of interconnected capacity is determined by the relevant TSOs.

Energy retailers[8] are required to buy ROs adequate to meet their actual demand at times of scarcity:

- retailers may choose the level of capacity to buy, at their own risk; and
- this structure ensures that demand side response is implicitly included in the scheme, to the extent that the actual demand would be reduced at times of scarcity.

Capacity providers[9] (including weather variable providers) may sell ROs based on their actual contribution to system capacity, at times of scarcity:

- providers may choose the level of reliable capacity to sell, at their own risk; and
- to measure their contribution, capacity providers are committed to meet pre-agreed characteristics relating to physical "availability" of the contracted capacity.

---

8. Depending on the local arrangements, the obligation may potentially extend to large customers and distribution companies if they have to buy energy to cover losses.
9. Capacity providers may include generation, storage, demand side providers, and interconnected capacity.

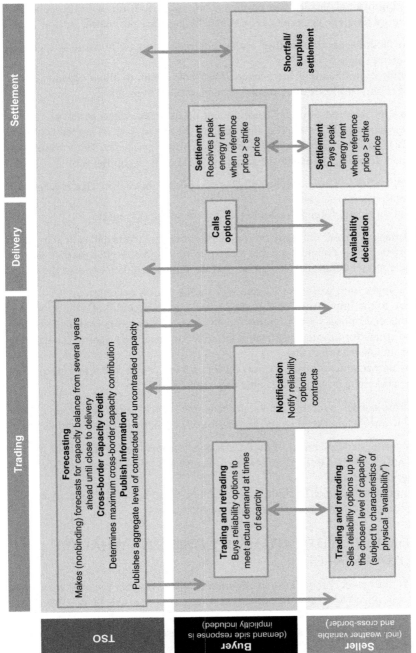

FIGURE 12.1 **Strawman design of decentralized ROs scheme.** (*Source: Pöyry Management (2015).*)

Administered penalties are applied at critical periods for under-procurement by energy retailers, or under-performance by capacity providers:

- to facilitate this, reliability option contracts must be notified to a central agency; and
- contract notifications are permitted after the event, to allow capacity shortages and surpluses to be resolved by market participants.

Aside from the physical commitment and its associated penalties, a reliability option includes a call option held by the buyer, with an agreed spot (reference) market and expiry time, an agreed strike price, and other terms (eg, contract duration). In exchange for an upfront payment (option fee):

- option holders are hedged against price spikes (above the strike price) in the reference market; and
- option sellers forego market revenue from such price spikes.

A decentralized reliability option market permits buyers and sellers to agree their own contract details; notably the expiry time and reference market, as well as the contract duration, strike price, and the time when they conduct the trade:

- an upper limit would be set centrally for the option strike price;
- due to the importance of day-ahead markets, ROs could take the form of a financial option settled against the day-ahead market; but options may also be struck for physical settlement intraday, or financially against a balancing or imbalance price; and
- by agreement, strike prices may either be fixed or indexed (eg, to some fuel or price indicator, or even to the day-ahead price).

Cross-border participation in the scheme (which could apply equally in other locational markets) would be possible by capacity providers, subject to:

- securing the agreement of the interconnector operator for the use of the capacity (whether through purchase of a forward transmission right or other agreement); and
- eligible capacity to be limited to the TSOs' estimates of the capacity contribution of the interconnector(s) at times of scarcity.

4 DECENTRALIZED RELIABILITY OPTIONS—WHAT ARE THE ADVANTAGES?

CRMs must meet both policy and commercial goals. In order to assess the merits of decentralized ROs, a set of evaluation criteria are proposed. Market design assessments are by definition subjective, but a set of issues have been selected to test the differences between schemes. Ultimately, these relate to the often-repeated objectives of achieving a "secure, affordable, and sustainable energy market."[10]

---

10. These objectives are often repeated, and were set out, for example, in European Commission (2010).

The European Union is building an integrated energy market across its 28 member countries, plus Norway and Switzerland, using a system of price zones and market coupling. There are strong provisions to prevent countries from closing markets or favoring their own producers, and any national CRMs will be required to be open to providers from outside the country.

The fundamental idea of the integrated energy market is to allow free trading of electricity across the region, reflecting congestion through locational spot pricing. There is an emphasis on market coupling, similar to the locational markets in Texas, Pennsylvania-Jersey-Maryland (PJM), and Singapore, but using price zones instead of nodes. The design addresses the need to integrate renewable generation to the wholesale markets through allowing trading across multiple timeframes: day-ahead, continuous intraday, and balancing.

The Target Model is the heart of a suite of codes designed to cement integration of the European electricity markets. The European Commission has set out the overall intention and allocated responsibilities to develop a series of Network Codes. ACER (the European regulatory coordination agency) created the "Framework Guidelines" for the content of these Codes (ACER, 2011), and ENTSO-E (the TSOs organization) is drafting the detail (ENTSO-E, 2013). Ultimately, these will become EU law (European Commission, 2014a). The core of the design is day-ahead market coupling, continuous intraday trading, and balancing and imbalance arrangements based on marginal pricing. Day-ahead market coupling is well under way; the other elements are lagging behind. A set of reference documents is provided in the references.

The European Commission has put in place State Aid guidelines covering capacity adequacy mechanisms[11], which have legal force from 2014 to 2020. The requirements may be summarized as dealing with the efficiency of any scheme (related to the underlying objective of affordability), within the context of the Internal Market for Electricity. As well as a scheme being efficient, it needs to be effective in delivering the objective of security.

The requirements of the EU are not unique, although they are inspired by the desire to create a common market from the 28 member states. The economics of the intention are applicable in any market where there is a desire to trade between price areas (nodal or zonal), and to allow trading of electricity over different timeframes up to real time.

From these high level principles, a set of evaluation criteria are proposed, as presented in Fig. 12.2, against which to compare designs of CRM.

Different scheme designs fit different circumstances, but the design of decentralized ROs is intended to be a market-wide, quantity-based scheme and,

---

11. The European State Aid Guidelines are part of an initiative which "aims to create a framework for policies to support the shift toward a resource-efficient and low-carbon economy" that helps to: (1) boost economic performance while reducing use of resources; (2) identify and create new opportunities for economic growth and greater innovation and boost the Union's competitiveness; (3) ensure security of supply of essential resources; (4) fight against climate change and limit the environmental impacts of the use resources (European Commission, 2011).

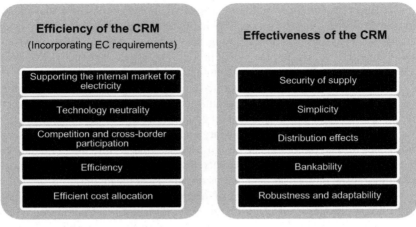

FIGURE 12.2 Evaluation criteria for assessing different CRM designs. *(Source: Pöyry Management (2015).)*

in this chapter, the concept is evaluated against similar designs. Two comparisons are made; a capacity ticket (eg, a capacity auction or capacity obligation) against a reliability option, and a centralized against a decentralized scheme.

ROs deliver security of supply, protect consumers and can help avoid energy price distortions. Most market-wide CRMs are intended to supplement "missing money" without addressing the underlying causes, which limit or prevent scarcity prices. These limits—where they exist—are generally measures to protect consumers from price shocks or poor reliability.

A reliability option is a hybrid between a physical commitment and a commercial option. The physical commitment is intended to deliver security of supply. It creates a supplementary revenue stream to deliver missing money (as for other market-wide CRMs), but the inclusion of the commercial option has an important influence:

- customers are protected from scarcity prices in the spot market; and
- spot price volatility can be hedged by the seller through the sale of the option in a "fixed-for-floating" swap of revenues, lowering the risks (and cost of capital) for investment in capacity.

These two effects mean that a reliability option scheme can reduce missing money from the energy market both indirectly and directly. ROs put in place the customer protection, which permits the regulators to remove any underlying distortions to energy price formation. If this is done, price volatility will reveal the value of demand side management, interconnection, and intraday flexibility.

As a result, the reliance on the physical commitment could be made transitional, leading to an improved version of the energy-only market in which investment risks could be managed through sale of a combination of forward sales of call options, and fixed volume energy contracts.

From a consumer perspective, ROs remove any incentive for generators to exercise market power over periods of scarcity; they also offer a hedge to consumers through direct compensation over periods of short-term price spikes. Capacity tickets, on the other hand, present the risk of overcompensation for generators at the expense of consumers as, in the absence of regulatory measures to limit price spikes, generators may attempt to exercise market power over periods of scarcity, in addition to receiving the upfront capacity payment.[12]

On the other hand, ROs present a more complex solution when compared to capacity tickets, and may be perceived as "riskier" by investors, as both a penalty and a commercial incentive for performance are in place.

Ultimately, the benefits of avoiding distortion of competition and trade, protecting consumers and facilitating innovative technologies better may outweigh any downsides of a ROs scheme. Table 12.1 shows the appraisal of capacity tickets against ROs, beginning with the European Commission's requirements which are likely to act as a constraint on the designs, which national governments may adopt.

Decentralizing ROs promotes an "active" role for market participants, and allows for the value of different types of capacity to be revealed based on its flexibility.

The underlying intent of a decentralized CRM (whether for capacity tickets or ROs) is to minimize the importance of central decisions and design parameters, and thereby reduce regulatory risk. Gains can potentially be realized from a decentralized approach in setting the capacity requirement, and the terms of procurement.

A centralized approach can accommodate the introduction of long-term contracts for new generating units, providing for greater investment certainty, and resulting in a lower cost of capital. However, a central agency is more likely to over-procure capacity, when compared to market participants, meaning that security of supply is better guaranteed, but that the outcome may be less efficient (with the associated cost borne by customers).

In terms of competition, a centralized platform for selling capacity means there is a common route to the market for all capacity providers, and a simpler product design promotes liquidity. A decentralized approach, on the other hand, adds complexity, and challenges liquidity as the number of products traded increases.

The decentralized model places greater responsibility on market participants, and allows them to optimize their own portfolios better. Demand side response is implicitly included, whether it participates directly in the scheme or not.

---

12. In a possible future world, the use of dynamic spot energy prices could be combined with the trading options to hedge volatility. This could be the basis for new set of customer offers (eg, as described in the chapter: Network Pricing for the Prosumer Future: Demand-Based Tariffs or Locational Marginal Pricing?) which would allow customers to form a full role in price formation and reduce market power directly.

TABLE 12.1 Comparison Between Capacity Tickets and ROs

EC key CRM requirements	Criteria	Capacity tickets	ROs	Comments
	Security of supply	✓	✓	ROs provide for stronger incentives for capacity providers to perform, as both a penalty and a commercial incentive exist
Competition and trade/Cross-border participation	Internal Market for Electricity	✓	✓✓	Capacity tickets risk damaging the underlying energy price signals at times of scarcity, limiting effectiveness of demand side and interconnection. ROs allow for the removal of regulatory interventions, which could result in energy market price distortions, while protecting consumers. Both options could provide for cross-border participation
Technology neutrality and decarbonization	Technology neutrality	✓	✓	ROs protect the underlying energy price signals and avoid price distortions, facilitating DSR better. ROs are more easily adapted to reward flexible capacity appropriately
Competition and trade	Competition	✓	✓	Both schemes allow for competition within the scheme. ROs, however, facilitate competition in the energy market better, through limiting energy price distortions over scarcity periods
Competition and trade/Time-bound intervention	Efficiency	✓	✓✓	ROs have the potential to deliver a more efficient outcome in terms of capacity on the system by allowing option contracts with different parameters (strike price, duration, and expiry time). ROs protect consumers, making explicit regulatory set price caps redundant
Allocation of costs	Efficient cost allocation	✓	✓	Both schemes aim at targeting costs associated with funding capacity contracts over periods of scarcity, and in proportion to the consumers contribution to demand over peak periods
	Simplicity	✓	✗	ROs are more complex than tickets, as option settlement has to be considered
	Distributional effects	✗	✓	With capacity tickets there is a risk of overcompensation toward generators (paid by consumers), limited in the RO scheme as there is direct compensation for short-term price spikes
	Bankability	✓✓	✓	In both schemes, penalties should be strong enough to incentivise performance but should also be manageable. The presence of both a penalty and a commercial incentive under ROs may present additional risk for investors
	Robustness and adaptability	✓	✓✓	Both schemes require regulatory intervention and centrally determined parameterization. ROs provide for flexibility to be adapted to reward capability more appropriately and can more easily be adapted to meet national needs.

Source: Pöyry Management (2015).

Decentralized ROs have further advantages over a centralized reliability option scheme. The use of ROs tends to fit with centralized energy markets, with a "spot" price that represents the value of energy.

However, under the European Target Model, with day-ahead, continuous intraday trading, and balancing energy markets,[13] there is no single "spot" price. The most convenient reference market for a centralized reliability option would be the day-ahead market, which will generally have good liquidity. However, the day-ahead market is too early for real scarcity to be revealed, and the use of day-ahead as the sole choice of reference price would mean that the reliability option does not distinguish between flexible and less flexible capacity. This seems to lose one of the advantages of trading capacity in the form of an option.

The principal advantage of decentralized ROs is that the options can be struck against different markets, including intraday and imbalance. Therefore, investments in flexibility will also benefit from being able to lock in fixed revenue streams, as well as investments in capacity. Participants may also choose the timing and duration of their contracts and the level of the strike prices contracted, making the trading of ROs a part of the portfolio of traded products. This freedom will allow value to be revealed for different types of capacity, while allowing the value to adapt to changing system requirements.

Overall, it can be argued that centralized ROs naturally fit better with more centralized energy trading arrangements, whereas decentralized ROs are more in line with most European electricity markets that value bilateral trading, and place greater responsibility on market participants.

Table 12.2 shows the appraisal of centralized and decentralized ROs, again beginning with the European Commission's requirements that—as described earlier—could be applied to any markets with a desire to trade between price areas, and to permit trading in a range of timeframes up to real time.

5 DECENTRALIZED RELIABILITY OPTIONS—THE WAY FORWARD

The convention design of ROs has been adapted to the circumstances in EU electricity markets; with continuous traded markets, growing needs for flexibility, and increasing shares of nonconventional capacity. Crucially, decentralized ROs meet the EC's stated requirements under the State Aid guidelines (which ultimately require efficient operation in the context of the Internal Market for Electricity), while also delivering capacity effectively.

By decentralizing the design, the arrangements place less reliance on administrative and policy decisions (with the regulatory risk that this brings), and greater reliance on the decisions of market actors, more in line with the spirit of the EU's Target Model for electricity trading.

---

13. The Target Model relates the cross-border trading arrangements, which are being implemented across the EU, described earlier in Section 4.

TABLE 12.2 Comparison Between Centralized and Decentralized ROs

EC key CRM requirements	Criteria	Capacity tickets	ROs	Comments
	Security of supply	✓✓	✓	A central agency is more likely to over-procure capacity, thus providing for greater security of supply; this may however mean overcapacity, and a less efficient outcome, in contrast to a decentralized approach
Competition and trade/Cross-border participation	Internal Market for Electricity	✓	✓✓	A decentralized approach is more in line with the thinking of the EU Target Model of placing increased responsibility on market participants, whilst allowing them to hedge their position through traded instruments
Technology neutrality and decarbonization	Technology neutrality	✓	✓✓	A decentralized approach may prove better at facilitating demand side response as retailers have better information regarding their customers' demand elasticity
Competition and trade	Competition	✓✓	✓✓	Centralized procurement provides common route to market for all capacity providers, and a simpler product design promotes liquidity. Demand side response is implicitly included (and better facilitated) with a decentralized approach
Competition and trade/Time-bound intervention	Efficiency	✓	✓✓	Decentralized procurement allowing different strike prices, contract duration and expiry of options should allocate resources more efficiently, and reward capacity for its value to the system more appropriately
Allocation of costs	Efficient cost allocation	✓	✓	Both options aim at targeting costs associated with funding capacity contracts over periods of scarcity, and in proportion to the consumers contribution to demand over peak periods
	Simplicity	✓	✗	(Potential) additional complexity in decentralized option, as there may be a variety of contract types
	Distributional effects	✓	✓✓	In a decentralized environment, assuming the development of more than one product, capacity will be rewarded for its real value to market participants and, by extension, to the system
	Bankability	✓✓	✓	Longer-term signals to investors with centralized procurement. More difficult to impose long-term obligations on retailers in a decentralized option. However, regulatory risk increases in a centralized scheme due to the importance of centrally determined parameters
	Robustness and adaptability	✗	✓	Decentralized procurement provides for a more flexible framework, able to adapt to evolving market conditions

Source: Pöyry Management (2015).

Capacity gives an option to deliver energy. By basing the scheme around the pricing of options, the scheme reflects the underlying economics of different types of capacity. As markets change, the value of flexibility inherent in the options will change, without the need for clumsy regulatory intervention.

ROs are a hybrid, containing commercial and physical (administered) obligations. This hybrid nature is a strength. It makes the scheme suitable for markets with different degrees of sophistication, and allows a transition toward more market based arrangements, moving away from reliance on the physical commitments and penalties, toward reliance on the commercial incentives, which are more consistent with the operation of the underlying energy markets.

The underlying principle of decentralization is that market actors will use innovative means to deliver the necessary level of system reliability, whereas a centralized system will tend to act conservatively and underwrite overcapacity at the expense of consumers. Demand-side response is fully facilitated (both implicitly and explicitly)—a crucial step toward a mature energy market with a fully active demand-side.

In the initial design, an appropriate set of penalties is required to ensure that the market actors meet their obligations, but there is scope to vary the design to strengthen the physical commitment, if it is deemed necessary. In time, the penalties may be lifted and the commercial incentives, which are inherently part of the option contracts, may suffice for delivering the required amount of reliable capacity.

To underpin these proposals, it is essential that any other distorting features of the electricity markets are removed, in particular balance responsibility for all participants, marginal pricing for balancing energy and imbalance, effective intraday markets, and the relaxation of controls or TSO policies which constrain the ability of the market to reveal scarcity. In a further phase of development, the markets would be strengthened by the use of shorter settlement periods and gate closure, by moving the "main" traded market closer to real-time, and by improving the performance of intraday markets.

6   CONCLUSIONS

This chapter outlines a strawman design for decentralized ROs, building on the centralized schemes, which have been implemented elsewhere.

Ultimately, this blueprint can be applied to all European countries (or regional markets), or just a subset of these. It creates a framework where different countries (or regional markets) can adopt this blueprint, or continue with an energy-only market without distorting trade and competition in the underlying energy markets.

REFERENCES

ACER, 2011. Framework Guidelines on Capacity Allocation and Congestion Management for Electricity. ACER, Ljubljana, http://www.acer.europa.eu/Electricity/FG_and_network_codes/ Electricity%20FG%20%20network%20codes/FG-2011-E-002.pdf.

ENTSO-E, 2013. An Introduction to Network Codes and the Links Between Codes. ENTSO-E, http://www.slideshare.net/ENTSO-E/130328-introduction-to-network-codes.

The European Commission, 2010. Energy 2020 – A Strategy for Competitive, Sustainable and Secure Energy. The European Commission, Brussels, ref COM (2010) 0021, Jan. 26.

The European Commission, 2011. A Resource-Efficient Europe – Flagship Initiative Under the Europe 2020 Strategy. The European Commission, Brussels, ref COM (2011) 639, Nov. 10.

The European Commission, 2013. Generation Adequacy in the Internal Electricity Market – Guidance on Public Interventions. The European Commission, Brussels, European Commission Staff Working Document.

The European Commission, 2014a. Commission Regulation Establishing a Guideline on Capacity Allocation and Congestion Management. The European Commission, Brussels, http://ec.europa.eu/energy/sites/ener/files/documents/cacm_final_provisional.pdf.

The European Commission, 2014b. Guidelines on State Aid for Environmental Protection and Energy 2014–2020. The European Commission, Brussels, June 28, ref. 2014/C 200/01. http://eur-lex.europa.eu/legal-content/EN/TXT/?qid=1448582033762&uri=CELEX:52014XC0628(01).

Pöyry Management Consulting, 2015. Decentralised Reliability Options – Securing Energy Markets. Pöyry Management Consulting, Londonwww.poyry.co.uk/news/dro.

Chapter 13

Network Pricing for the Prosumer Future: Demand-Based Tariffs or Locational Marginal Pricing?

Darryl Biggar*, Andrew Reeves†

*Australian Competition and Consumer Commission, Melbourne, VIC, Australia;
†Former Chairman, Australian Energy Regulator, Hobart, Tasmania, Australia

1 INTRODUCTION

How should the utility of the future charge for the use of the electricity distribution network?

This question of optimal pricing of electricity distribution networks is part of the broader question of optimal wholesale electricity pricing—that is, the wholesale pricing of energy, and the wholesale pricing of network services. Ultimately, the question of optimal wholesale pricing depends primarily on the impact of that wholesale pricing on outcomes in the retail electricity market. In the Australian context, and in other markets with retail competition, policymakers do not directly determine the retail market outcomes—rather, these are determined by competition between retailers. But the policy settings in the wholesale market, including the structure of the wholesale prices, affect those retail market offerings.

To answer the question of optimal wholesale pricing, we must first answer the following question: what should policymakers be seeking to achieve in the retail market? The answer this chapter offers is as follows: end-users should make efficient decisions regarding the usage of, and investment in, local devices and appliances, and be efficiently insured against risk.

As discussed in detail below, efficient usage and investment decisions could be achieved either by (1) exposing the end-customer to efficient price signals, or (2) through control, by the retailer, of the end-user's devices and appliances.[1]

---

1. The role of efficient wholesale pricing and direct control of customer devices is also discussed in the chapters by Covina et al., Gimon, and by Knieps, in this book.

Future of Utilities - Utilities of the Future. http://dx.doi.org/10.1016/B978-0-12-804249-6.00013-0

But this is not the end of the story. If end-users are risk averse, they would also like to be insured against price or bill risks. If end-customers vary in their preferences, each end-customer will prefer a different combination of price signals, insurance, and direct control over local devices and appliances. Some end-customers will prefer a retail contract that insulates them from the volatile wholesale prices, possibly combined with some control by the retailer over their devices and appliances. Other end-customers will be prepared to take some price risk, and may themselves set up their devices to respond automatically to those prices. The optimal outcome in the retail market is not a "one-size-fits-all" retail contract; rather, the optimal retail market outcome will involve customers choosing from a range of contracts, varying on these dimensions, ideally all efficiently priced.

Given this objective for the retail market, what should policymakers focus on achieving at the wholesale level? Policymakers should focus on two objectives: (1) ensuring that network tariffs, including distribution network tariffs, are set efficiently; and (2) ensuring that retailers have access to the hedging instruments they require. As long as the wholesale prices reflect the efficient price of electricity at the location of the customer, retailers will have an incentive to induce the end-customer, directly or indirectly, to make efficient usage and investment decisions. As long as retailers have access to the hedging instruments they require, they will be able to interface effectively between the volatile wholesale energy and network prices, and the smoothed retail contracts that end-customers desire.

The policy debate in Australia has focused on moving distribution networks to more "cost reflective" tariffs. This chapter argues that, as Australia moves to more cost-reflective network tariffs, policymakers should simultaneously ensure that retailers have access to the hedging instruments that they need to convert dynamic or time-varying wholesale charges into the range of retail tariffs that customers desire.

As discussed in more detail in the chapter by Nelson and McNeil, several commentators in Australia and around the world have argued for a move to "demand-based" distribution network tariffs, with a component that depends on each customer's maximum consumption during each billing period. This chapter notes that simplified demand-based charges risk muting the price signals that provide incentives for retailers to induce efficient usage and investment decisions amongst end-customers (or, for end-customers exposed to such charges, to make efficient usage and investment decisions). In addition, retailers should have access to the instruments that they need to hedge the network pricing risks arising from a dynamic demand-based charge. In the absence of such hedges retailers may choose to pass these charges directly on to customers, whether or not customers prefer this charging structure. Alternatively, customers will be charged an unnecessarily high price for charging structures that depart from the demand-based structure.

This chapter is divided into six sections. Section 2 sets out a vision for overall efficiency in retail markets. This vision involves customers choosing from a range of retail contracts with some combination of price variability, direct load control, and insurance or smoothing by the retailer. Section 3 discusses how competition between retailers will drive retailers to offer a range of end-user contracts. But, for those contracts to be priced efficiently, retailers must have access to hedging instruments. Section 4 sets out an approach to hedging network pricing risk that ensures retailers have the hedging instruments they require. Given that retailers are able to hedge their price risk, Section 5 explores the nature of efficient wholesale pricing, including efficient pricing of the distribution network. Section 6 explores the implications of these ideas for the current debate on network pricing in Australia.

2 EFFICIENCY IN RETAIL CONTRACTS

As many chapters of this book have noted, the electricity industry is on the cusp of a fundamental transformation.[2] This transformation relates to the way that smaller customers pay for electricity and interact with the electricity market.

Historically, small customers were passive consumers of electric power, and were charged a simple fee based on their usage. But, in the utility of the future, the role of small customers will be quite different. Small customers now have access to a wide range of controllable devices that impact substantially on the level of net electricity consumption at each point in time, known throughout this volume as Distributed Energy Resources (DER).[3] Regulatory requirements for zero net-energy buildings will further accelerate these developments.[4] In the prosumer future electricity customers will no longer be passive, but will face dynamic decisions as to the amount of electricity to produce or consume at each point in time, and in which devices and appliances to invest.

In this new world, what do economically-efficient outcomes for small consumers look like? Overall economic efficiency will involve a combination of both (1) efficient decisions regarding usage and investment, on the one hand, and (2) efficient smoothing of risks, on the other. Both aspects of overall efficiency are important.

Efficient decision-making has two dimensions:

- Efficient *usage* decisions: that is, the decision to produce or consume electricity locally up to the point where the marginal value of the last unit

---

2. See, for example, chapter by Sioshansi in this book, RMI (2014), page 6, and Reeves (2014).
3. Such devices include local generation (solar PV, fuel cells, or small-scale wind turbines), electric vehicles, battery storage, or smart appliances such as controllable air-conditioning loads or pool pumps.
4. Discussed further in the chapter by Sioshansi, in this book.

produced or consumed locally is equal to the marginal cost of producing and transporting an additional unit of electricity over the network to that location.

- Efficient *investment* decisions: that is, the decision to invest in an electricity producing or consuming device only if the (forward looking) additional economic surplus created at the times when the device is producing or consuming exceeds the fixed cost of the device.[5]

But efficient decision-making is not the only aspect of overall efficiency. In practice, most small customers are risk averse. Most would like some degree of protection from volatility in their electricity bill, or at least protection from volatility in the price they pay for electricity. Furthermore, the degree of risk tolerance of small consumers varies. While some may be prepared to pay a retail price that varies dynamically with wholesale market conditions, others would be prepared to pay a premium in exchange for insurance against volatile wholesale prices. If we are to achieve overall efficient outcomes, it is not enough for consumers to face efficient price signals; end-customers must also receive the degree of insurance or risk-sharing they desire.

Of course, there may be a trade-off between these objectives. As discussed further in this chapter, one way to achieve efficient decisions is by exposing the end-customer to a retail price that reflects the marginal cost of producing electricity and delivering it to the location of the customer. But, as long as there are periodic episodes of congestion arising on the distribution network, such a retail price would likely be highly volatile, and would expose the end-customer to substantial risk (Lineweber, 2013).[6] On the other hand, a retail tariff that smooths that price signal in some way inevitably distorts the signals for usage and investment. The objectives of efficient decisions and efficient risk sharing are, to an extent, in conflict.

This conflict can be mitigated if the customer is prepared to cede some control over his/her local devices and appliances to the retailer. In this case, the end-customer may pay a smoothed retail tariff for his/her net consumption, while the retailer uses its direct control to ensure that the customer's devices and appliances are used when and only when it is efficient to do so. Put simply, a customer facing a flat retail tariff has no incentive to charge his/her electric vehicle, or to use his/her air-conditioner when it is efficient to do so. But the

---

5. In the Australian National Electricity Market there is a legislative statement that is a valuable summary of the overall objective for public policy in the electricity market. The National Electricity Objective is to: "promote efficient investment in, and efficient operation and use of, electricity services for the long term interests of consumers of electricity with respect to: (1) price, quality, safety, reliability and security of supply of electricity; and (2) the reliability, safety, and security of the national electricity system."

6. Lineweber (2013) points out that most retail customers do not want dynamic prices—in fact he argues that there are "deep-seated customer preferences for predictability and consistency."

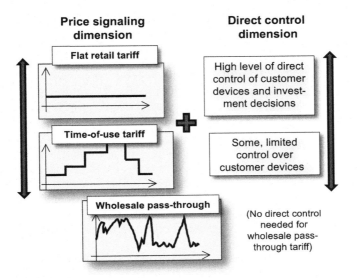

FIGURE 13.1 Achieving efficiency in the retail market involves customers choosing from a range of retail contracts which vary over the following two dimensions.

retailer, facing the volatile wholesale price, has an incentive to do so on behalf of the customer, and, in exchange for taking over control of these devices, can share some of the resulting economic benefits with the end-customer.

To summarize, overall efficient outcomes for small customers will involve retail contracts that vary along two dimensions:

1. Efficient use of prices signaling usage and investment decisions, balanced with the efficient level of smoothing and insurance desired by the customer; and
2. Efficient use of direct control of the customer's devices and appliances, from no direct control at all, to substantial control over the use of the customer's devices and appliances, together with decisions to invest in those devices and appliances.

These two dimensions of retail contracts are illustrated in Fig. 13.1.

Importantly, customers will differ from each other in the trade-offs they are prepared to make. Some customers may prefer less insurance, and may be less prepared to cede control over local DER to another party, but will be prepared to make active production and consumption decisions in response to changing wholesale market conditions. Such customers would probably choose a retail tariff with a high degree of wholesale price pass-through. Other customers will prefer a stable retail bill, and will be prepared, in return, to cede a high degree of control over local resources to the retailer. The overall efficient outcome for retail customers will likely involve different customers making different choices along these two continua. There will not be one-size-fits-all.

3   RETAIL COMPETITION AND RETAIL TARIFFS

For many years, researchers and policymakers in the electricity industry have taken a keen interest in the design of retail electricity tariffs. Hundreds of studies have been conducted, and thousands of pages of reports have been written comparing the merits of, say, inclining block tariffs (IBT) or time-of-use (ToU) pricing, against more dynamic approaches such as critical-peak pricing (CPP), variable-peak pricing (VPP) or real-time pricing (RTP) (see, eg, Faruqui and Lessem, 2012; Electricity Expert Panel, 2014).

But this focus on price structures is only part of the story—price signals are just one aspect of overall economic efficiency. There will not be one single retail tariff that suits all end-user customers. Rather, the best retail contract for a customer will depend on factors unique to that customer, such as the customer's preferences for risk, the nature of their local controllable devices and appliances, and their willingness to cede a degree of control over those resources.

The question for policymakers is not which retail tariff should be imposed. Rather, the question for policymakers is how to design a framework that allows each retail customer to be matched to the retail contract that, for that customer, best balances the trade-offs set out above?

Fortunately, there is a straightforward answer to this question: in an electricity industry with competition between retailers, provided the conditions set out below are satisfied, market forces will drive retailers to offer customers a range of retail contracts from which customers can choose the contract that meets their needs best, including the best balance on the two dimensions mentioned previously.

This is an important observation. It suggests that policymakers need not be directly concerned about the retail market. Instead, policymakers should focus on getting the conditions in the wholesale market right, so that, through the actions of retailers, end-customers are faced with an efficient range of contracts from which to choose.

But what are the necessary conditions on the wholesale market to achieve efficient retail market outcomes? There are two such conditions:

- First, the retailers must have access to hedging instruments that allow them to hedge the temporal and locational wholesale price risks they face when interfacing between the wholesale market and the retail contracts that customers desire;
- Second, the wholesale prices faced by a retailer in serving a customer must reflect the cost of producing and delivering electricity to the location of the customer. These costs must reflect the temporal and locational variation in market conditions in the wholesale market.

The next two sections address each of these issues in turn.

4 HEDGING TEMPORAL AND LOCATIONAL PRICING RISK

This chapter has argued that, in order to achieve efficient outcomes overall, retailers should be able to provide a range of retail contracts to end-customers. Some of those retail contracts will inevitably involve some form of smoothing or insurance (such as the common flat retail tariff). As long as the wholesale price is volatile, a retailer purchasing at the volatile wholesale price, while selling to the end-customer at a smoothed price, will inevitably be exposed to risk. If retailers are risk averse, they will seek to manage that risk through a range of risk management and hedging instruments.

But what if retailers do not have access to the risk management and hedging instruments that they require? In this case, there will be adverse outcomes for end-customers. Specifically, retailers will either:

- Include a risk premium in the retail contract, as compensation for the risk they are forced to bear; or
- Restrict the range of contracts they are prepared to offer to those retail contracts that limit their exposure to risk—namely, retail contracts that match the structure of the wholesale prices they face.

In short, either end-customers will be paying more than necessary for the retail contracts they desire, or the range of contracts available to end-customers will be restricted. Either way, end-customers are worse off, and overall economic efficiency is undermined.

Let's look at these effects more closely. Informally, the wholesale price for electricity can be viewed as comprising two components: a component that reflects the cost of producing electricity (the "energy" component), and a component that reflects the cost of transporting electricity to the location of the customer (the "transport" or "network" component).[7] In principle, each of these components can be, and should be, volatile. Retailers would like to hedge both of these components.

Let's focus first on energy component of the wholesale price. In the Australian National Electricity Market, the wholesale spot price in each pricing region reflects, more-or-less accurately, the marginal cost of producing electricity at the designated regional reference node. This price is determined every five minutes and can be highly volatile. Most retailers seek to offer smoothed (flat) retail tariffs to their end-customers. As a result, in the absence of hedging, retailers would incur substantial risk. Fortunately, there is a large range of hedging instruments available to allow retailers to manage these risks. The same is

---

7. This separation between the energy component and the network component is somewhat artificial. Nodal prices reflect both the marginal cost of production and the marginal cost of congestion. However, this distinction is useful for the presentation in this chapter, and partially reflects the situation in the wholesale market in Australia.

true in many other liberalized wholesale electricity markets around the world: a wholesale spot price (often differentiated by location) is determined through a market process, and a range of hedging instruments are available to allow retailers to hedge the resulting risks.

In the case of the energy component of the wholesale price, generators and retailers face roughly equal and opposite exposure to the price risk and, therefore, are natural counterparties in the trading of hedging instruments. In Australia, as in many other liberalized wholesale markets, there is an active market in these hedging instruments. The instruments themselves are varied, and range from bespoke and tailored products to standardized swaps and caps. On the whole, retailers do not seem to have too much difficulty acquiring the portfolio of hedge products necessary to provide end-customers the (mostly flat) retail contracts that end-customers seem to prefer.

But what if these hedge products were not available? In this case, as noted above, retailers who sought to offer smoothed or flat retail contracts would incur substantial risk. Such retailers might choose to self-insure this risk, or to pass this risk on to other parties, such as financial intermediaries, or insurance companies. But, either way, directly or indirectly, such retailers would be forced to include a risk premium in the retail price. Alternatively, if they are unwilling or unable to manage this risk, retailers might choose to restrict the range of contracts they offer.

Retailers can reduce their exposure to risk by offering retail customers a retail contract with a structure that matches the structure of the wholesale prices (such as a wholesale-pass-through contract). The problem here is that many end-customers do not want to face a time-varying network price. If they are forced to take such a tariff, the customers will be disadvantaged and there may arise a political backlash against the electricity reform process. Access to adequate hedging instruments is essential for retailers to be able to provide efficiently the range of retail contracts that end-customers desire.

A similar argument applies, of course, to the network or transport component of the wholesale price. Recent rule changes in Australia[8] require distribution networks to charge cost-reflective tariffs.[9] It is not yet clear precisely what this will mean in practice. However, the intent of the rules is that distribution networks would move away from the simple flat network tariffs they have used in the past, to more sophisticated ToU or dynamic tariffs, subject to certain constraints (AEMC, 2014). The same issues arise in many countries. As policymakers seek to move toward the utility of the future, they are seeking more sophisticated ways of charging for the distribution network.

Let's suppose that a distribution network offers a new, more sophisticated tariff at the wholesale level, involving some temporal or spatial variation in the network charges. As before, the particular retail contract that is best for each

---

8. See AEMC, http://www.aemc.gov.au/Rule-Changes/Distribution-Network-Pricing-Arrangements.
9. National Electricity Rules, Clause 6.18.5 (f).

customer will vary from customer to customer, but some customers will likely prefer a flat retail contract. A retailer who purchases at a time-varying wholesale price and sells at a flat retail price incurs risk.[10] To date, little consideration has been given to developing mechanisms that would allow the retailer to hedge this distribution network pricing risk. As a result, the problems identified above may arise—retailers will be forced to include a risk premium when they offer retail contracts with a flat network component. Alternatively, retailers will restrict the range of retail contracts they offer in order to reduce their risk. For example, if the network component of the wholesale price is time-varying, retailers will restrict their offers to a time-varying retail contract, even if customers desire a flat tariff.

To summarize, retail competition on its own is not sufficient to ensure that customers face a range of efficiently priced retail contracts. In addition, retailers must also have access to instruments that allow them to hedge the risks they face. In most liberalized wholesale markets, such hedging instruments already exist for the energy component of the wholesale price. But no such instruments exist for hedging the network component of pricing risk. As Australia moves to more sophisticated distribution network tariffs, new instruments will need to be developed to hedge network pricing risk.

4.1   An Approach to Hedging Network Pricing Risk

What will these new network hedging instruments look like? Surprisingly, there has been relatively little academic attention to the question of hedging network pricing risks. The work that has been done has been focused on the properties of Financial Transmission Rights (FTRs), which are unfortunately not able to hedge many of the risks transacting parties face (Biggar and Hesamzadeh, 2013).

This chapter sets out an alternative approach, which requires network businesses to offer a bundled pricing-plus-hedging product on request. Under this approach, each network business would be required to offer, for each end-customer, a wholesale network tariff structure requested by a retailer, for a retailer-specified profile of consumption. The network business would determine a price (a quote) for the delivery of the nominated profile of electricity to the specified end-customer at the nominated structure of prices.

This approach allows each retailer to match the structure of the network tariff to the structure of the retail tariff the retailer desires to offer the end-user. If the retailer wishes to offer a flat retail tariff, the retailer would request (and the distribution network would be required to offer) a flat network tariff for the forecast consumption profile of the customer. If the retailer wishes to offer a ToU tariff, the retailer would request (and the distribution network would offer)

---

10. Even if there is no uncertainty in the network price (as in a ToU network tariff), the retailer still faces uncertainty about the volume of electricity transported to each customer and therefore still faces risk.

a matching ToU tariff, for the forecast consumption of the customer. As noted further, as long as the retailer pays the correct efficient network charge at the margin, this approach combines hedging with efficient price signals, allowing the retailer to offer a range of contracts to its end-users.

The justification for this approach is as follows: generators are the natural counterparty to retailers for hedging the energy price risk, and therefore are the natural provider of energy price hedges. In the same way, networks are the natural counterparty to retailers for hedging network price risk and, therefore, are the natural provider of network price hedges. Each network effectively provides two services: (1) the real (physical) service of transporting electricity, charged at a potentially dynamic network tariff; and (2) the service of hedging network pricing risk. This approach combines these two services into a bundled product.

The hedging instrument that a retailer needs depends on the retail contract it intends to offer to its end-customer. Let's suppose, for example, that the customer desires a flat retail tariff, but the retailer faces a dynamic wholesale network tariff at the location of that customer. In this case, the retailer would like to construct a hedging instrument that hedges the risk arising from purchasing network services at the dynamic network tariff, and on-selling them to the end-customer at a flat rate. The volume of this hedging instrument that is required is a volume equal to the forecast net consumption of the end-customer at each point in time.[11]

Exactly the same principles apply if the retailer seeks to offer some other form of retail contract to the end-customer. For example, let's suppose the end-customer prefers a ToU retail tariff. In this case, the retailer requires a hedging instrument that hedges the risk arising from purchasing network services at the dynamic wholesale network tariff, and on-selling to the customer at the ToU retail rate. Again, the volume required is equal to the forecast net consumption of the end-customer at each point in time.

In fact, the same principle applies no matter what structure of retail contract the retailer seeks to offer. For every possible retail contract, there will be some corresponding network-price hedging instrument that the retailer requires. Mathematically, let's suppose the (possibly time-varying) retail price is P_t^R, the dynamic wholesale network price is P_t^W, and the forecast consumption of the end-customer is Q_t^F —the hedging instrument the retailer requires would then pay out the following amount each period:

$$\left(P_t^R - P_t^W\right)Q_t^F$$

Now, let's combine the hedging service with the underlying transport service. The distribution network now provides a bundled service (transport + hedging), which is potentially tailored for each end-customer. Combining the underlying

---

11. This hedging instrument would pay out, at each point in time, the difference between the flat retail price and the wholesale network price, multiplied by the forecast net consumption of the end-customer at that point in time.

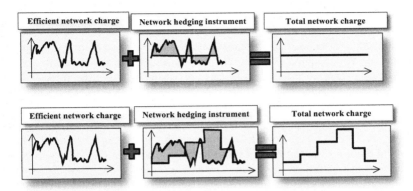

FIGURE 13.2 Combined network price plus hedging matches the structure of the retail contract. Note: These charts refer to the charge for the forecast consumption of each customer. The difference between actual and forecast consumption is charged (or paid) at the dynamic network price.

price for transport with the hedging instruments set out above, we find that the total payment from the retailer to the network, when the end-customer consumes at the rate Q_t, is equal to the flat retail price P_t^R for the forecast consumption volume Q_t^F of the customer, plus the dynamic network charge P_t^W for any variation in the actual consumption from the forecast consumption:

$$P_t^W Q_t + \left(P_t^R - P_t^W\right) Q_t^F = P_t^R Q_t^F + \left(Q_t - Q_t^F\right) P_t^W$$

This approach could be used for any retail contract the retailer desires to offer. In each case, the distribution network would be required to provide a network tariff that matches the structure of the retail tariff for the forecast consumption profile of the customer. The total price the retailer pays for that contract would depend on that consumption profile. Furthermore, the retailer would pay the dynamic spot network tariff for any deviation of actual consumption from forecast consumption.[12]

Fig. 13.2 illustrates how the hedging product of the distribution network allows the retailer to convert a volatile wholesale network price into the retail price that the end-customer desires.

4.2 An Example

An example may make this proposal clearer. Let's suppose that the distribution network is priced using a dynamic efficient tariff, such as locational marginal

---

12. We could also envisage hedging arrangements in which the distribution business also takes on volume risk, such as a "load following hedge." However, this is not necessary for our purposes and shifts responsibility for control of end-customer volumes to the distributor. We prefer a framework in which responsibility for control of the customer devices and appliances rests with the retailer (a competitive business) rather than the distributor (a monopoly).

pricing. Let's suppose a retailer wishes to offer its end-customers a retail tariff that incorporates, say, a ToU tariff structure with, say, the retailer controlling the air-conditioning load of the customer at peak times. Under the proposal above, the retailer would nominate the tariff structure (in this case, the timing and duration for each price band of the ToU structure), and the forecast load profile of the customer. The distribution business would then be required to provide and price a hedging instrument that allows the retailer to hedge between the dynamic wholesale network charges and the desired ToU retail tariff. As noted above, this could be achieved if the distribution business offers a tariff with a structure that matches the desired retail tariff, corresponding to a volume given by the forecast load of the customer. The network tariff in each ToU price-band would correspond to the volume-weighted average of the network tariff in that price band.

This is illustrated below. Let's suppose that there are 7 periods in each day. The probability of network congestion in each period is set out in the table. The uncongested network price is 2 ¢/kWh, whereas the congested network price is 1000 ¢/kWh.

Under this proposal, the retailer would announce the tariff structure that it desires, and the forecast load profile of the customer. In this case, the retailer seeks to offer a three-part retail tariff with an off-peak component (tariff A), a shoulder component (tariff B), and a peak component (tariff C). The duration of each of these components and the forecast load profile of the customer are set out in Table 13.1.

Given this information, the distribution business would be required to offer a matching network tariff. From the information given previously, we can compute that the distribution business should offer a three-part tariff, with tariff A equal to 14 ¢/kWh, tariff B equal to 29 ¢/kWh, and tariff C equal to 62 ¢/kWh. These tariffs yield the same revenue to the distribution business as the distribution business would receive if the customer consumed exactly its forecast load profile. In addition, the retailer will be charged the dynamic price for any excess consumption above the forecast load profile (and will be paid the dynamic price for any shortfall of consumption below the forecast load profile). This approach allows the retailer to offer the end-customer the ToU retail tariff that the retailer desires. At the same time, the retailer retains the incentive to engage in direct control of the customer's devices and appliances, such as controlling the air-conditioning load at times of network congestion. Competition between retailers will ensure that the benefits of such direct load control are passed on to end-customers.

In summary, as noted above, as the utility of the future moves to more sophisticated distribution network tariffs, new instruments will need to be developed to allow retailers to hedge network pricing risk. This section sets out one such approach, which requires distribution networks to provide a bundled transport + hedging service. The hedging component of that service depends on the structure of tariffs and the profile of consumption of the end-user and, therefore, must be tailored for each end-customer.

TABLE 13.1 Illustration of the Pricing of a Network Hedge to Allow a Retailer to Offer a Simple ToU Tariff to a Specific Customer

Period	0	1	2	3	4	5	6
Probability of congestion	0.10%	1%	2%	3%	6%	2%	1%
Uncongested network price ($/kWh)	0.02	0.02	0.02	0.02	0.02	0.02	0.02
Congested network price ($/kWh)	10	10	10	10	10	10	10
Expected network charge ($/kWh)	$0.03	$0.12	$0.22	$0.32	$0.62	$0.22	$0.12
Forecast consumption profile (kW)	1	2	2	2	1	1	1
Tariff	A	A	A	B	C	B	A

5   EFFICIENT NETWORK PRICING

Achieving overall efficiency is not just a matter of ensuring that retailers have access to the hedging instruments they require. In addition, of course, the wholesale network tariffs must induce end-customers (or retailers, through direct load control) to make efficient decisions—efficient production and consumption decisions, and efficient decisions regarding investment in devices and appliances. How should network tariffs be designed in order to induce efficient usage and investment decisions?

The answer to this question is straightforward, in part due to the separation of roles discussed previously. Once the retailer has access to a range of hedging instruments, such as those discussed above, the retailer has both the incentive and the ability to provide the degree of insurance or smoothing that the end-customer desires. Policy-makers no longer need be concerned with insurance or risk-management in the wholesale tariffs. The only question remaining is how to achieve efficient decision-making.

A retailer will only have an incentive to ensure a customer makes efficient consumption and investment decisions if that retailer faces the correct efficient price signals. In the context of the electricity industry, the correct efficient price signals are the prices that emerge from what is known as a "smart market" or, in the language of wholesale electricity markets, as "security constrained optimal dispatch." The resulting prices are known as nodal prices, or locational marginal prices.[13]

Many liberalized wholesale electricity markets around the world already compute nodal prices for major points of connection on the transmission network. These same principles and mechanisms currently used to efficiently price transmission networks can also be applied to distribution networks.

The answer to the question: "how should we efficiently price distribution networks?" is, in principle, straightforward: using the same mechanisms that are currently used for pricing transmission networks.[14] Yes, distribution networks are somewhat different to transmission networks—distribution networks are often less meshed than transmission networks, and the number of assets on distribution networks is at least an order or magnitude larger than on transmission networks. But there is no in principle reason why the same mechanisms that are currently used for pricing transmission networks should not be applied to distribution networks. Yet, there is currently very little attention being paid to issues of nodal pricing of distribution networks.[15] There remain important questions as to the level of granularity at which distribution marginal pricing should be applied—should it be down to the zone substation, the feeder level, or the street level? These questions remain to be resolved, but do not, in our view, undermine the case for locational marginal pricing at the distribution network level.

The combination of locational marginal pricing of electricity and the provision of a range of hedging instruments (for hedging both energy and network pricing risk), combined with retail competition, will ensure that end-customers are offered a range of retail contracts with the mix of insurance and direct load control they prefer, and those contracts will induce the end-customer (or retailers acting on their behalf) to make appropriate usage and investment decisions

---

13. See the chapters by Knieps and Gimon, in this book.
14. There are technical differences of transmission and distribution networks which give rise to different engineering concerns. For example, maintaining voltage levels may be a more important consideration at the distribution network level. However, these issues do not preclude the determination of nodal prices for distribution networks.
15. One of the limited exceptions in the academic literature is Sotkiewicz and Vignolo (2006). Some important recent work in this area has been carried out by a group at the Lawrence Berkeley National Laboratory, see De Martini and Kristov (2015). Borenstein et al. (2002) advocate dynamic pricing (but not necessarily full nodal pricing) for retail customers. The RMI (2014) argues for "deliberately and incrementally increasing rate sophistication" for residential and small commercial customers, by increasing the temporal and spatial granularity of pricing. Allcott (2011) reports the results of a Real-Time Pricing trial in the Chicago area.

to the extent it is efficient to do so. It seems to us that this is the best possible outcome.[16]

6 THE NETWORK PRICING DEBATE IN AUSTRALIA

In Australia, as in other liberalized wholesale electricity markets around the world, prices at the wholesale transmission network level are determined through a bid-based security-constrained optimal dispatch mechanism.[17] In contrast, in Australia, as in other countries, distribution network prices have historically not been determined through a market process, but through a regulatory process. Distribution network operators in Australia have used a range of tariff structures. However, in practice, the smaller (mass market) customers have almost uniformly been charged simple pricing schemes—such as a simple two-part tariff with a fixed fee per day of connection and a fee per kWh of consumption.[18]

Unsurprisingly, these arrangements have given rise to concerns about inefficient usage and investment decisions by retail customers, including inefficient decisions regarding the use of air-conditioning at times of network congestion, and inefficient decisions to invest in solar PV (Wood and Carter, 2014; Ergon Energy, 2015; Wood and Blowers, 2015).

As noted in the chapters by Sioshansi, and Nelson and McNeill, in recent years, several countries have experienced low or no growth in electricity demand. This has also been a concern in Australia. The decline in demand for energy, coupled with the simple volumetric tariff, has led to a decline in revenue for network businesses. Under a revenue cap arrangement, a decline in revenue allows an increase in regulated network charges. As a result, customers are further encouraged to switch to alternatives, further eroding network revenues. In order to reduce the risk of a further erosion of revenues, network businesses in Australia have been eager to move toward network tariffs that are less dependent on throughput (kWh). In particular, several commentators have called for a move to a demand-based tariff.

A "demand-based" tariff includes a component based on the historic peak demand of a customer (typically measured in kW) during a specified billing period. For example, a demand-based tariff might specify that the customer

---

16. We have focused here on efficient pricing of distribution networks. In practice, another legitimate objective for network tariffs is ensuring sufficient revenue on average to cover the prudent costs of the distribution business. We will not focus on questions relating to revenue adequacy and the setting of this fixed charge in this chapter.

17. The Australian NEM does not currently use full nodal pricing at the transmission network level—instead the NEM uses a form of regional pricing. Nevertheless the process for pricing the wholesale transmission network has been superior to the process for pricing the wholesale distribution network.

18. The per-kilowatt-hour fee may vary with the amount of consumption of the customer (known as an inclining or declining block tariff).

must pay an amount based on the customer's peak demand between 4 and 8 pm on weekdays, during the previous month. Alternatively, the tariff might specify that the customer must pay an amount based on the average of the five highest demands between 6 and 10 pm over the previous quarter or year.

Several commentators in Australia and overseas have argued for the widespread adoption of a demand-based tariff (Wood and Carter, 2014, p. 13; Simshauser, 2014; Electricity Expert Panel, 2014; Ergon Energy, 2015; Geode, 2013). Some distribution businesses in the state of Victoria have recently proposed rolling out demand-based charges for all residential and small business customers.[19] In the United States, the Rocky Mountain Institute asks whether demand charges are the "next big thing" in distribution pricing, and notes that "more than a dozen utility companies across the country have implemented or are currently considering residential demand charges" (RMI, 2015. See also RMI, 2014).

It is straightforward to demonstrate that a simple demand-based tariff is equivalent to a two-price charging system.[20] The customer faces one price (equal to the usage charge component) at times when demand is not at that customer's peak for that measurement period (or outside the measurement period), and it faces a second price (equal to the usage charge plus the demand charge) at that time when demand is at its peak during the measurement period.[21,22]

A demand-based tariff has some features that are similar to a locational marginal price: it is dynamic (in the sense that the peak price can occur at any time in response to "market" conditions), and it involves prices that are materially higher at those peak times.

But a demand-based tariff also has many features that are quite different to a locational marginal price. In particular, the prices implicit in a demand-based tariff depend on each customer's own time of peak demand. They do not necessarily correspond to times when the local network is congested. This gives rise to inefficient consumption and investment decisions. In addition, since the

---

19. See, for example, the Tariff Structure Statements submitted to the AER by the Victorian distribution network service providers on 25 September 2015.

20. Let's suppose we have a simple demand tariff with a demand charge D ($/kW) and a usage charge P ($/kWh). Let's suppose that a customer consumes at the rate Q_t during each half-hour period during a month. The total payment from the customer to the network at the end of each month is then $\max\{Q_t\}D + P\Sigma_t Q_t$. The price is the slope of this function and is equal to P at time periods when the consumption is below the maximum and equal to $P + D$ at the time when consumption of the customer is at the maximum.

21. A more complex demand-based tariff, such as where the peak demand charge is based on a weighted average of the highest rates of consumption, will result in more than just two effective prices.

22. Some commentators (chapter by Rowe, Sayeef, and Platt) have argued for combining the demand charge with a time-varying (eg, ToU or CPP) usage charge. In this situation the demand charge effectively adds a dynamic component to the underlying (ToU or CPP) usage charge, which has the effect of significantly increasing the price the customer pays at times of his/her peak consumption during the billing period.

prices are unique to each customer, incentives arise for neighbors to negotiate mutually-advantageous arrangements.

Consider the case, for example, of a customer whose peak demand occurs at night, at times when the local network is uncongested (this could be the case, eg, for a restaurant or nightclub). Such a customer has a strong private incentive to reduce consumption at that customer's peak time, even though the social benefits to the industry are small. In fact, that customer has a strong incentive to install battery storage, and to charge those batteries during the day and discharge the batteries at night, at the time of the customer's peak consumption. Perversely, this may exacerbate network congestion during the day. In short, due to the difference between the timing of the customer peak and the network congestion, the demand-tariff may lead to inefficient consumption and investment decisions.

In principle, the timing problem could be overcome if the measurement period could be set, in advance, to match the time of network congestion. But network congestion depends on a range of network, supply, and demand conditions. To forecast reliably when congestion will occur in advance is difficult to impossible. Furthermore, a demand charge based on the customer's peak demand each month gives rise to a peak price, once and only once each month. This could only correspond to nodal pricing if there was one, and only one, episode of network congestion each month.

Another observation is that, since the timing of the individual peak of customers varies between customers, under a demand charge, neighboring customers will face quite different effective prices for electricity at the same time. This gives rise to an incentive for customers to arbitrage these price differences through shifting the metering of electricity from one customer to another. In particular, neighboring customers with different load profiles will have a strong incentive to take electricity supply at a single point, with a single meter, and to share the single electricity bill according to their own consumption. In the case of a multi-tenant building, there may arise economic incentives, for example, to install a single electricity meter, and pay a tariff based on the peak demand of all of the occupants, rather than metering each apartment separately. Developers of new residential or industrial parks may seek to aggregate demands for new housing estates, rather than installing separate meters for each dwelling, and so on.

In addition, it is not clear that demand charges are suited for embedded generation. In principle, a customer who is exporting power to the grid at the time of network congestion should be *paid* for the network capacity it is freeing up. It is not clear that advocates of demand charges envisage that the demand charge will be paid to embedded generators.[23] Moreover, it is at least theoretically possible that, in the presence of embedded generation, network constraints may bind in the direction of flow *away* from the end-customer. In this case, customers should

---

23. Furthermore, the demand charge inefficiently creates incentives to invest in embedded generation that can produce throughout the measurement period—not just the times of network congestion.

be paid for increasing their consumption at such times. It is not clear that this possibility has been contemplated in designs for demand tariffs.

As noted earlier, several networks in Australia have announced that they intend to adopt a demand-based tariff for all small business and residential customers in the future. The Australian Energy Regulator will consider whether such tariffs are consistent with the National Electricity Rules. A demand-based tariff, with its dynamic structure, is arguably an improvement on a flat network charge (and possibly an improvement on ToU or CPP). While simple demand-based tariffs have some shortcomings, they may be an interim or practical step, or a complementary transition, to fully cost-reflective tariffs.

In any case, little attention has been paid, to date, as to how retailers will hedge the pricing risks created by a dynamic demand-based charge. As emphasized earlier, in the absence of such tools, it is likely that retailers would choose to simply pass this price structure on to end-users, or will charge a higher-than-necessary price for tariffs which smooth this risk. Either way, end-customers are not well served, or there is a risk that support for needed reforms will be undermined.

7 CONCLUSIONS

As the other chapters in this book have made clear, the electricity industry is undergoing a fundamental transformation. Small customers will no longer be passive but will, directly or indirectly, be active participants in the wholesale electricity market. There is no one single retail tariff that will suit all retail customers. Instead, the ideal tariff for each customer will differ in the degree of exposure to price signals they are prepared to tolerate, together with the degree of remote (retailer) control of their local devices and appliances. The goal for policymakers is not to determine the best network tariff, but to create a regulatory framework in which customers are offered a range of retail contracts from which customers can choose the retail contract that best meets their needs, and which promotes overall economic efficiency.

Competition between retailers will drive retailers to offer end-users a range of contracts. The range of contracts will be economically efficient provided two conditions are satisfied: first, retailers must have access to a full range of hedging products that allow them to convert the wholesale prices into the retail contracts that customers desire; second, the wholesale price associated with serving a particular customer must reflect the short run marginal cost of producing electricity and delivering that electricity to that customer. When these conditions are satisfied, end-users will face the correct trade-offs between price signals and insurance or smoothing, and between local and remote control of devices and appliances.

Recent reforms in Australia require distribution network tariffs to be more cost-reflective, and to be based on the long-run marginal cost of distribution services. There has been considerable interest in a move to demand-based tariffs as a possible improvement on the status quo. In our view, a combination of dynamic prices and hedging instruments could yield efficient outcomes and for customers, a choice of services to meet each customer's preference for cost, control, and risk.

ACKNOWLEDGMENT

The views expressed here are those of the authors and do not reflect the views of the Australian Competition and Consumer Commission or the Australian Energy Regulator.

REFERENCES

AEMC, 2014. "Distribution Network Pricing Arrangements". Rule Determination, November 27, 2014.

Allcott, Hunt, 2011. Rethinking real-time electricity pricing. Resour. Energ. Econ. 33, 820–842.

Biggar, D., Hesamzadeh, M., 2013. "Designing Transmission Rights to Facilitate Hedging". mimeo, May 2013.

Borenstein, S., Jaske, M., Rosenfeld, A., 2002. Dynamic Pricing, Advanced Metering, and Demand Response to Electricity Markets. Centre for the Study of Energy Markets, Berkeley, CA, CSEM WP 105, October 2002.

De Martini, Paul and Lorenzo Kristov, 2015. "Distribution Systems in a High Distributed Energy Resources Future: Planning, Market Design, Operation and Oversight", Lawrence Berkeley Laboratory, Future Electric Utility Regulation, Report No. 2, October 2015.

Electricity Expert Panel (QLD), 2014. Network Tariff Stabilisation Review: Project 1 Network Tariff Design for Residential Consumers. Electricity Expert Panel, Brisbane, December 5.

Ergon Energy, 2015. Consultation Paper: the Case for Demand Based Tariffs. Ergon Energy, Townsville, March.

Faruqui, A., Lessem, N., 2012. "Managing the Benefits and Costs of Dynamic Pricing in Australia". Report prepared for the AEMC. The Brattle Group, September 2012.

Geode, 2013. "Geode Position Paper on the Development of the DSO's Tariff Structure". GEODE Working Group Tariffs, September 2013.

Lineweber, D., 2013. "Few Residential Customers Want Dynamic Prices Yet", January 2013.

Reeves, A., 2014. "Perspectives on Regulation in a Changing Environment: What Does Success Look Like in Energy Regulation", Speech to Energy Network's Association 2014 Regulation Seminar, Brisbane, August 6, 2014.

Rocky Mountain Institute (RMI), 2014. Rate Design for the Distribution Edge: Electricity Pricing for a Distributed Resource Future, August 2014.

Rocky Mountain Institute, 2015. "Are Residential Demand Charges The Next Big Thing in Electricity Rate Design?", Blog Post, RMIOutlet, May 21, 2015.

Simshauser, P., 2014. "Network Tariffs: Resolving Rate Instability and Hidden Subsidies". AGL Working Paper, No. 45, October 2014.

Sotkiewicz, P.M., Vignolo, J.M., 2006. Nodal pricing for distribution networks: efficient pricing for efficiency enhancing DG. IEEE Trans. Power Syst. 21 (2), 1013–1014.

Wood, T., Blowers, D., 2015. Sundown, Sunrise: How Australia can Finally get Solar Power Right. Grattan Institute, Melbourne, May.

Wood, T., Lucy, C., 2014. Fair Pricing for Power. Grattan Institute, Melbourne, July.

Chapter 14

The Evolution of Smart Grids Begs Disaggregated Nodal Pricing

Günter Knieps
University of Freiburg, Freiburg, Germany

1 INTRODUCTION

In the current debate on the reform of electricity markets, in addition to "stand-alone" generators of renewable energy, the evolution of smart grids, with particular focus on microgrids, gains increasing relevancy (eg, chapter by Marnay). Smart grids enable local platforms (microgrids) integrating locally and real-time based generation and consumption of renewable energy. The roles of consumers and producers of electricity are no longer clearly distinguished, merging into "prosumers." Nevertheless, microgrids cannot be considered as isolated commons, but still need to be interconnected with the traditional electricity networks. On the one hand, there are time intervals when local demand exceeds locally generated electricity, and electricity has to be "imported" from the electricity distribution network; on the other hand, there may also be time intervals when local generation exceeds local demand, and electricity can be "exported" to the electricity distribution network. Therefore, the evolution of microgrids poses important challenges to the traditional "top down" electricity networks, and thereby challenges the business models of utilities of the future. Injections from smart microgrids may even lead to temporary changes in the direction of power flows from local grids toward distribution networks. Moreover, day-ahead markets selling the right and obligation to inject or extract electricity at different nodes of the electricity networks are challenged by the increasing role of real-time markets taking into account the specific real-time import/export conditions of microgrids.

Unfortunately, the role of intelligent grid management of complementary electricity networks has been seriously neglected within European countries (Bieser, 2014). After all, the potentials of smart grids are not only limited to local microgrids, but challenge the whole architecture of traditional electricity systems (Fang et al., 2012). On the one hand, the decentralized injection of

Future of Utilities - Utilities of the Future. http://dx.doi.org/10.1016/B978-0-12-804249-6.00014-2

renewable energy, either via isolated stand-alone generation or via microgrids, increases the requirements of node-aware active traffic management; on the other hand, the increasing role of "intelligent" real-time managed electricity transmission and distribution networks increases the potentials for implementing time sensitive node-dependent pricing systems. Up until now, pricing the opportunity costs of network access due to network constraints has not been implemented in most European electricity market designs. As a consequence, microgrid operators lack incentives to take into account the network constraints, and resulting positive or negative externalities of their injection and extraction schedules into the electricity distribution network. Therefore, an efficient allocation of network capacities gains particular importance in solving the incentive problems of interaction between microgrids and distribution networks.

There is already a substantial body of literature investigating the technical and organizational peculiarities of smart grids, with particular focus on distributed generation and the subsequent challenges of the value chain of the traditional top-down organized electricity systems (Fang et al., 2012). The lessons from the GridWise Transactive Energy Framework (GridWise Architecture Council, 2015) are pointing out the necessity of considering end nodes on the physical power grid no longer as simple "dumb" loads, but as being increasingly interactive with the grid. The provision of economically founded incentives for participation in actively managed systems in the generation, transmission, and use of electric power seems to be an important gap in the current reform debate.

The aim of this chapter is to provide a critical appraisal of the evolution of microgrids, from the perspective of their integration into, and compatibility with, the complementary electricity networks, from a network economic point of view. The focus is on the proper economic incentives for all relevant actors involved. In particular, the analysis should not be limited to the participants of the microgrid platforms, but also take into account the day-ahead and intraday markets for injection into, and extraction from, complementary electricity networks. Make or buy decisions to build and operate a microgrid platform, as well as decisions to interact with complementary electricity networks, should take into account the node-specific opportunity costs of changing power loss and network scarcity on all lines of an electricity network which are caused by injection or extraction. Hence, from the perspective of the usage of scarce capacities of distribution networks, microgrids play the role of (net) extraction nodes, or (net) injection nodes. Since smart grids emphasize the increasing importance of adding information and communication technologies (ICT) to electricity networks and complementary parts (eg, smart metering at customer side), decisions of microgrids on intraday injection and extraction into the electricity distribution network become increasingly relevant, with particular focus on real-time responsiveness.

Inefficient network management instruments have to be replaced by a consistent market framework taking into account the pricing of the node-specific opportunity costs of network access. Within liberalized electricity markets, a disaggregated framework of nodal pricing is required, separating between decentralized generation and extraction decisions based on local and time differentiated network constraints. Therefore, network operators need to provide disaggregated nodal pricing signals, in order to guarantee an optimal allocation of scarce network capacities. Injection and extraction prices should reflect the node-specific costs of network access, resulting in efficient decentralized injection and extraction decisions by market participants. Implementing disaggregated nodal pricing can have a strong impact on the efficiency of electricity markets. This holds not only within networks, but also regarding the incentive compatible integration of microgrids into electricity networks. The potentials of microgrids and prosumer activities create major challenges for the utilities of the future electricity power sector. There seems to be no doubt that the evolutionary path toward distributed energy and smart grids is irreversible. In order to survive, future electricity utilities should be enabled to exploit the new business opportunities, not obstructed by regulatory policies (chapter by Hanser and Van Horn; Nillesen et al., 2014; chapter by Sioshansi).

The chapter is organized as follows: In Section 2, the smart grid innovation potentials within electricity systems are characterized, focusing on the traditional value chain and the evolution of microgrids and distributed generation. In Section 3, the changing allocation problems of electricity networks driven by prosumer activities within microgrids are analyzed. For this purpose, the disaggregated nodal pricing model focusing on a uniform transmission network and day-ahead allocations (Knieps, 2013) is extended, taking into account flexible flow patterns in smart grids, and decentralized interaction between different network levels. In order to analyze the interaction of microgrids with the distribution networks, the differentiation between different network levels (high voltage, HV; medium voltage, MV; and low voltage, LV) is required. Moreover, to take into account the potentials of smart grids for real-time allocations, the day-ahead perspective is extended to include intraday allocation problems. In Section 4, the interaction incentives of microgrids for distributing energy into (medium voltage) distribution networks are analyzed. Full incentives for real-time microgrids require real-time distribution grid nodes. In contrast, as pointed out in the conclusions, ignorance of node dependency for injection or extraction of microgrids, as well as subsidized injection of renewable energy, cause disincentives for microgrids and welfare inefficient prosumer behavior.

2 SMART GRIDS FROM THE PERSPECTIVE OF TRANSACTIVE ENERGY FRAMEWORKS

Smart grids and transactive energy (TE) frameworks are both open concepts, without a precise overall accepted formal definition. However, in contrast to traditional electricity systems, smart grids are based on modern information and communication infrastructure, supporting two-way flows of electricity and information, capable of delivering power in more efficient ways, and responding to events that may occur everywhere within electricity systems (Fang et al., 2012, pp. 944ff.; Barrager and Cazalet, 2014).

From a narrow perspective, the smart grid approach could be applied to the traditional value chain of electricity systems, improving efficiency of high voltage transmission networks, substations, and medium voltage distribution networks and low voltage local networks. Smart management systems, based on smart infrastructure systems, improve the efficiency of electricity transmission under network constraints by taking into account real-time traffic management. From a broader forward-looking perspective, smart grids are also enabling distributed generation, and thereby supporting the goals of a TE framework. The GridWise Architecture Council defines TE as "A system of economic and control mechanisms that allows the dynamic balance of supply and demand across the entire electrical infrastructure using value as a key operational parameter" (GridWise Architecture Council, 2015, p. 11).

The concept of TE shifts the focus of smart grids to the active role of prosumers, not only consuming, but also generating energy based on a low-voltage distribution grid similar to microgrids. The traditional top-down architecture of electricity grids may become challenged by the evolution of microgrids and virtual power plants. As a consequence, real-time allocation and related price signals become increasingly important to optimize allocation of electricity, subject to the constraints of the grid (CPUC, 2014, pp. 4 ff.). It is important to differentiate between a sequence of balancing transactions from day-ahead bids, intraday hour-ahead prices, and prices for additional resources transacted for the final matching. The additional resources (every 4 s) are transacted at the true real-time prices (King, 2014, p. 201).

Distributed generation via microgrids requires node-aware time sensitive distribution grids, as increasing congestion problems in distribution networks arise (CPUC, 2014, p. 15). Real-time relevancy increases due to short-term generation of renewables, and short-term response of demand. The role of interaction of retail markets and wholesale markets is of particular interest. Prosumers are active on the retail market, the microgrid node serves as the interface to the distribution wholesale market. Not every household is bargaining with the distribution grid (either for extraction or injection). Instead, an aggregator plays the role of agent for the prosumers and other consumers within a microgrid, thereby extending the retailer functions of traditional utilities. Meanwhile, the changing role of distribution utilities gradually turning into distributed system platform providers is increasingly being discussed (CPUC, 2014, pp. 14, 16).

Prosumer activities within microgrids, and TE concepts, require a differentiation of allocation problems at different voltage levels, as well as a differentiation regarding different time perspectives. The opposite drivers of network hierarchies within smart grids, compared to traditional electricity systems, may be summarized as follows:

- Local initiatives to build microgrids and provide TE: software defined low voltage distribution grid acting both as supplier and load (Coll-Mayor et al., 2007, p. 2458). Microgrids are characterized by exchange of power via low voltage powerlines between renewable energy sources (RES) generation (wind energy generators, solar panel generators), energy storages and loads (from consumers). Exchange of information may be carried out via wireless access points-based wireless networks (Fang et al., 2012, p. 951).

- Islanded microgrids, in which distributed generators power the users within the microgrid, without obtaining power from the distribution grid, do not represent the typical scenario. Instead, there is usually a connection to the traditional power grid, and the possibility of obtaining power from the electricity utility exists (Fang et al., 2012, pp. 951 ff.). The interaction of microgrids with the distribution system causes positive and negative network externalities (Felder, 2014, p. 414). Microgrids possess the characteristics of acting as a single node from the perspective of a distribution grid. Microgrids may be run semi-autonomously, either in the grid-connected or in the island mode, not connected to the traditional top-down "bulk power" electricity system consisting of generation and transmission (Felder, 2014, p. 400).
- The role of aggregators at the interface between the suppliers of decentralized energy resources (DERs) and the wholesale market becomes relevant. Demand-side aggregators (eg, large industrial customers) may sell their demand response services to the system, multiple wind or solar generators may use aggregators to sell their energy to the system (Keay et al., 2014, pp. 184f.). Aggregators may become active to organize smart grid prosumer activities and the interaction with distribution networks, they may also become active to enable direct injection of RES into distribution networks. Due to the flexible flow pattern in smart grids, injection of RES may also take place at the distribution grid (Fang et al., 2012, p. 948).
- Network intelligence with application of ICT, Distributed Generation (DG), and the increasing role of prosumers, enable two-ways power flows, and faster exchange of information, including short-term price signals and demand responsiveness (Barrager and Cazalet, 2014; Cazalet, 2014; CPUC, 2014; King, 2014, pp. 189ff.).

3 THE CHANGE OF ELECTRICITY SYSTEMS ARCHITECTURE AND ITS IMPACT ON DISAGGREGATED NODAL PRICING

The transition to local grids and TE shifts the focus to an active role of medium voltage distribution networks. The proper incentives for generation at the right place (node) and the right time become highly relevant, no longer only for transmission networks, but also for distribution networks.

3.1 The Basic Principle of Disaggregated Nodal Pricing

The principle of nodal pricing was initially developed in an aggregated manner, for the context of an integrated electricity system, with end-to-end centralized control. Optimal spatial spot pricing for electricity indicates the value of energy at a given time and location. Prices at each node reflect the integrated locational value of electricity, including the costs of electricity generation and the transmission costs within the network (Bohn et al., 1984, p. 364). Decentralized injection decisions in liberalized electricity systems require the application of the principle of nodal pricing in a disaggregated manner, separating between the node-dependent marginal costs of generation, and pricing signals for network constraints provided by the network operators.

The traditional hierarchy of electricity systems differentiates between high voltage (long distance) transmission networks, medium voltage (regional) distribution networks, and low voltage (local) consumer networks. In top-down electricity systems, electricity is generated and injected into the high voltage transmission network. Distribution networks are focused on the passive transition (forwarding) of electricity to the local low voltage consumer networks. Therefore, the relevant wholesale market for generation decisions is on the high voltage transmission level. In Knieps (2013), the framework of disaggregated nodal pricing has been developed. A single electricity transmission network has been considered, and day-ahead allocation has been analyzed in order to derive the day-ahead generalized merit order rule, as described in Box 14.1.

The interaction of the different actors involved in implementing the concept of disaggregated nodal pricing is schematically illustrated in Fig. 14.1.

3.2 Transactive Energy and the Changing Market Architecture

Within traditional power grids (Fang et al., 2012, Fig. 4, p. 948; Felder, 2014, p. 399), power is stepped up to high voltage transmission level by means of generation step up transformers, and injected into the high voltage transmission grid. At transmission substations, power is stepped down to distribution voltage, entering

---

BOX 14.1   The Generalized Merit Order Rule

Disaggregated nodal pricing can be characterized as follows:
- Electricity transmission prices raised by the network carrier consist of node-dependent injection and extraction prices based on system externalities. System externalities—changing power loss and network scarcity on all lines of an electricity network with a given number of interconnected nodes—are the opportunity costs of electricity injection or extraction depending on the node (location) where generation or extraction takes place.
- The generalized merit order indicating at which nodes injection is incentive compatible because generation costs and injection price do not exceed marginal willingness to pay on the wholesale market. Market value, according to the shadow value of the system balance equation, reflects the opportunity costs of the last unit of demand, which is still served on the wholesale market, knowable to the different generators at the different nodes.
- Nodal prices at extraction nodes are equal to the sum of the (uniform) wholesale price, and the node-dependent extraction price.

Source: *Knieps, 2013, pp. 156ff.*

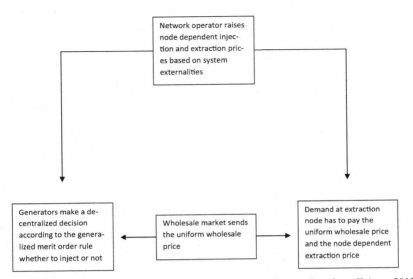

FIGURE 14.1 Disaggregated nodal pricing framework. *(Source: Based on Knieps, 2013, pp. 156ff.)*

the distribution network. Finally, at service location, power is again stepped down at distribution substations to the service voltage. Transmission grid substations transform electricity down from high voltage to medium voltage, whereas distribution grid substations transform electricity down from medium voltage to low voltage levels. Smart transmission grids and distribution grids (smart control centers, smart power networks, smart substations) enable more flexible control and operation information on the status of capacity usage. As a consequence, increasing potentials arise for implementing pricing signals reflecting node-dependent scarcity, and the opportunity costs of network usage (Fang et al., 2012).

Network constraints and the resulting system externalities of injection and extraction at different nodes can be characterized for high voltage transmission networks, as well as for medium voltage distribution networks separately. In contrast, at the local retail level, low voltage power lines are used by generation or consumption, irrespective of the individual location within the microgrid; only the aggregated net consumption or net production is exchanged at the microgrid node. The location where consumption or generation occurs is of no relevance within a microgrid. Therefore, only the intersection node to and from the distribution network (vertical) or to other microgrids (horizontal) are considered, nodes within a microgrid are irrelevant for influencing locational decisions; instead, prosumer activities are characterized by different generators, storage and loads, which are aggregated by concentrators, and interconnected to the microgrid node at the distribution grid.

However, the change of power flows at distribution substations, and the subsequent need for step up transformers, becomes relevant for injecting power into the distribution grid at the microgrid node. Step up transformers may also become relevant at other injection nodes of the distribution grid, or for injecting power from distribution networks to high voltage transmission networks. Driven by bottom up injection from microgrid nodes, active management of all injection sources within the medium voltage distribution networks, including stand-alone renewable resources and conventional energy sources, becomes necessary. The wholesale market for electricity injected into the high voltage transmission network has to be differentiated from the wholesale market for electricity injected into the medium voltage network. From the perspective of distribution network operators import from high voltage transmission networks may be beneficial, depending on the generation and transmission charges for high voltage electricity, compared to the generation and transmission charges for medium voltage electricity. On the other hand, it may also become beneficial to export electricity from the distribution network to the transmission network.

Decentralized injection of renewable energy, either into transmission or into distribution networks, requires an upgrade of network intelligence to take into account node dependency, but also for real-time awareness. Day-ahead and intraday markets are interrelated, the scope of the markets for balancing energy depends on the organization of the day-ahead and intraday markets (King, 2014, p. 201). The time perspective of allocations differentiating between day-ahead, intraday (hourly, minutes) ahead, and real-time may play an important role.

The disaggregated nodal pricing framework described in Fig. 14.1 must be extended to take into account different levels of electricity transmission networks, and real-time allocations. The analysis of microgrids and distributed generation shifts the focus to the relation between distribution networks and local networks/consumers and prosumers. Changes in the architecture of smart electricity systems encompass the growing importance of distribution networks, the interaction between different distribution networks by (horizontal) import/export agreements, and the interaction of a distribution network with microgrids.

Transmission network nodes, distribution network nodes, and local loads within local networks are to be differentiated, reflecting different levels of network usage with different opportunity costs. This enables a differentiation between injection of (stand-alone) renewable energy into the transmission networks, and into the distribution networks. Moreover, microgrids may also inject electricity at the microgrid node into the distribution network.

According to the generalized merit order principle, not only the generation costs, but also the node-dependent injection costs, characterized by the system externalities of the usage of a distribution network or a transmission network, respectively, are to be taken into account. It is assumed that there exists a competitive wholesale market for high voltage electricity, transmitted on a transmission network, as well as a competitive wholesale market for medium voltage electricity transmitted on a distribution network. Market values result according to the shadow values of the relevant system balance equations of each network, equal to the opportunity costs of the last unit of demand, which is still served.

TABLE 14.1 The Principle of Decentralized Interaction Between Different Network Levels

Injection on different levels		Wholesale markets	Retail market
→	Transmission network	High voltage electricity	
↓↑			
→	Distribution network	Medium voltage electricity	
↓↑			
→	Microgrid/local network		Low voltage electricity

The principle of the generalized merit order is extended, taking into account not only injection by generation within one network level, but also import/export from other network levels.

3.3 Distributed Generation and Disaggregated Nodal Pricing

3.3.1 Market Driven Incentives for Power Flows

In contrast to the traditional hierarchical organization of electricity systems, within smart grids power flows can be organized by market transactions between transmission network owners and distribution network owners, as well as between microgrids and distribution networks, as illustrated in Table 14.1. Power flows may be bidirectional between transmission and distribution networks and microgrids. The relative importance of the different network levels is influenced by the evolutionary change of grid usage, depending on market-driven incentives, rather than historical top-down organization principles. Prosumers and other participants of a microgrid decide on the microgrid level for each time interval how much energy they buy or sell. Net import or net export takes place at the microgrid node of the distribution network.

3.3.2 The Generalized Merit Order Reconsidered at Different Network Levels

1. *High voltage: system externalities among the high voltage nodes of a transmission network*
   The transmission network provider signals the opportunity costs of injection to the different generators located at the different injection nodes. Injection takes place either via import from other transmission networks, import from a distribution network, or at generation locations (nodes) within the network. The sum of generation costs (or import price) and injection charge

(according to the system externalities at that node) must not exceed the wholesale price equal to the shadow value of the system balance equation of the transmission network.

2. *Medium voltage: system externalities among the medium voltage nodes of a distribution network*
Each distribution network has its own network constraints, and system externalities at the different nodes. Injection takes place either via import from a transmission network or another distribution network, or via injection by a generator at the distribution level, or via a microgrid or Virtual Power Plant. In each injection, a node of the distribution network is used, causing opportunity costs of distribution network usage, according to the system externalities and network loss within the distribution network. The sum of generation costs (or import) and injection charge (according to system externalities at that node) must not exceed the distribution wholesale price, equal to the shadow value of the system balance equation of the distribution network.

3.3.3 Outside Options for Microgrids

The opportunity costs of injection and extraction at a microgrid node cannot be considered in isolation; instead, the system externalities of injection or extraction due to the change of scarcity within the whole distribution network have to be taken into account. In contrast, within retail markets on low voltage local networks, no system externalities are relevant, therefore the individual locations of the prosumers within a microgrid is not relevant. Instead, only the total net consumption or net production measured by the aggregator and forwarded to the microgrid node at specific time intervals are relevant. The basic outside option for prosumer and consumer only activities within microgrids is the real-time generalized merit order of the distribution network containing the microgrid node. The following cases may be differentiated.

- If willingness to pay within a microgrid is lower than the wholesale market price of the distribution network, it is worthwhile for the microgrid to sell electricity to the wholesale market.
- If microgrid demand exceeds microgrid generation, it is worthwhile to import electricity from the distribution wholesale market, as long as willingness to pay exceeds the distribution wholesale market price, plus the injection charge at the microgrid node.
- Import into the microgrid only takes place if microgrid demand exceeds microgrid generation. Otherwise, due to arbitrage conditions, it is possible to use microgrid generated electricity at opportunity costs equal to the distribution wholesale price, thereby saving the injection charge at the microgrid node.

If real-time price signals are provided by the distribution network operator, the price signals for prosumer activities are sufficient because all relevant decisions of prosumers can be based on them. At time intervals when the import price at the microgrid node is high (generalized merit order of distribution network plus extraction charge at microgrid node) incentives may arise for microgrid users to reduce their consumption, or to expand their own generation, maybe even by additional gas generators.

4 MICROGRIDS NECESSITATE REAL-TIME NODAL PRICING WITHIN DISTRIBUTION NETWORKS

The organization of microgrids creates challenges not only from the perspective of technical and organizational issues (eg, the chapter by Marnay), but also from a network economic point of view. In particular, the incentives for local decision makers within the microgrids should become compatible with the incentives of distribution operators.

4.1 Alternative Objectives for Microgrids

1. *Self-sufficient commons*
   Island solutions combine smart small-scale generation (solar panels, smart wind turbines), applying curtailing with demand responsive smart metering and local incentive pricing to balance demand and supply locally. Concentrators organize the minimum necessary interaction with the microgrid distribution node. Welfare effects depend on the opportunity costs of injection or extraction at the relevant node within the distribution network. Similar to cogeneration, microgrids change the overall level of capacity usage of the distribution network. From the perspective of an overall electricity distribution system, a self-sufficient commons goal need not be welfare maximizing. This is the case, if the willingness to pay for additional electricity is higher than the generalized merit order wholesale price of the distribution grid, plus the extraction charge at the microgrid node, and additional demand should be fulfilled by importing electricity, given that prosumer generation is not sufficiently high.

2. *Welfare maximizing incentives for microgrids*
   In order to provide the proper economic incentives for prosumers and other consumers (given the required microgrid infrastructure), the opportunity costs of extraction from, or injection into, the microgrid node within the distribution network are relevant. Opportunity costs vary between day-ahead and real-time. In order to provide the proper economic incentives within microgrids, real-time nodal pricing should be applied because only real-time nodal pricing ensures that the real-time opportunity costs of production and consumption within a microgrid are sufficiently signaled, and no additional pricing instruments are required. Similar to stand-alone renewables, the investment decisions of generators can be financed via subsidies or alternative financial arrangements, so that the danger of interfering with network usage decisions is avoided. However, for the case that only day-ahead nodal prices are applied, real-time allocation still requires additional pricing instruments (eg, peak load pricing) within the microgrid community.

4.2 The Relevancy of the Real-Time Generalized Merit Order for Wholesale Markets on the Distribution Networks

Meanwhile, innovations in smart metering technologies also enable real-time pricing which is already part of the smart grid developments. Within traditional electricity systems, time sensitive pricing may reduce electricity wholesale consumption from extraction nodes. However, interactive prosumer behavior with a time-responsive component requires TE concepts. Increasing need for real-time nodal pricing arises in order to enable customer responsiveness, or, more generally, innovative smart prosumer activities.

Whereas in the traditional top-down organization of electricity systems day-ahead allocation played the key role, and the real-time operating reserve was considered residual, within TE systems this time perspective is reversed. Real-time allocation signals within microgrids are forwarded via the microgrid node toward the distribution network, and thus cannot be ignored by distribution operators without running into serious disincentive problems for local microgrid prosumers. Real-time allocations bundled by an aggregator are online interrelated with the distribution network, thereby causing positive or negative real-time system externalities to the distribution network. System externalities from a day-ahead perspective of distribution network capacities are different from real-time system externalities, and no longer decision relevant within the short-term real-time interval.

The real-time allocation of microgrid energy requires real-time node-dependent pricing, not only at the microgrid node, but within the whole distribution network. The opportunity costs of prosumer activities are determined by the opportunity costs of outside conditions at the microgrid node of the distribution network. Real-time pricing signals for generation and consumption within a microgrid reflecting the real-time disaggregated microgrid nodal price are sufficient for real-time allocation of electricity within the microgrid. Further pricing instruments (eg, peak load pricing) become superfluous.

There is no doubt that the increasing role of renewable energy generation strongly increases the volatility of wholesale prices, challenging the traditional approach to dispatch and electricity pricing (chapter by Woodhouse; Keay et al., 2014, pp. 166ff.). Efficient price signals on the distribution wholesale markets provided by real-time based disaggregated nodal pricing create the proper economic incentives for prosumer activities within microgrids. Nevertheless, the question arises whether the political goal of reducing the risks from volatility in electricity bills on the retail markets, due to volatile wholesale prices, would conflict with the objectives of efficient decision making for prosumer activities.

The necessity of implementing nodal pricing in distribution networks, taking into account the marginal costs of generation, and the marginal costs of congestion, has also been pointed out in chapter by Biggar and Reeves. According to these authors, policy-makers should not be concerned with regulatory micromanagement focusing on insurance or risk management for volatile wholesale tariffs. Instead, the focus should be on the role of competitive retailers offering consumers a range of optional contracts, hedging the temporal and locational wholesale price risks.

However, within microgrids, the role of the retailer merges with the role of the aggregator, bundling the supply of prosumers and providing the interface between microgrids and the distribution network. Interfering with or ignoring the price signals at the microgrid node seems, therefore, to be a costly strategy for providing insurance against volatile wholesale market prices. Instead, risk management should better take place within the fixed part of nonlinear retail tariffs, leaving the wholesale market variable price signals based on the opportunity costs of network usage unchanged.

5 CONCLUSIONS

The current policies in European countries often prohibit pricing for injection of electricity at generation nodes; moreover, priority access and guaranteed access for electricity from renewable energy sources are the rule. To the extent that microgrids are exporting electricity to the distribution grids, similar incentives arise as with the generation of stand-alone renewable energy. Microgrid owners may maximize generation and revenues from injection into distribution networks. In their demand function, prosumers take into account the opportunity costs of exporting electricity at high renewable energy prices without reflecting the opportunity costs of electricity injection at the microgrid node. The welfare effect of reduced consumption and exporting can be negative, due to negative system externalities, overload of the network, and subsequent requirements to redispatch or injection of additional generation capacity available within a short interval of time (operating reserve).

As a consequence of an ignorance of locational and time-specific differences within the relevant electricity networks, the wrong incentives for generation and consumption decisions are created. For the case that export from microgrids (via the microgrid node) into the distribution network is carried out under the same conditions as for stand-alone renewable energy, and import into a microgrid takes place at electricity generation tariffs only, the real-time opportunity costs of injection or extraction from distribution networks are not taken into account. Microgrid owners are then incentivized to maximize revenue from export, ignoring the generalized merit order effects of real-time transactions. Capacity limits of microgrids are fully exhausted, even if system externalities at the microgrid node are highly negative. Moreover, incentives for reduced consumption are provided due to the high opportunity costs of electricity export, even if increased consumption would reduce system externalities at the microgrid node. Incentives for import reductions will arise, even at time intervals when positive system externalities at the microgrid node would make import socially beneficial. As more and more local communities are jumping on the environmental bandwagon of microgrids, there seems to be no alternatives for electricity utilities of the future to incentivize prosumer activities by proper pricing signals. As has been shown in this chapter, the time is right to introduce disaggregated nodal pricing, taking into account flexible flow patterns in smart grids, and decentralized interaction between different network levels.

REFERENCES

Barrager, S., Cazalet, E., 2014. Transactive Energy: A Sustainable Business and Regulatory Model for Electricity. Baker Street Publishing, San Francisco, CA, eBook.

Bieser, G., 2014. Smart grids in the European energy sector. Int. Econ. Econ Policy 11 (1–2), 251–259.

Bohn, R.E., Caramanis, M.C., Schweppe, F.C., 1984. Optimal pricing in electrical networks over space and time. Rand J. Econ. 15 (3), 360–376.

Cazalet, E.G., 2014. Transactive energy: interoperable transactive retail tariffs. In: Sioshansi, F.P. (Ed.), Distributed Generation and its Implications for the Utility Industry. Elsevier, Amsterdam, pp. 205–229.

Coll-Mayor, D., Paget, M., Lightner, E., 2007. Future intelligent power grids: analysis of the vision in the European Union and the United States. Energy Policy 35, 2453–2465.

California Public Utilities Commission (CPUC), 2014. Transactive Energy: A Surreal Vision or a Necessary and Feasible Solution to Grid Problems? Policy & Planning Division, October.

Fang, X., Misra, S., Xue, G., Yang, D., 2012. Smart grid – the new and improved power grid: a survey. IEEE Commun. Surv. Tut. 14 (4), 944–980.

Felder, F.A., 2014. What future for the grid operator? In: Sioshansi, F.P. (Ed.), Distributed Generation and its Implications for the Utility Industry. Elsevier, Amsterdam, pp. 399–415.

GridWise Architecture Council, 2015. GridWise Transactive Energy Framework, Version 1.0, January, www.gridwiseac.org

Keay, M., Rhys, J., Robinson, D., 2014. Electricity markets and pricing for the distributed generation era. In: Sioshansi, F.P. (Ed.), Distributed Generation and its Implications for the Utility Industry. Elsevier, Amsterdam, pp. 165–185.

King, C., 2014. Transactive energy: linking supply and demand through price signals. In: Sioshansi, F.P. (Ed.), Distributed Generation and its Implications for the Utility Industry. Elsevier, Amsterdam, pp. 189–204.

Knieps, G., 2013. Renewable energy, efficient electricity networks and sector-specific market power regulation. In: Sioshansi, F.P. (Ed.), Evolution of Global Electricity Markets: New Paradigms, New Challenges, New Approaches. Elsevier, Amsterdam, pp. 147–168.

Nillesen, P., Pollitt, M., Witteler, E., 2014. New utility business model: a global view. In: Sioshansi, F.P. (Ed.), Distributed Generation and its Implications for the Utility Industry. Elsevier, Amsterdam, pp. 33–47.

Part III

Utilities of the Future:
Future of Utilities

Chapter 15

New Business Models for Utilities to Meet the Challenge of the Energy Transition

Paul Nillesen*, Michael Pollitt[†]
*PwC, Amsterdam, The Netherlands; [†]Cambridge University, United Kingdom

1 INTRODUCTION

The energy transformation is being driven by five global megatrends interacting with a set of shifts taking place within the power sector. The five megatrends[1]—technological breakthroughs, climate change and resource scarcity, demographic and social change, a shift in global economic power, and rapid urbanization—are challenges for all businesses. Technological innovation is at the heart of the shifts that are occurring in the power sector. Advances are happening in many parts of the sector—for example, in large-scale technologies, such as offshore wind and high-voltage DC transmission, in distributed and smaller-scale customer-based energy systems with the internet-of-things, and on the load side with demand declining and shifting.

The energy sector is on the frontline of concerns about climate change. The sector as a whole accounts for more than two-thirds of global greenhouse-gas emissions (OECD/IEA, 2013), with just over 40% of this stemming from power generation. Resource scarcity or availability, and the associated geopolitics and economics of gas, oil and coal supply, are key factors shaping power market policy.

By 2025 another billion people will have been added to the global population; this will reach about eight billion. Explosive population growth in some areas set against declines in others makes for very different power market growth potential in different parts of the world. Africa's population is projected to double by 2050, while Europe's is expected to shrink. Over the next 2 decades, nearly all of the world's net population growth is expected to occur in urban areas, with about 1.4 million people—close to the population of Stockholm—added each week (Seto and Dhakal, 2014).

---

1. See http://www.pwc.com/gx/en/issues/megatrends/index.jhtml.

Future of Utilities - Utilities of the Future. http://dx.doi.org/10.1016/B978-0-12-804249-6.00015-4

As echoed in other chapters of this book, in the power sector, these trends are leading to a number of simultaneous disruptions, involving customer behavior, competition, the production service model, distribution channels, and government policy and regulation. The extent and nature of these disruptions varies from market to market. But, in many markets, their intensity leads to the impact being transformational, rather than incremental. It is not a question of whether new market models will take shape, as this is already happening, but which new business models will be pursued in the sector, and what countries and regulators will do to increase access to reliable electricity supply, and what existing power utilities will do to keep up with the change and alter their course.

In this chapter a set of business models is examined that are nascent or likely to emerge as companies take advantage of the changes and exploit new value pools, or are forced to adapt to defend existing profitable revenue. Section 2 examines which of these megatrends will have the most impact for traditional power companies. Section 3 describes eight possible business models and their likely viability in the different market environments. The chapter's conclusions include an assessment of the probability that existing players will be able to adapt and retain their position, or whether new entrants are likely to capture the new emerging value pools, leaving the traditional players in a situation where they "lose the winners, and win the losers."

2 AREAS OF DISRUPTION

The disruptions taking hold in the power sector may be just the start of an energy transformation. It is not a question of whether the business models pursued in the sector will change, but rather what new forms they will take and how rapidly companies will have to alter course. The pace of change will be different in each market and each specific situation. The implication for companies is that they should assess their strategy and implement the changes they need to make in time or, even better, ahead of time. Already, of course, many have reset their compasses with a switch in priorities and emphasis. In what follows, the five areas in which disruption is having an impact are discussed (Fig. 15.1) and where it will be important for companies to assess their strategies. Together they form the context in which future market and business models will be framed.

2.1 Customer Behavior

There is evidence of a gradual erosion of power utility company revenues (net of fossil fuel costs) as distributed energy gains an increasing foothold. Some commentators go so far as to predict that customers will be saying "goodbye to the grid" in the future. In some places, this is already happening. For example, in Germany, almost half of the installed renewable energy capacity is owned by private citizens and farmers, rather than the traditional utilities (Agora Energiewende, 2015). Significant changes in the economics and practicalities of self-generation and storage are needed for such a scenario to occur, on any

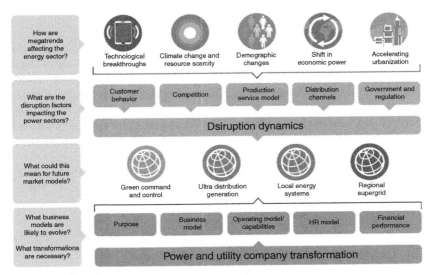

| How are megatrends affecting the energy sector? | Technological breakthroughs | Climate change and resource scarcity | Demographic changes | Shift in economic power | Accelerating urbanization |

| What are the disruption factors impacting the power sectors? | Customer behavior | Competition | Production service model | Distribution channels | Government and regulation |

Dsiruption dynamics

| What could this mean for future market models? | Green command and control | Ultra distribution generation | Local energy systems | Regional supergrid |

| What business models are likely to evolve? | Purpose | Business model | Operating model/ capabilities | HR model | Financial performance |

| What transformations are necessary? | Power and utility company transformation |

FIGURE 15.1 The impact of megatrends on the power sector and disruption dynamics. *(Source: PwC, 2014, p. 3.)*

kind of scale. But even if customers do not literally say *goodbye to the grid*, power utility companies face the prospect of playing the role of being providers of secondary or back-up power to customers.[2]

Instead, the traditional utilities could become part of the change by being more active participants in the self-generation market, providing advice on equipment, metering, and using the opportunity to secure more of the home and business services space. Some regulators, such as in the State of New York, are actively promoting this vision of the future by creating distribution service platform providers (DSPs) to facilitate and seek out distributed energy resources on both the demand and supply side (State of NY Department of Public Service, 2014).[3]

The growth in self-generation can create a reinforcing dynamic. As well as the decline in revenues to decentralized sources, there is the impact of cost pressures on the centralized system which, in turn, reinforces the movement to decentralization. The reaction of some in the industry has been to press for new regulatory policies to allow for some form of cost recovery, in recognition of utilities being left with the fixed cost of the grid but a shrinking revenue base.[4] However, as one academic study points out: "In the short run, these steps very well could insulate the utility from solar PV competition but at the same time create substantial medium- and long-term risks, including those of customer

---

2. See chapter by Burger and Weinmann in this volume for a further discussion of the developments in Germany.
3. See also the contribution by Jones et al. in this volume.
4. http://www.forbes.com/sites/williampentland/2013/05/30/electric-power-politics-net-metering-gets-nasty/.

backlash, deferral of adaption, and stimulation of enhanced competition" (Grapy and Kihm, 2014, p. 31.).

The net metering of domestic PV is one such problem area. Under many domestic PV subsidy regimes, this means that those who self-generate can avoid most or all of any contribution to the joint grid costs of the network, or of the subsidy regime. This means that higher PV penetration not only raises total subsidy costs, but is also associated with a diminishing revenue base from which to recover system costs. Recently, Germany has recognized this problem with the introduction of a €4.4 cent/kWh charge on own consumption of domestic PV, so that even a net zero household would contribute, somewhat, to shared system costs.

Both in terms of regulatory relations and customer relations, utility companies need to align their ambitions with those of their customers in a new energy future, ensuring their services are relevant to and cost-effective for as many customer situations as possible.

2.2 Competition

Energy transformation is shifting the opportunity for good margins into new parts of the value chain. But lower barriers to entry in these areas of the value chain and the need for new capabilities mean there is the prospect of existing companies being outflanked and outpaced as more nimble and able competitors seize key revenue segments.

New roles for companies come into view. In a distributed energy community with its own microgrid, players other than power utilities can play an energy management role. This could be for local systems such as transport networks, residential communities, or industrial communities. For example, distributed energy is a key focus both for incumbent power utility companies and new entrants. It is a big market space, potentially worth tens of billions globally. It covers a wide spectrum of opportunities, from energy controls and demand management activities that save energy, to local generation, both small- and larger-scale, embedded in own use or local networks, through to distributed storage that can shift loads or, ultimately, end grid dependency.

Engineering and technology companies such as GE, Siemens, and Schneider Electric have long been important players as equipment providers in larger-scale segments of the distributed energy market. The growth and extension of distributed energy is likely to blur the boundaries between such companies and the power utility sector, both at the individual customer and community levels.[5]

Demand management services are another key area and, already, we see companies such as Kiwi Power in the United Kingdom providing services to industrial and commercial clients, offering demand reduction strategies that they claim might typically see larger businesses reduce their electricity bills by around £100,000.[6]

---

5. See, for example, chapter by Cooper in this volume.
6. www.kiwipowered.com.

In addition, there is considerable interest from companies seeking to explore the opportunities that come from existing home and online services, as well as future smart grid and distributed energy provision. "The battleground over the next five years in electricity will be at the house," David Crane, CEO of NRG Energy, told Bloomberg Businessweek. "When we think of who our competitors or partners will be, it will be the Googles, Comcasts, AT&Ts who are already inside the meter. We aren't worried about the utilities, because they have no clue how to get beyond the meter, to be inside the house" (Goossens et al., 2014).

Competition is also being enhanced by smart meter technology. In New Zealand the move to Advanced Meter Infrastructure (AMI) has resulted in a dramatic reduction in switching times for residential consumers since 2004, from around 40 days to less than 5. As AMI meter penetration has increased— it is now more than 50%—this has seen switching rates more than double since 2008, and increase in the number of new retail competitors in the market (Beatty, 2014). The US market also offers significant opportunities, with more than 52 million AMI meters in 2013, of which 90% are residential.[7]

2.3 Production Service Model

The production service model of centralized generation and grids is being joined by a much more disintermediated and distributed model. New supply sources requiring centralized infrastructure, such as offshore wind, are coming on stream but the danger for utilities is that other assets and infrastructure are left stranded. The centralized infrastructure that has long been a source of strength of the industry can be a source of weakness vulnerable to market, policy, or disaster risk.

In Europe, the changing economics of generation brought about by a combination of the rise of renewables, the collapse in the carbon market and cheaper international coal prices; this has left much gas generation out of the market. Even modern plants, completed as recently as 2013, have had to be temporarily mothballed, and many others have been taken out of the market more permanently. In total, over the course of 2012–13, 10 major EU utilities announced the mothballing or closure of over 22 GW of combined cycle gas turbine (CCGT) capacity in response to persistently low or negative clean spark spreads, of which 8.8 GW was either built or acquired within the last 10 years (Caldecott and McDaniels, 2014).[8]

Disaster risk led to all of Japan's nuclear reactors being gradually taken offline after the 2011 Fukushima disaster and they remained offline 3 years later. In Germany, the reaction to Fukushima was to begin to phase out nuclear

---

7. See EIA (http://www.eia.gov/tools/faqs/faq.cfm?id=108&t=3).
8. See also the contributions by Woodhouse, Boscan and Poudineh, and Burger and Weinmann in this volume.

power altogether. Official policy in Japan is to bring plants back into operation, as and when the atomic regulator deems new, stricter safety standards are being complied with. But opinion polls have consistently shown that a majority of Japanese are opposed to restarting reactors, and nuclear assets are unlikely to regain the same role in Japan's energy system as they had before Fukushima. By contrast, PV and wind power have extremely high reliability factors—although the availability of space remains an issue for Japan. The availability factor for wind turbines is 97%; this contrasts with long average global nuclear fleet availability of around 75%. Thus, even though solar and wind are intermittent sources of energy, they are not subject to large, and often covarying equipment failures.[9]

In the United States, one can draw a direct line from environmental policy to the stranded asset risk faced by many of the country's coal generation plants. Coal-fired power plants are subject to the Mercury and Air Toxics Standards (MATS), which require significant reductions in emissions of mercury, acid gases, and toxic metals. The standards are scheduled to take effect in 2015 and 2016, with generators needing to install costly pollution-control equipment, if they want to keep their coal plants running. The US Energy Information Administration expects about 60 GW of coal generation to shut down between 2012 and 2018, a reduction of about a fifth (EIA, 2014). A further threat to coal comes in the form of the proposed Clean Power Plan, which will require carbon emissions from the power sector to be cut by 30% nationwide below 2005 levels by 2030.

There are also significant problems with siting electricity transmission assets and expanding both large scale power transfers and new power plants. This has led to long delays in getting even short lengths of new transmission capacity built, and there are ongoing problems of expanding large scale generation in particular countries such as Italy and parts of the United States (such as California). By contrast, it has proved possible to expand rapidly small scale PV and wind projects. Gigawatts of both solar and wind have been installed relatively quickly closer to demand centers in many European countries, including Germany and the United Kingdom.[10]

These developments highlight the risk of overreliance on a concentrated centralized power generation asset mix. The wrong type of asset mix can leave companies vulnerable to rapid transformation, arising from market or policy forces or the forces of events, in the case of nuclear. Such forces provide a wake-up call, which is likely to accelerate the move to alternative decentralized power systems.

---

9. See for example also the contribution by Lobbe & Jochum in this volume.
10. Germany added 7 GW of solar PV per year in each of 2010, 2011, and 2012, see http://www.solarwirtschaft.de/fileadmin/media/pdf/2013_2_BSW-Solar_fact_sheet_solar_power.pdf. Even the United Kingdom has added 7 GW of solar PV since 2011, see https://www.gov.uk/government/statistics/solar-photovoltaics-deployment.

2.4   Distribution Channels

In a digital-based smart energy era, the expectation is that the main distribution channel will be online, and the energy retailing prize will hinge on innovative digital platforms (refer chapter by Cooper in this volume) in order to secure the energy automation, own generation, and energy efficiency customer space. Already, many companies are shifting their positioning to cluster energy management offerings around a central energy efficiency and energy saving proposition, and using new channels, such as social media to engage with customers, such as E.ON.[11]

A risk for traditional energy companies is that their distribution channel to end customers becomes disintermediated in ways that are not dissimilar to what has happened to incumbent publishers and booksellers with the advent of Amazon. Not only is the channel to market for incumbents dominated by the new platform, but the actual demand for product is eroded as the platform acts as an aggregator for self-publishing and second-hand sales. And, of course, the offering is now much wider than just books, with the combination of a trusted brand, and sheer presence, providing a marketplace joining consumers to a wide range of product providers. There is some evidence that this is happening in New Zealand, where it is possible for households to be directly exposed to the wholesale market (eg, via the retailer, Flick Electric). The next stop is for individual households to trade energy with each other across the AMI technology, which some companies are experimenting with already (such as Reposit Power in Australia)[12] and the ability to do this via online communities is being actively researched (Bourazeri et al., 2012).[13]

Smart grids, microgrids, local generation, and local storage all create opportunities to engage customers in new ways—for example, see the discussion of mircogrids (chapter by Marnay in this volume). Increasingly, we are seeing interest in the power sector from companies in the online, digital, and data management world, who are looking at media and entertainment, home automation, energy saving, and data aggregation opportunities. In a grid-connected but distributed power system, there are roles for intermediaries that can match supply and demand, rather than meet demand itself.

A key consideration for incumbent power utilities is if their brands are perceived as being part of the past that is being broken away from, rather than the future for customers. An energy saving or demand management proposition may be perceived as more credible, coming from a new entrant rather than an incumbent, so use of the brand needs to be carefully considered. At the same time, old brands could become valuable again as a cacophony of new brands emerges and causes confusion.

Another important challenge for companies arises from the need to be an expert in managing data in a smart home, smart city, and smart company

---

11. https://www.eonenergy.com/for-your-home/saving-energy.
12. http://reneweconomy.com.au/2014/households-to-use-solar-storage-to-trade-electricity-with-grid-99142.
13. See also the contribution by Nelson and McNeill in this volume.

environment. As well as data from smart devices and the grid, additional layers of information about demographics, behavior, customer characteristics, and other factors will often be required to exploit the data opportunity best. Many power utility companies already use sophisticated data analytics for customer segmentation purposes, and this can be built on and supplemented by enhanced analytics, big data from social media, and learning from other industries.

However, as consumers become more conscious of the value and potential data security issues associated with AMI technologies, there may be pressure to have more control of their own data, and who has access to it. Active consent for data collection and analysis may need to be sought, and consumers may be able to access and sell their data on to new entrants offering new services reducing the natural data analysis advantages of current utilities.

2.5 Government and Regulation

Energy is, by its nature, a key economic and political issue. More than in many other sectors, firms in the power sector depend on the political context for their license to operate and public trust in their activities is a big factor. The cost of power is an important element in household budgets, as well as business and industrial competitiveness. The availability of power is a "make or break" matter for everyone. So it is inevitable that the activities of power utility companies are never far from the center of the public and political spotlight. Recent and current events in different countries discussed in the earlier section (and in other contributions in this book) on the production service model highlight the potential for the public and political will to alter the nature of the business.

The political context shapes the utility business model. Changes in that context can impact utilities dramatically. This has always been the case but, in a more dynamic energy transformation context, political and regulatory decisions become even more significant. The different political approaches to energy transformation in different countries are key to explaining why the impact on fossil and nuclear generation has been faster and more dramatic in Europe, compared to elsewhere—with Germany showing the most noticeable and dramatic impact.

Political and regulatory concerns seem likely to impact the energy transformation in three important ways. There will be concern about data privacy and security, impacts on vulnerable customers, and financial regulation of entities selling energy services (Pollitt, 2016). These will necessarily limit the scope of action of all players in the energy sector and will offer the potential for negative technological impacts to result in regulatory intervention, which affects substantially the pathways of economic development followed in individual countries.

2.6 The Need for Innovation

Incumbent companies that do not innovate could risk seeing themselves succumbing to the pressure points, and being eclipsed in the same way that incumbents like Kodak, Blockbuster video stores, and high street booksellers were in other sectors. Certainly, sector transformation could shrink the role of some power utility companies to providers of back-up power. The energy transformation will gather pace, and growth will become more innovation-dependent, with success coming to those companies that use innovative technologies, products, services, processes, and business models to gain competitive advantage, to stay ahead of change, and create new markets for their products and services. In Africa, but also in many other places, innovation will be driven, in part, by the fact that power utilities will not be able to support the increasing demand for electricity supply, and businesses will evaluate other solutions, such as looking at different means of cogeneration.

Business model innovation is just one element of the innovation required, but is likely to be a key part. The difficulty is that business history tells us that the majority of business model innovations are introduced by newcomers, and incumbents often find it hard to respond successfully. Incumbent companies sometimes try to hold on to the existing model for too long, or fall between two stools as they try to manage two competing business models at the same time—the original business model and the new model (Nillesen et al., 2014).

2.7 Avoiding a Capital Crunch

As well as innovating, utilities need to make sure that a weakened investment case does not close off growth routes. Some utilities have suffered significant rating downgrades and have seen their value diminish substantially. In 2014, 3% of the regulated utilities in the USA had an S&P rating of A or higher, whereas in 2007 that was still 13% (Edison Electric Institute, 2014). In 2013 Moody's, for example, revised its rating methodology to address explicitly the implications of distributed generation.[14] Others have had to deleverage, reducing debt relative to cash flow, in order to maintain credit ratings. These developments, primarily affecting companies in Europe and some in Africa, come at a time when they also face major capital investment challenges to replace aging infrastructure, as well as make energy transformation investments, such as smart grids. In parallel, many such utilities need to deploy capital to pursue diversification away from mature, low, or flat-growth markets, toward fast-growth regions. This leads to a stretch between shrinking margins and stranded conventional assets, and a need to reinvest in new technologies, infrastructure, and renewable energy.[15]

In the United States the challenge for power companies has been to convince investors that peak stock valuations can be maintained, a key part of which will be to demonstrate that they can negotiate the challenges of energy transformation without facing the kind of conditions that have engulfed their European

---

14. See https://www.moodys.com/research/Moodys-Warnings-of-a-utility-death-spiral-from-distributed-generation--PR_312101.
15. See for example, The Economist (2013), How to lose half a trillion euros.

peers. Innovative and alternative approaches to financing are becoming more commonplace in the sector. Partnerships and strategic tie-ups with sovereign wealth funds, insurance and pension funds, already becoming more numerous, are likely to increase in importance.

3 FUTURE UTILITY BUSINESS MODELS

In defining future business models, in the world outlined earlier, companies need to first understand and challenge their company purpose and positioning in the markets of the future. This is called "blueprinting the future" (Fig. 15.2) and consists of several fundamental steps, starting with defining "where to play" in terms of business segments, markets, products, and services. Core, adjacent, and growth market participation areas are assessed based on attractiveness, capability to compete, and potential for profitable success. Next comes assessing "how to play" in these selected areas, defining the go-to-market strategies to be adopted by participants in pursuing their market aspirations, for example, new products, innovative unbundled pricing. Subsequently, the focus is on the most important dimension of the blueprint, "how to win." This element defines the particular tailored approach that is most appropriate for a company in order to achieve competitive market success, such as partnering or channel expansion.

To evaluate fully the aforementioned choices, companies need to examine their current core capabilities against the type and level necessary to compete effectively, and prosper in a more decentralized and disaggregated marketplace. In particular, incumbents and new entrants need to take stock of which capabilities are distinctive and differentiable, such as asset management or regulatory prowess, and which may need to be developed or strengthened, for example,

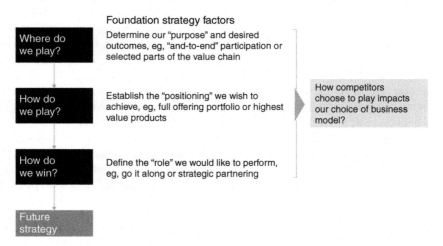

FIGURE 15.2 Blueprinting the future. *(Source: PwC, 2014, p. 17.)*

innovation or commerciality. This is referred to as "Capabilities-Driven Strategy" (Leinward and Mainardi, 2010).

The energy value chain of the future will be more interconnected than ever before.[16] This value chain forms an integrated ecosystem of unique elements that are highly interrelated. Incumbents will need to focus on extending beyond independent views of each value chain element into a more integrated view of how these elements can interact with each other in the future, for example, how the benefits of increased knowledge about system performance can bridge the gap to enhance the customer experience. Nontraditional entrants will need to determine how they interact between incumbents and customers, in a manner that does not "island" assets or "diminish" customer relationships.

3.1 Traditional Core Business

Alternative business models of the future may be very different from the traditional model that dominated power and gas delivery for decades. In the past, operating an integrated utility from generation through to customer supply was well understood, because the utility controlled the entire value chain. However, this model has been supplanted in many countries through market restructuring, and may be rendered further obsolete through the convergence of distributed technology and customer engagement.

In the traditional model, both tangible assets and franchise customers were considered important to preserve the benefits of physical integration, economies of scale, and access simplicity. Often regulated prices and monopoly service franchises guaranteed revenue collection. As policies encouraging competition emerge, to take advantage of market options or regulatory mandates, specific segments of the value chain became available for specialization and for new entry. Now, unbundling opportunities are starting to extend deeper into the value chain, and enable more specialist participation.

In emerging business models, a greater emphasis will be placed on obtaining higher margin from prices/revenues, rather than cost reduction, to get higher earnings and profit growth. Depending on how a traditional utility thinks the electricity industry may evolve in its country/region, and what market models may emerge, it needs to evaluate where to play across the value chain. Should a traditional utility leverage multiple business models? And if so, which ones, and how should they transform their business to be successful?

The succeeding sections identify eight alternative business models (Fig. 15.3), which are described with respect to their scope, rationale, basis for competition, and source of earnings. This should help utilities think through which business model options might be right for them, and the key decisions required to enable them to develop their new market position in sufficient time.

---

16. This process has already developed massively in the telecoms sector. See Li and Whalley (2002).

	Description	Capabilities required		Description	Capabilities required
Gentailer	Own generation assets and sell retail energy to customers in a competitive market	• Market insights • Project development • Energy trading/hedging • Origination, product development and pricing • Customer management	Product innovator	Offers electricity and behind-the-meter products such as solar, fuel cells, EV chargers, and smart devices	• Customer acquisition, retention, cross-selling • Pricing and bundling • Energy sourcing • Customer service
Pure play merchant	Owns generation assets and sells power into competitive wholesale markets or through bilateral contracts	• Market insights • Project development and finance • Origination • Energy trading/hedging • Risk management	"Partner of Partners"	Offers standard power and gas plus a range of energy services using high quality, branded partnerships	• Customer acquisition, retention, cross-selling • Partner management and service delivery • Innovative financing • Customer service
Grid developer	Acquires/develops, owns and maintains transmission assets connecting generators to distribution systems	• High voltage line design and operation expertise • Real-time electricity supply/load management • Stakeholder management • Investor relationships	Value added enabler	Use score "big data" capabilities to provide enhanced energy services to customers not wanting to actively manage their energy use	• Big data management, analytics, and security • Ability to educate and guide customers and deliver value
Network manager	Operates transmission and distribution assets and provides network access to generators and retail service providers	• T&D line design and operation expertise • Real-time supply/load management and central/DG resource integration • System data analytics	"Virtual" utility	Aggregates generation from distributed systems and acts as intermediary between/with energy markets without owning generation/T&D assets	• Energy sourcing and real-time balancing • Customer management • Partner development and management • Demand management

FIGURE 15.3   Overview of emerging business models across the value chain. *(Source: PwC (2014).)*

3.2   Gentailer Model

The gentailer model is typically applicable in markets where the generation and retail portions of the value chain are competitive and the transmission and distribution companies operate as a regulated monopoly.

Australia, the United Kingdom, and New Zealand are countries that have deployed this model successfully. In New Zealand, the five major generators are also the top five retailers. This type of market development often has regulatory implications, which are likely to influence future development of this business model. This model has also developed in areas where traditional IPPs have moved into retail energy sales, as energy markets have deregulated. NRG Energy and NextEra Energy are two examples of utilities operating in this model in the United States that developed or acquired retail capabilities to complement their generation positions.

Companies that want to be successful gentailers will need to do a number of things:

● Monitor and understand the way different customer segments use smart technology and assess how to harness these preferences into mobile and/or tariff solutions.

● Monitor regulatory change enabling transmission and distribution operators to develop, own, and operate distributed energy resources.

● Monitor new entrants offering products/services that increase customer engagement in managing their energy needs through mobile tools or social media.

● Identify potential partners to help develop behind-the-meter distributed energy resource business plans.

- Review make/buy decisions to support their asset position and investment requirements under alternative market scenarios to determine what generation products to offer in the future.
- Invest in understanding customer segmentation and what that means for switching rates and retaining the most profitable customers.
- Develop a behind-the-meter distributed energy resource business, organically or through networks partnerships, or acquisitions.
- Increase customer engagement via new energy management solution offerings and intelligent tools.
- Develop alternative pricing packages and approaches that provide greater optionality or risk/reward tariffs to customers.

3.3   Pure Play Merchant Model

In the Pure play merchant model, players prefer liquid markets with rising and/or high peak wholesale energy prices and high price volatility. Examples include deregulated regions like Texas, California, and New England in the United States, and countries in emerging markets like Chile.

Areas with low natural gas and coal prices are typically not well suited for merchant players because these low-cost inputs often depress wholesale energy prices and do not provide significant "spread" for merchants to leverage.

Companies that want to be successful pure play merchants will need to do a number of things:

- Implement world-class operational procedures to minimize costs of operations, manage price and volume risk exposure.
- Develop robust investment plans to create a balanced generation portfolio, either across technologies or markets.
- Investigate alternative products to offer from the generation portfolio in order to mitigate against market change or merit order structure.
- Assess feasibility of expanding capacity of traditional fossil fuel plants with solar or energy storage in order to expand ability to play using same grid interconnection.
- Explore options for developing distribution level or behind-the-meter projects (eg, solar, charging infrastructure).
- Evaluate investment to enhance plant operational capabilities to capture value of flexible thermal capacity.
- Development of disruptive grid-level solutions, such as energy storage that compete with merchant generators.
- Development of ultra-efficient generation and excellence in operations.
- Identification of profitable regions for future merchant potential, and key early investments (eg, land).
- Ability to cost-effectively plan, develop, finance, construct, and operate the right asset mix.
- Strong analytics and energy trading capabilities to hedge business risk.

3.4 Grid Developer Model

Grid developers are typically established by regulation in areas with existing infrastructure. Examples of this model include transmission system operators (TSOs) in Europe and independent system operators (ISOs) in the United States. Additionally, new grid developers may be formed in areas that lack sufficient transmission infrastructure between generation and load centers.

For example, a grid developer may be created to build and operate transmission lines between remote generation assets like a hydropower facility or wind farm and a distant urban area, or to provide new transmission infrastructure where there are transmission constraints. Examples of the newer grid developers in the United States include Electric Transmission Texas (ETT) and Clean Line Energy Partners (CLEP).

A successful grid developer has a very strong operating track record and excellent capabilities in designing, operating, and maintaining high-voltage transmission lines, and supporting infrastructure.

Companies that want to be successful grid developers will need to do a number of things:

- Identify new locations (within their own market or in new markets) for large-scale renewable generation, flexible thermal generation, and associated transmission build.
- Work with alternative owner classes, such as financial sponsors, to shape market bidding processes where competitive transmission protocols will exist in the future.
- Review existing contracting and procurement procedures to assess whether they are maximizing value for money, risk allocation, and whether they reflect regulatory settlements.
- Streamline grid connection processes to improve resource productivity and lower operating costs.
- Consider whether alliances with DSOs may provide economics of scale and increased scope for new investment.
- Develop close relationships with distribution system operators and regulators to support large-scale generation growth plans.
- Convince investment partners to invest in large scale, low-cost renewable project developments that require new transmission to urban areas.
- Implement world-class operational procedures to improve cost-effectiveness, and develop new forms of contracts with generators, load managers, and the supply chain that incentivize strong performance and cost management.

3.5 Network Manager Model

The network manager models typically exist in regions where the generation and retail portions of the value chain are competitive, and the transmission and distribution companies operate as a regulated monopoly. Australia, the

United Kingdom, and New Zealand are countries that have successfully deployed this model, although, with the expansion of distributed generation and microgrids, we expect this role to expand and become more relevant in many global regions.

Companies that want to be successful network managers will need to do a number of things:

- Work with regulators to approve investments focused on advancing system capabilities (smart grid).
- Harden the infrastructure through targeted investment.
- Advance reliability standards and awareness.
- Develop partnerships with distributed generation developers to install, operate, and manage network connections.

3.6   Product Innovator Model

The product innovator model will be most relevant in markets where the regulatory framework allows choice and the level of customer acceptance of new technologies and products is high. A market with a high penetration of distributed energy will be attractive for a product innovator who can help provide products that enable or complement distributed energy. Examples of product innovators that have moved beyond pure energy supply include Direct Energy and TXU Energy in the United States and Powershop in New Zealand.

It is expected that companies will enter into discussions with providers from other markets, for example, Google Nest, to discuss how to partner around customer product development and provisioning, rather than default to automatic competition.

Companies that want to be successful product innovators will need to do a number of things:

- Develop a compelling product suite, including everything from the commodity to the home device.
- Develop opportunities for superior products that integrate with other needs, such as home or building security or automation.
- Invest in data analytics tools to develop excellence in customer acquisition and retention.
- Enhance brand reputation to support product expansion.
- Keep customer acquisition costs low, and operating costs overall low (eg, attractive energy supply contracts or own energy supply).
- Diversify the product portfolio beyond commodity products to preserve margin.
- Establish partnerships to enhance product range.
- Focus on retention of profitable and high-volume customers, and minimize switching.

3.7 "Partner of Partners" Model

The "partner of partners" model is most relevant in markets where there is a high proliferation of energy technology and choice, and customers are seeking ways to simplify their lifestyle while lowering upfront costs. A market with a high penetration of distributed energy is attractive for a "partner of partners" who can help provide simple and innovative service-based solutions. One example of a "partner of partners" model is NRG Energy in the United States, with its eVgo and Sunora offerings. Few utilities have embraced this model, as it involves nontraditional partnership arrangements.

Companies that want to be successful partner of partners will need to do a number of things:

- Develop a compelling suite of services, and identify the right solution provider partners.
- Create a range of relationships with solution partners.
- Expand the range of channels to market that can be leveraged.
- Develop bundles of offerings targeted at the connected customer.
- Enhance their brand value.
- Keep cost of service low, such as cross-selling multiple products to a single customer.
- Keep customer satisfaction high by establishing clear customer service standards, pinpointing and addressing customer pain points quickly, and offering a wide range of innovative yet convenient services.

3.8 Value-Added Enabler Model

Most utilities have performed value-added roles that are knowledge-based in the past, specifically around energy efficiency programs or energy management in industrial processes. The level of knowledge-based energy management anticipated in this model, however, extends the scope and scale of these activities dramatically into the mass market, to a much deeper level.

Manufacturers such as Honeywell and Mitsubishi are focused on addressing certain control elements, like power monitors and smart thermostats. Others, such as Google Nest, are playing in a similar vein with the objective of providing customers a "set and forget" experience, leveraging their massive data centers to provide real-time and predictive energy consumption data. This space is relatively wide open to utilities, particularly if they understand how to leverage system and customer data to provide premise and action-based insights and solutions. Utilities may also benefit from constraints on third party data usage, under data privacy laws, enabling them to offer solutions or to become a partner of choice for solution providers.

Companies that want to be successful value-added enablers will need to do a number of things:

- Build-out "big data" platforms.
- Educate customers on energy management options.

- Design innovative methods for sharing customer benefit.
- Deploy data synthesis methods and tools.
- Develop data security protocols.
- Enter into partnerships with data managers to improve competitive position.
- Robust data analytics systems.
- Pricing mechanisms for performance.
- Decision tools for customer adoption.
- High quality data security management systems and protocols.

3.9 Virtual Utility Model

Markets with high penetration of generation connected at distribution level (Germany, US states of Hawaii and California) or a regulatory setting that offers a high degree of freedom for customer choice (US states of New York or Texas, the United Kingdom, and Australia) are ideal for the virtual utility model. Island systems and remote systems are also ideal markets for this business model. Utilities may combine distributed generation with their own generation, providing a route to market for independent generators, and expanding their own asset portfolios.

Companies that want to be successful virtual utilities will need to do a number of things:

- Participate in distributed generation and storage projects.
- Develop partnerships with technology-based providers.
- Increase customer engagement via intelligent tools.
- Define future roles for incumbents in concert with regulators.
- Offer structure and pricing of a wide range of energy products (low cost, high renewables, local generation etc.).
- Ability to trade energy (sourcing and consumption).
- Partnering with key players across the energy value chain.

The different business models described earlier have a different business focus, business alignment, and profitability basis. A summary of the models is given in Fig. 15.4.

4 CONCLUSIONS

The five global megatrends introduced at the start of this chapter are currently working their way through the energy sector. At the heart of this disruption is new technology and new applications of existing technology. The changes bring existential challenges for incumbent utilities, but also opportunities for the traditional players and new entrants alike.

Utilities may choose from a range of paths to move forward from where they are today. But business model clarity may be difficult to achieve, as a lot of uncertainty exists on how future markets may develop and mature. Thus, multiple

Business models	Business focus	Business alignment	Profitability basis
Traditional core business	Assets—customers	Generation—T&D—retail	ROIC
Gentailer	Assets—customers	Generation—retail	Competitive margin
Pure play merchant	Assets	Generation	Competitive margin
Grid developer	Assets	Transmission	Regulated ROIC
Network manager	Assets	Transmission—distribution	Regulated ROIC
Product innovator	Customers	Retail	Competitive margin
"Partner of Partner"	Customers	Retail	Competitive margin
Value-added enabler	Customers	Retail	Competitive margin
"Virtual utility"	Customers	Distribution—retail	Competitive margin

FIGURE 15.4 **Summary of business models and profitability basis.** *(Source: PwC, 2014, p. 19.)*

models may need to be deployed to meet diverse market needs or specific regulatory structures in various countries, or even jurisdictions.

While not always obvious, companies have several levers that can be used to advance their readiness for the future and position themselves for success, such as ownership of customer and grid data, policy influence and regulatory know-how, long-term existing relationships across the sector, and large, though damaged, balance sheets and cashflows. These can help leverage the successful development of their future business models.

At the same time, it is important for companies to recognize that future markets are likely to create networks of participants in new partnerships and collaborations that become a norm of the go-to-market models. Companies can take advantage of these levers to strengthen their starting point for defining their future roles, and for subsequent market participation. For example, utilities already hold large quantities of data that have not been utilized effectively, giving them the opportunity to add value through better utilization and communication of this data.

Similarly, utilities are a natural collaborator with regulators in the shaping of responsive policies to accomplish public interest objectives, including how to enable customers to achieve greater control and choice. Utilities are also attractive partners to new entrants that wish to offer high-value products, but do not wish to support them in the manner customers are used to from their utility. Utilities will need to determine "where" it makes sense for them to participate in the future energy market, and "how" they can best position themselves for success.

There is no single business model that will be the panacea for every utility. Rather, they will have to be adaptive to the development of the marketplace, and the evolution of the connected customer. Just as utilities are unsure of market direction, customers are equally uncertain of what really matters to them in energy decision making.

These gaps between foresight and expectations provide the "open seas" where utilities can forge new business models that fundamentally reshape the historical relationship with customers, and position incumbents for a broader and more value-creating future.

REFERENCES

Agora Energiewende, 2015. Report on the German Power System. http://www.agora-energiewende.de/fileadmin/downloads/publikationen/CountryProfiles/Agora_CP_Germany_web.pdf

Beatty, R., 2014. NZEM—Retail Transformation, Presentation to Australian Utility Week, 19 November.

Bourazeri, A., Pitt, J., Almajano, P., Rodriguez, I., Lopez-Sanchez, M., et al., 2012. Meet the meter: visualising smartgrids using self-organising electronic institutions and serious games. Sixth IEEE International Conference on Self-Adaptive and Self-Organizing Systems Workshops (SASO). IEEE, Lyon, pp. 145–150.

Caldecott, B., McDaniels, J., 2014. Stranded Generation Assets: Implications for European Capacity Mechanisms, Energy Markets and Climate Policy. Working Paper, January 2014. Smith School of Enterprise and the Environment, University of Oxford.

Edison Electric Institute, 2014. Edison Electric Institute Credit rating analysis Q2 2014. http://www.eei.org/resourcesandmedia/industrydataanalysis/industryfinancialanalysis/QtrlyFinancialUpdates/Documents/QFU_Credit/2014_Q2_Credit_Ratings.pdf

US Energy Information Administration (EIA), 2014. Annual Energy Outlook 2014 (AEO2014) reference case.

Goossens, E., Chediak, M., Polson, J., 2014. Why Google, Comcast, and AT&T are Making Power Utilities Nervous. Bloomberg Businessweek, May 29, 2014.

Grapy, E., Kihm, S., 2014. Does disruptive competition mean a death spiral for electric utilities? Energy Law J. 31 (1), 1–44.

Leinward, P., Mainardi, C.R., 2010. The Essential Advantage: How to Win With a Capabilities-Driven Strategy. Harvard Business Review Press, Harvard, NJ.

Li, F., Whalley, J., 2002. Deconstruction of the telecommunications industry: from value chains to value networks. Telecommunications Policy 26 (9–10), 451–472.

Nillesen, P., Pollitt, M., Witteler, E., 2014. New utility business model: a global view. In: Sioshansi, F.P. (Ed.), Distributed Generation and its Implications for the Utility Industry. Elsevier, Amsterdam.

OECD/IEA, 2013. Redrawing the Energy-Climate Map: World Energy Outlook Special Report. OECD/IEA, Paris.

Pollitt, M., 2016. The future of electricity network regulation: the policy perspective. In: Finger, M., Jaag, C. (Eds.), The Routledge Companion to Network Industries. Oxford, Routledge, pp. 169–182.

PwC, 2014. The road ahead—Gaining momentum from energy transformation. http://www.pwc.com/gx/en/utilities/publications/road-ahead-gaining-momentum-energy-transformation.jhtml

Seto, K.C., Dhakal, S., 2014. Human settlements, infrastructure, and spatial planning. In: Climate Change 2014: Mitigation of Climate Change. Contribution of Working Group III to the Fifth Assessment Report of the Intergovernmental Panel on Climate Change (Chapter 12).

State of New York Department of Public Service, 2014. Developing the REV Market in New York: DPS Staff Straw Proposal on Track One Issues, CASE 14-M-0101, August 22, 2014.

Chapter 16

European Utilities: Strategic Choices and Cultural Prerequisites for the Future

Christoph Burger, Jens Weinmann
European School of Management and Technology (ESMT), Berlin, Germany

1 INTRODUCTION

In the introductory chapter of this book, the question emerges how (and when) energy incumbents will react and reposition themselves to tackle the challenges that have emerged over the last couple of years. This chapter is an attempt to provide an answer for European energy utilities.

For those utilities, it is not the first fundamental change that they have experienced and overcome. In the early 2000s, European players successfully managed the transition from a regulated, state-controlled regime to liberalization. The creation of wholesale and retail markets injected competitive momentum in the electricity and natural gas supply industry. The new system did not hamper revenue streams of most players.

Since then, the competitive landscape has changed: The International Energy Agency (IEA) estimates that by 2020, grid parity, which means photovoltaic power will have the same price level as electricity retail prices, will be reached in many markets across the globe. Renewable generation may increase from 20% to a quarter of global gross power generation between 2011 and 2018 (IEA, 2015).

In Germany, for example, more than 1.5 million private, nonutility owners of energy-generating assets, privileged by priority feed-in, participate in the market (also see chapter by Löbbe and Jochum). At zero marginal cost, renewable energies like wind or solar exert a measurable downward pressure on wholesale market prices. Generation with conventional power plants does not yield the revenues the utilities were used to, a "perfect storm," as Peter Terium, the CEO of German utility RWE, notes in the Preface of this book. Some companies have even started to mothball newly constructed and highly efficient gas-fired power plants.

Future of Utilities - Utilities of the Future. http://dx.doi.org/10.1016/B978-0-12-804249-6.00016-6

Energy companies in continental Europe have realized that their chances to prosper or at least maintain their current status are shrinking rapidly. Given limited financial resources, utilities face the choice whether they want to expand their existing business model into growth markets beyond their current geographical extension, or whether they want to transform themselves into Utility 2.0, which implies becoming a provider of service solutions.

Most major utilities in Europe attempt to pursue both strategies simultaneously, but differ in which path they emphasize, both in terms of actual financial commitment and corporate communications.

In Sections 2 and 3 of this chapter, diverging strategies of European energy incumbents are identified. Both internationalization strategies and the move toward Utility 2.0 are analyzed, using a case-based, qualitative approach that highlights priorities of selected European energy companies. In Section 4, a categorization of European energy utilities, according to their dominant strategies, is established. Examples of niche players and strategies of incumbents from other industries, which enter the new markets of Utility 2.0 are presented in Section 5, and key organizational prerequisites for utilities of the future are deducted. The chapter concludes with a long-term perspective on the future of energy markets.

2 INTERNATIONALIZATION STRATEGIES

2.1 Differences in Internationalization Strategies

The system of independent nation states that characterized the decades after World War II in Europe was reflected in electricity supply industries. Cross-border trade occurred in limited instances, expansion into adjacent markets was no strategic objective. The situation fundamentally changed with the advent of liberalization in the 1990s. Utilities were given the choice, partially even encouraged by governments, to acquire assets abroad.

The internationalization of European utilities occurred in two waves:

The first wave started in the late 1990s and early 2000s. During that decade, major European utilities used their financial endowment and credibility on stock markets to do large-scale acquisitions. Privatizations became a major opportunity for European companies to extend their reach. First, they focused on Latin America and the Caribbean, where in particular Spanish utilities, but also companies like EDF, acquired conventional generation and distribution assets. By the mid-2000s, privatizations in Eastern Europe and Central Asia gained momentum. The largest investments were undertaken by E.ON with US-$6.2 billion and Enel with US-$3.8 billion, whereas the most active utilities in terms of numbers of deals were EDF and RWE, with seven and six acquisitions, respectively. Meanwhile, some consolidation took place in the European market, for example, when Italian utility Enel became the major stakeholder of Spanish power company Endesa, in February 2008.

Fig. 16.1 shows the number of acquisition deals and the total value of their acquisitions of selected European players during the privatizations in Eastern Europe and Central Asia.

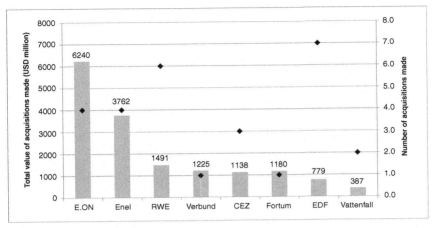

FIGURE 16.1 **Acquisitions of major European energy incumbents during the privatizations in Eastern Europe and Central Asia.** *(Source: Authors, based on data gathered by analyst reports and utilities' annual reports.)*

The second wave of internationalization started in the later part of the 2000s and early 2010s, albeit by a slightly different set of players. Companies like Iberdrola, EDP, or E.ON, specifically targeted renewable energy companies abroad. Examples are Iberdrola's acquisition of Rokas Group (Greece) in 2004; EDP's acquisition of wind power capacity from Hydra Energy (USA) and EOLE 76 Group (France) in 2008; and E.ON's acquisition of Energi E2 Renovables Ibéricas and Airtricity (USA and Canada) in 2007. M&A value of wind, solar, and renewable enablers in Europe grew from below US-$2 billion in the first quarter of 2009 to about US-$8 billion in the fourth quarter of 2014, the number of deals from less than 20 to about 80. Germany, France, and the United Kingdom were leading the development, often with nonutility players establishing their presence in the market. Assets went from developers and aggregators to "destination owners," such as pension funds (Impax Asset Management, 2015). In addition, electric utilities became targets of companies from outside Europe. Most prominently, China Three Gorges acquired a 21% stake of Portuguese utility EDP for US-$3.5 billion, in 2011, beating the bids of E.ON and Brazil's incumbent utility Eletrobras (Bugge, 2011).

Over the last 20 years, many European utilities have reached a high degree of internationalization. Even smaller players like Swiss utility Axpo are present in more than 20 markets (based on company data from 2015). An analysis of the origin of revenues of 16 major players confirms that half of them earned more than 50% of their revenues outside their country of origin. Dong, Vattenfall, GDF-Suez (Engie), E.ON, and Enel lead the batch in terms of internationalization with more than 60% of revenues generated abroad. Only a quarter of the utilities in the sample generated a significant majority of revenues domestically, namely Fortum, CEZ, EnBW, and SSE. Fig. 16.2 shows the ranking of the sample in descending order.

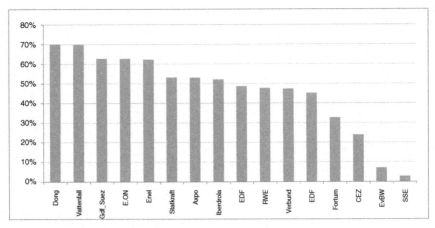

FIGURE 16.2   Foreign (nondomestic) revenues as percentage of total, by customer location (2014). *(Source: Burger and Pandit (2014).)*

A closer analysis of the geographic distribution of nondomestic revenues reveals, though, that most utilities have strong positions in countries adjacent to their home market. Swedish utility Vattenfall, for example, focuses on three main markets: the Nordics, Germany, and the Netherlands, according to company statements: "Vattenfall has a top three position in all of these markets, and together they account for around 95 per cent of our total cash flow. Aside from Vattenfall's three main markets, the UK is a growth market, primarily in wind power" (Vattenfall, 2014). Similar to Vattenfall, German utility RWE comments: "Europe is and remains the focus of our electricity and gas business. Our most important markets are Germany, the Netherlands, the United Kingdom as well as Central Eastern Europe. In the field of electricity generation from renewables, we are also active outside our core markets, for instance in Spain and in Italy" (RWE, 2015).

This regional focus is also described in Kolk et al. (2014). The authors identify "policy harmonization and market integration at the regional level" as drivers for the core market strategies of many European utilities, and detect an "incomplete home-region internationalization" (Kolk et al., 2014, pp. S90/S91).

However, a small number of European utilities pursue biregionalization or internationalization strategies, which means that they have major operations in at least one more region beside their home region. The following two examples highlight two different strategies, one based on fossil, the other one on renewable energy generation.

- Engie (GDF-Suez): Internationalization based on conventional (fossil) assets
  GDF-Suez, in early 2015 rebranded as Engie, was the result of the merger of Gaz de France and Suez in 2008. Two years later, it acquired UK-based energy company International Power and integrated its own operations of the United Kingdom and Turkey into the new subsidiary.

International Power's history begins as early as 1993, when its predecessor, National Power, started to invest abroad. In 2000 domestic and international operations of National Power were split into the UK business, renamed Innogy/npower, and international business, renamed International Power, the legal successor to National Power. Over the next 10 years, International Power expanded its generation portfolio with investments in fast-growing markets such as South America, the Middle East, South-East Asia, and Australia. When GDF-Suez took control of International Power in 2011, it added around 35 GW of generation capacity to its own 82 GW (Jacobs and Nair, 2012). The acquisition created the world's largest independent power producer. International Power got rebranded as GDF-Suez Energy International in 2012.

- Iberdrola: Internationalization based on green (renewable) assets
  Spanish utility Iberdrola has successfully established itself as international player in the field of renewable energies. Being already a privately owned company, even before the European liberalization of energy markets at the end of the 1990s, Iberdrola started its green internationalization strategy at the beginning of the 2000s with the advent of a new CEO, Ignacio Galan. After having acquired stakes in several Spanish wind power companies, Iberdrola bought the largest wind power producer in Greece and started installing wind farms in Morocco. In 2006 it acquired US wind power producer Community Energy, and started to expand into the US market. Its acquisition activities temporarily increased the debt burden, but after some divestments, rating agencies and shareholders regained confidence in the corporate strategy of the utility.

  At the beginning of 2015, its renewables business had an installed capacity of more than 14.5 GW, of which onshore wind power accounted for 96%. While the company still has the highest amount of installed onshore capacity in its domestic market in Spain, with 5.8 GW, investments in the USA follow closely, with 5.6 GW. Behind NextEra Energy Resources, Iberdrola Renewables is the second-largest wind operator in the USA. Its other main markets are the United Kingdom and Ireland with 1.4 GW and Latin America with 0.4 GW (Iberdrola, 2015).

  Similarly, Portuguese utility EDP mirrors Iberdrola's green internationalization strategy.

2.2 Observations on Internationalization Strategies

Despite differences among the internationalization strategies of European players, some similarities can be identified:

- From acquisitions to alliances
  The cases presented here suggest that most internationalization strategies have rather been undertaken by acquisitions, than by cooperations with local partners. Especially in the field of renewable energies, utilities benefit from the fact

that markets are often fragmented, with a range of smaller private players, rather than large energy companies, as a dominating force. As acquisition targets, these smaller players also serve as channels of information on the regulatory idiosyncrasies and the competitive environment of the host country or state.

However, over the last couple of years utilities started searching local firms for cooperation, when they expanded into new territories or technologies. For example, EDF Energies Nouvelles teamed up with wpd offshore for wind energy projects off the French coast, and Iberdrola with SunPower to build a solar power plant in Arizona. The increasing tendency to enter cooperations reflects a change in the mindset of utilities, which often tried to develop inhouse solutions, rather than joint products and services. While acquisitions have the advantage that market entry is fast and (relatively) efficient, cooperations often serve as a platform for exploiting synergies in complex technology fields (Musiolik and Markard, 2011; Powell and Grodal, 2005). They are more pronounced in strategies to become a Utility 2.0, which will be discussed Section 3.

- Diversification of regulatory risk
  One of the major uncertainties of investments in renewable energies is regulatory risk. Prominent cases of governments altering or even abandoning subsidy schemes have occurred. For example, the Spanish government's feed-in system was introduced in 2007, but sharply reduced in Sep. 2008 because it proved too successful with private investors, and completely suspended in 2012. The strategy that companies such as Iberdrola pursue, when they invest abroad, is geographic portfolio diversification to reduce overall regulatory risk. The fragmented regulatory structure of, for example, the USA makes it possible to hedge against unpredictable state interventions successfully. For example, Iberdrola's North American subsidiary operates in 24 US states, from New England to the West Coast.

  As many other utilities, Iberdrola considers the low-margin business of running the grid a secure and reliable source of income—and a further effort of risk reduction. In 2013, Iberdrola generated 76% of its core earnings from grids and wind. News agency Reuters quotes CEO Ignacio Galan: "We have much more transmission and distribution than power generation. We are more like National Grid than like E.ON" (Clercq, 2013).

- State ownership versus internationalization
  Governments may doubt whether risky investments abroad are justifiable. As major shareholders, they may prefer a concentration on the domestic market. RWE and Germany's third largest electricity supplier, EnBW, are majority-owned by municipalities and the government of the South German state of Baden-Württemberg, respectively. Compared to other European players, they only show cautious investments in markets outside Europe, where both companies have operations only in Turkey.

  State-owned utilities may also exhibit less cautious strategies, though, as Swedish company Vattenfall shows. In 1996, it started to expand into other European countries, first Finland, then Germany, Poland, and later the Netherlands. Over the last couple of years, announcements of planned divestments in its main foreign markets hint that a strategic shift toward concentrating on the safe harbor of its home territory may occur.

Both extremes—RWE and EnBW on the one hand, Vattenfall on the other—show that public ownership exerts some influence on corporate decision-making with respect to internationalization.

- Postacquisition integration—the lack of mobility of the workforce

As in any industry sector, a successful internationalization strategy of an energy company is affected by cultural differences between acquirer and target, such as working habits, dealing with hierarchies, orders, and the legal system. Many privatizations of European players in Eastern Europe and Central Asia needed several years before their endeavors became commercially successful.

In addition, recruitment of employees who are willing to leave their country, in order to work abroad for a couple of years, may seem problematic among utilities, often rooted in the cultural footprint of their traditional demarcations during times of state regulation, according to personal communications with executives from European energy companies. This phenomenon may add to the focus on expansion into adjacent territories, or regionalization, described in Section 2.1.

As the preceding discussion illustrates, internationalization strategies of major European electricity companies started long before the crisis provoked by the emergence of decentralized, renewable energy. With stagnating or even shrinking domestic markets, in terms of energy demand, utilities have to explore new sources of revenues. Applying the same, proven business model abroad is, according to the data on nondomestic revenues in Fig. 16.2, an attractive path for many European players in the electricity sector.

3   STRATEGIES TOWARD UTILITY 2.0

Developing new streams of revenues in the domestic market appears to be an alternative—or complementary—path that a number of major European utilities have chosen. Fundamental changes in the business model are a recurring motive in corporate strategy. For example, IBM reinvented itself from being a computer manufacturer to a company that offers complex service solutions and IT consulting, while Finnish company Nokia switched from producing snow boots to cell phones. These changes are difficult to imagine in the power sector, where long product cycles prevail, where planning horizons often stretch over more than 25 years, and demand for the commodity electricity is relatively inelastic.

However, digitalization transforms all industries, changes the way humans interact, and blurs the line between final consumers and producers, persons and machines. The decentralized energy revolution would not have been possible without digitalization, and its consequences affect directly the business model of utilities: where can money be earned when electricity is produced in autonomous island grids, or even on a household-scale? What is the role of a utility in this new energy world?

One answer that the electricity supply industry currently explores is the transformation into Utility 2.0. According to Farrell (2014, p. 5), Utility 2.0 can be defined as "second generation electric company that can accommodate and thrive alongside distributed clean power generation, energy storage, and advanced energy management." Other definitions include a decoupling of economic prosperity from resource use, that is, enhancing energy efficiency instead of selling kilowatt-hours, customer engagement, and acting as an incubator (Vanamali, 2015). All definitions have in common that Utility 2.0 alters incrementally or radically the business principles of utilities as they exist today.

As a broad categorization, two trends in the Utility 2.0 strategy can be observed: the management of *assets*, often associated with the operation, and sometimes even ownership, of decentralized generation units, and the management of *information*, such as the smart grid, smart meters, and the like. Most European energy companies would claim that they are active in both segments, but their emphasis differs.

3.1 Management of (Decentralized) Assets

Of these two strategies, the management of (decentralized) assets is more closely linked to the traditional business model of utilities than the management of information, because electricity companies have always been handling the financing, construction (or supervision of the construction), and operation of power plants, transmission and distribution lines, and other capital-intensive assets. With decentralized energy, the scale changes, and consequently the qualifications of the workforce and the structure of revenues. Setting up small energy-generating units is more labor-intensive than building and operating large plants, and requires a higher degree of individualization and flexibility.

Two European players pursue a strategy to become leaders in the management of assets, German utility E.ON and French utility Engie (GDF-Suez).

- E.ON: Utility 2.0 based on the management of assets
  German energy utility E.ON has experience of more than 25 years in managing distributed energy assets. The company claims to have around 1 GW of distributed power and 3 GW of distributed heat in operation across almost 4000 sites in Europe (E.ON, 2015). In 2012 the company bundled its activities in managing energy-related consumption and production units at their customers' facilities or buildings, in a new business unit called "E.ON Connecting Energies." The new entity provides its services for companies such as Tesco, Marks & Spencer, or Metro (E.ON, 2015). Leonhard Birnbaum, head of the unit and member of the E.ON board, stated that, in 2013, the company's earnings stemming from decentralized units amounted to €400 million, and revenues were around €1 billion (Flauger, 2013).

- Engie (GDF-Suez): Utility 2.0 based on the management of assets
  Other companies like Engie (GDF-Suez) combine the management of assets with complex service solutions. In its "positive energy territory" strategy, the utility supports local authorities in initiatives to increase energy efficiency, lower energy use, installation of decentralized generation of renewable energy, and the optimization of local resources (GDF-Suez, 2015). In Jan. 2015, GDF-Suez announced the creation of a joint venture with the Chinese company SCEI DES (Sichuan Energy Investment Distributed Energy Systems), where the first natural gas distributed energy system in an industry park in Sichuan shall be established (GDF-Suez, 2015).

The examples of E.ON and Engie suggest that management of distributed generation assets targets a market where traditional utilities may benefit from a competitive advantage due to their technical competencies and brand recognition. It is a more cumbersome and fragmented way of revenue generation than running large power plants, though. E.ON's reported earnings in decentralized energies, in 2013, are dwarfed by earnings of the traditional business units.

3.2 Management of Information

Large amounts of data can be handled with greater ease than ever before. Where standard load curves provided approximations of the energy consumption of most users in the past, mass rollout of smart meters allows for insights and personalized data analytics, down to the level of each residential customer. The local optimization of energy flows in distribution grids becomes a necessary prerequisite to deploy an increasing amount of photovoltaics. In an IBM white paper, an electric utility representative is quoted as saying, "meter data management database holds 30 terabytes of information, including meter event data, alerts, health status updates and over 75 billion meter reads per year" (IBM, 2013). A subset of the management of information is the Internet of Things, including Machine-to-machine (M2M) communication (see chapter by Cooper). The Swedish communications technology company Ericsson predicts that "the global number of devices being managed by utility companies is projected to grow from 485 million in 2013 to 1.53 billion in 2020" (Ericsson, 2014).

In contrast to the management of assets, management of information represents a utility's move into a sphere where numerous multinational enterprises and competitors are already active, most of them from outside the sector. Motivations of these new entrants may include gaining insights about customer behavior, or exploiting new marketing channels. If new entrants do not compete with utilities, they may offer their expertise to those utilities, which are ill-prepared to handle and analyze the data, in particular smaller players like municipal utilities or regional distribution grid operators (see chapter by Hanser and van Horn). The income from these additional services is then inevitably lost for the energy companies.

Although most utilities in the European context have a relatively cautious stance toward innovation, some incumbents have developed smart solutions in-house or in close cooperation and interaction with OEMs:

- Enel: Utility 2.0 based on the management of information
  Enel's move toward Utility 2.0 received a major push by Telegestore, the deployment of smart meters across Italian households, which started already in 2001. The reasons for that move were manifold, but Enel quotes automatic reading of the meters, the reduction of nontechnical losses, that is, energy theft, and the possibility of switching off the electricity for nonpaying customers, as well as field operations, purchasing, and logistics as reasons. The utility reports investments of around €2.1 billion into the installation of around 31 million smart meters until 2008, and that the switch to smart metering led to a reduction of losses of around €500 million per year (Borghese, 2010).

  Since its experience in Italy, the utility has built on its experience in the new smart energy world: "The Group intends to confirm its position as a world leader in the development of smart distribution networks and will consolidate its key operations at global level. The objective is to increase both the number of customers served by around 6% by 2018, and the number of smart meters installed by around 30% up to approximately 50 million devices" (Enel, 2014).

  Enel actively uses its expertise abroad. The company started in 2010 to replace conventional meters in its Spanish distribution network that belongs to Endesa, 13 million meters in total. In Spain, it uses a more advanced technology than in its domestic market. It also pushes pilot projects outside Europe: in 2013 the company signed a "Memorandum of Understanding" for Smart Grids with ICT Europe and Advanced Electronic Company (AEC) from Saudi Arabia; in 2014 a similar memorandum followed with the Mexican Instituto de Investigaciones Eléctricas. Especially in emerging markets that suffer from high nontechnical losses, mass deployment of smart meters may occur despite low savings potentials—which provides multiple opportunities for Enel to build on its experience.
- RWE: Utility 2.0 based on the management of information
  German utility RWE, whose CEO contributes a Preface to this book, has been identified as a research leader among European utilities (Burger and Weinmann, 2012, 2014). The company pioneered public charging technology for electric vehicles. It actively participated in standardization committees that determined the configuration of the charging stations and the protocols deployed for roaming. With this decisive move toward the new technology, the company was able to become the European leader in the installation of public charging stations and has now 2700 charging stations in operation across 16 European countries and the USA, according to company reports (RWE, 2014). Although the business model behind the charging infrastructure is not yet clear, RWE has the advantage of building upon its technological expertise, which it can sell to third-party suppliers, such as municipal utilities

or other players in the electricity supply industry that have an interest in facilitating the mass rollout of electric vehicles, in particular in urban areas. For example, in a contract with the city of Amsterdam, the company's Dutch subsidiary Essent installed more than 260 charging points in the Dutch capital.

RWE also developed an integrated solution for smart homes. The system runs on an independent protocol. This technical choice has the disadvantage that only certified products can be integrated into the system, but it reduces the risk of any undesired intrusion by hackers, thus making it less vulnerable to cyber-attacks. The company does not release numbers how many of its smart home systems have been sold, though.

The preceding two examples show that power companies are capable of developing and disseminating innovations linked to the management of information. They are not able to rely on proven business models, though, but have to accept that the phase of experimentation (and sometimes failure) is far from over.

3.3   Observations on Utility 2.0

In comparison with internationalization strategies, the move toward Utility 2.0 encompasses more fundamental organizational and behavioral changes. The following observations may be indicative for the disruption that an electricity supplier may undergo:

- Alliances as prerequisite
  Alliances and partnerships are an integral element of becoming a Utility 2.0. In the area of electric vehicle charging infrastructure solutions, RWE teamed up with French electronics company Schneider Electric in early 2014. While Schneider Electric contributes its experience in energy hardware and software components, RWE uses its expertise in smart charging products and IT systems. In the smart home market, RWE cooperates with companies like German washing machine manufacturer Miele or Dutch electronics company Philips. M2M communication is required in smart homes, as well as electric charging infrastructure, for example, in the automatic authentication of the vehicle. Telecommunications company Vodafone provides the expertise for the automated interaction as a service to RWE (Vodafone, 2015).
  But also nonutilities prepare for the market of Utility 2.0. In 2013 Siemens and Accenture entered into an agreement to form Omnetric, a joint venture that provides advanced smart grid solutions and services focused on data management and systems integration to improve energy efficiency, grid operations, and reliability. In eMeter, Siemens and IBM joined forces to provide tools for utilities to improve distribution planning, prevent outages, engage consumers, and increase profitability.
- Crossing the chasm/fail-fast
  The organizational implementation of business units that are oriented toward Utility 2.0 within the energy company poses major risks. Many new

products and services are not yet commercially viable or, if they are no longer loss-making, are still well below internal hurdle rates. Meanwhile, these hurdle rates are applied in, or imposed on, other units of the company, for example, when assessing domestic or international investments in new power plants. This double standard may be justified by long-term strategic considerations, but in day-to-day routines, frictions between business units may emerge. Many utilities have therefore chosen to bundle these new products and services in separate units. For example, RWE founded a subsidiary called RWE Effizienz in 2011, which contains its smart home, electric mobility, metering, and building efficiency services.

In addition, emerging technologies may temporarily rise to the peak of the hype cycle; this means they enjoy a favorable perception in media and in public opinion, as well as support by politicians, think-tanks, and corporate decision-makers. However, some products or services may not meet the high expectations. For example, savings potentials of smart meters hover typically around 3% (European Commission, 2015), mainly because customers' behavioral patterns bounce back into preinstallations routines. Similarly, market penetration of electric vehicles in most European countries is well below politicians' aspirations. Norway, considered one of the most successful markets for electric mobility, achieved a penetration rate of a 20% share of newly registered cars having an electric drivetrain in the first months of 2015 (smh/Reuters, 2015) because the Norwegian government introduced attractive conditions and subsidies to promote electric mobility. Luxury vehicle Tesla did not cost more than Volkswagen's compact car Golf, as German motor developer Fritz Indra comments (Meiners, 2015). Norwegians purchasing an electric vehicle did not only benefit from substantial tax reductions, they were also given nonmonetary incentives, such as the free use of bus lanes in Norway's capital, Oslo. In most other countries, the business case for public charging infrastructure remains a loss-making endeavor, and is typically justified by first-mover arguments or integrated services.

The organizational culture of energy utilities has been shaped by long-term planning horizons, continuity, safety, and reliability. Dealing with a whole spectrum of new products and services, of which a large fraction may remain loss-making over the next couple of years, or may become outright failures, is not necessarily engrained in the mindset of either the management or the employees (see the comments of RWE's CEO in the Preface of this book on changing attitudes of employees in his company, as well as chapter by Nillesen and Pollitt). Many utilities try to ignite a more entrepreneurial spirit by introducing new forms of innovation like incubators or accelerators, but they may have to keep these new ventures apart from more conventional business units, and equip them with a different workforce (see chapter by Löbbe and Jochum).

- Leveraging regulatory incentives
  The two energy utilities in Europe that have gathered the most extensive experience with smart meters are Enel in Italy and Vattenfall in Sweden. The large-scale deployment would possibly not have taken place without government intervention, or at least support. France, the United Kingdom, and most other European countries, follow the Swedish and Italian examples and aim to achieve a full smart meter rollout until 2022, as the plan of the European Commission foresees it. Irrespective of the primary motivation to install smart meters, be it actual savings, prevention of energy theft, or less costly meter reading, governments may act as a trigger to accelerate the move toward Utility 2.0. The European Commission expects that more than 70% of European consumers will have a smart meter for electricity by 2020 (European Commission, 2015).

One of the biggest challenges on the way to Utility 2.0 is the mental change process—from selling a commodity in large quantities on wholesale markets, or to indistinguishable final customers with a fairly inelastic demand, utilities have to adapt to customized, complex solutions for individual customers, customer segments, respectively. Especially in building efficiency and energy performance contracting, discussed in the context of Getec as a new entrant, in Section 5 of this chapter, each solution has to be individually developed and adapted to local conditions, in terms of energy consumption, behavioral patterns, and configuration of the building stock.

Typically, electric utilities have all the expertise necessary to succeed in the marketplace within their organizations, but experts may be scattered across different, or even internally competing, business units. But one of the greatest challenges is to develop an appropriate organizational restructuring of business units that enhances identification of inhouse specialists with their newly created business units.

In addition, value creation of the Utility 2.0 occurs in the context of integrated services. For example, the pure energy savings potential of a residential customer is so low that it hardly justifies any investment into smart devices. However, if the technology is combined with assisted living systems, for example, for elderly or disabled people, or with automated house protection systems against illegitimate intrusion and theft, it offers a benefit that becomes attractive for customer segments other than technology-affine, wealthy early adopters, who would be the primary target group.

4 A CATEGORIZATION OF STRATEGIES OF EUROPEAN ENERGY UTILITIES

The observations of Sections 2 and 3 suggest that European energy utilities have realized that the status quo that was established after liberalization does no longer guarantee corporate survival. Similar to national carriers in European

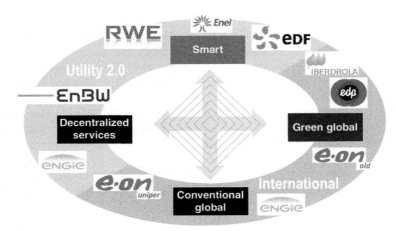

FIGURE 16.3 A categorization of strategies of European energy utilities. *(Source: Authors, based on annual reports and investor updates.)*

aviation, some utilities may disappear or get acquired, and will only remain as a brand, but under the sphere of influence of an outside player.

However, European energy incumbents choose diverging strategies to cope with that challenge. Based on the examples and narratives of utilities' strategies depicted in this chapter, Fig. 16.3 unites major European players in a two-dimensional sphere depicting either internationalization or the move toward Utility 2.0, as each utility's strategic focus. None of the utilities follows a monodimensional strategy, but, for example, players such as Iberdrola and EDP can be easily categorized as going "green and international," and Engie/GDF-Suez as "conventional and international," while RWE exhibits strong tendencies to become "smart."

The figure graphically underlines that no dominant strategy can be observed among major European utilities currently, but both internationalization and Utility 2.0 are options that are explored.

German energy company E.ON is represented twice, because its CEO Johannes Teyssen announced in late 2014 that the company would be split in two parts. One part would keep the E.ON branding and assemble its business units in renewables, distribution, and customer solutions, including operations in Germany, other EU countries and Turkey, where the company is mainly involved in retail operations. The other company, called Uniper, would obtain the upstream business, global commodities and trading, as well as conventional power generation, including the company's hydropower assets and its predominantly fossil power generation facilities in Brazil and Russia.

While some commentators claimed that E.ON intended to get rid of its nonprofitable conventional generation business, the company's CEO explained the division of assets in the following way: "These two missions are so fundamentally different that two separate, distinctly focused companies offer the best prospects for the future" (Drozdiak, 2014). While renewables-focused E.ON would benefit from a low volatility, Uniper might build on its natural gas business in Europe and Russia. It would focus on safeguarding the security of energy supply, according to Drozdiak (2014). This strategic move brings the new E.ON in a situation comparable to Iberdrola and EDP, which use regulated business to secure a lower-risk revenue stream. In terms of revenues, both companies are expected to yield approximately the same earnings, though, according to E.ON's management.

According to the categorization suggested in this chapter, other types of separation could have been imagined; for example, keeping all large scale operations, including renewable energies, in one company. With 782 MW of installed capacity, E.ON runs the world's second-largest onshore wind park in Texas, USA. These types of investments are more similar to conventional power generation projects than to small scale, complex service solutions in distributed assets or smart grids.

The example of E.ON shows that utilities may choose their strategies by concentrating on specific segments of the traditional energy value chain, meanwhile seizing synergies across countries or technologies. Of course, Fig. 16.3 does not imply that the trajectory of each utility is predetermined and fixed. Most European energy incumbents still have the financial endowment to redirect their course of action.

5 CULTURAL PREREQUISITES: LESSONS FROM NEW ENTRANTS

The corporate culture of traditional electricity supply companies may be the greatest hurdle in preparing a utility for current and future challenges. Some utilities may fail, where new entrants have a proven and successful track record. This part of the chapter sheds light on four key characteristics that distinguish new entrants in energy markets from classical energy companies: decentralized decision-making, customer centricity, flexibility, and new platforms. To illustrate each characteristic, an actual case of a new entrant is described and discussed.

- Decentralization: Getec
  The Getec Group belongs to a circle of companies that entered the energy performance contracting market. It was founded in 1993. With a turnover of almost €750 million across its business units, Getec is one of the largest nonutility energy solution providers, and a major player in the energy performance market in Germany. The company successfully builds, manages assets like cogeneration plants, purchases and sells electricity and natural gas on the wholesale market, operates wind and solar parks, and operates distribution grids.

  The handling of energy performance contracting is a complex service solution that requires expertise in buildings, energy technology and energy markets, commercial and legal matters. Since each building is unique, in terms of age, spaces, materials and user behavior, a one-size-fits-all solution is, by definition, not feasible in energy performance contracting. It is an expert-intensive but small-scale process that ends, compared with revenues as utilities used to earn them a couple of years ago, with modest margins and earnings.

Retrospectively, it seems clear that utilities did not enter this market aggressively. First of all, money could be earned on wholesale markets much more easily. Second, they may have faced a credibility problem with potential customers, because energy performance contracting has the objective of reducing energy consumption, whereas energy utilities would, at least traditionally in the old world, aim to increase revenues by increasing sales.

Getec's business model contains many elements that characterize the Utility 2.0 strategy. Its example points to one internal and one external factor that may be obstacles for the redefinition of an energy company's business model toward Utility 2.0.

First, the strategic shift implies a change in the corporate mentality from big to small, from commodity to service, from wholesale to customer orientation, and from long-term to flexible and adaptive. With a workforce that has all expertise for that type of business latently available inhouse, but dispersed across different business units, typically under no competitive pressures, and relatively expensive compared to outside firms, it requires utilities to initiate a structural reorganization to create dedicated groups with inhouse experts. Getec follows an organizational model of decentralization: each new business activity is implemented by a new legal entity. At a workforce of around 800, the Getec holding company owned direct and indirect interests in 46 companies, which were fully consolidated within the financial statements for the group, in 2013 (GETEC, 2014). Justifying high labor costs may prove to be equally challenging, especially if companies like Getec offer comparable services for less money.

Second, final customers have to believe in the authenticity of the strategic change, that the utility really aims to optimize energy expenditures of their customers. This may only work if the units operate under a different brand name or logo.

- Customer focus/lifestyle: Google and Nest
  Google acquired startup Nest for US-$3.2 billion. Nest produces an intelligent thermostat that integrates learning algorithms in its software. It allows for continuous improvements of heating or cooling of the house, hence energy savings and a pleasant room climate are the results of the learning process. Google's motivation to acquire Nest may stem from its ambition to provide a whole world that surrounds each Google user with Google products, to sell hardware to access information that enables it to provide intelligent data streams. On the move to a more transparent final customer, heating and cooling habits may be an interesting aspect.

  However, there are many competitors to Nest's learning algorithm. Why did Google choose Nest, and why were they willing to pay such a high price?

The design features of Nest show some similarities with Apple products. This is not astonishing, since one of Nest's founders worked for Apple beforehand, and designed the iPod (Wohlsen, 2014). As much as the functionality, Google realized that aesthetic features of a product weigh heavily in the purchase decision.

Utilities in liberalized markets across the globe have started realizing that winning and retaining customers requires a more customer-centric approach (see chapters by Gimon, and Smith and Macgill for discussions on customer centricity and customers of the future, respectively). Even though electricity and natural gas are commodities, understanding customer needs and tastes has become an important prerequisite for sales.

- Flexibility: LichtBlick

Hamburg-based eco-energy retailer LichtBlick was founded in 1998 as one of the first providers of green electricity for the newly liberalized mass market. In 2009 it expanded from sales of CO_2-neutral power and gas on the German retail market into a new business area: it entered an alliance with Germany's largest car manufacturer Volkswagen to sell so-called home generating units, residential micro-CHP (combined heat and power) plants (for a detailed description, see Burger and Weinmann, 2013). Until the end of the joint venture in May 2014, LichtBlick had sold around 1400 units (Stahl, 2014).

The lasting decline of wholesale market prices had negative repercussions on the underlying business model of the home generating units. As one of Germany's largest retailers of biogas, LichtBlick decided to include their micro-CHPs into the feed-in system, which provided subsidies for biogas. Whenever a home generating unit would be switched on, biogas would be fed into the natural gas grid. With this method, the company ensured financial viability of the units.

Together with Volkswagen, the company started offering energy tariffs for e-mobility for Volkswagen and Porsche cars in 2015, while abandoning the cooperation with Volkswagen with respect to the home generating units. In May 2015, LichtBlick announced to have entered an alliance with electric car and battery manufacturer Tesla Motors. While Tesla's new lithium-ion batteries should capture electricity during times of surplus renewable energies, and release it into the grid during scarcity, LichtBlick would be in charge to connect all batteries to form a virtual power plant, using software similar to the one developed for their home generating units (Schultz, 2015).

LichtBlick is an example of entrepreneurial spirit and managerial flexibility. It has a track record for successes, but also for failures. Since its existence, the company has not ceased experimenting and optimizing its strategy continuously—despite (or induced by) a very dynamic market environment.

- New platforms: Bankymoon

Digitalization and the Internet of Things open the door to consumers beyond the existing customer segments. Electrification and grid-based power supply will grow in many countries in Africa, Asia, and Latin America.

New customer segments will get access to basic infrastructure services. But 80% of the African population does not have a bank account, according to Lorien Gamaroff (2015), founder of startup company Bankymoon. Gamaroff uses the cryptocurrency Bitcoin to enable electronic payments for those who do not have access to traditional payment methods. Utilities may use Bankymoon's system to ensure that electricity bills are paid upfront, or to lower transaction costs of donations when, for example, they support charities or public institutions.

Bankymoon seizes one of the opportunities that emerge with digitalization and the Internet of Things. The startup combines elements of Utility 2.0 with internationalization. Most importantly, it uses a new platform, the cryptocurrency Bitcoin, for its transactions, thereby becoming a pioneer in innovation and in targeting potentially lucrative new customer segments.

6 CONCLUSIONS

As the analysis suggests, European power companies are on track to define their role in the new configuration of the power industry. A number of incumbents have initiated the transformation into a provider of complex service solutions, either specializing in the management of distributed generation facilities, or in the management of information and data analytics. However, in nascent business applications, such as public charging for electric vehicles, viable business models are still scarce, and require patience and endurance until markets mature. This may prove difficult, given the financial constraints that many utilities currently have to cope with.

Internationalization seems a viable alternative to Utility 2.0, with some European players successfully entering new markets while focusing on renewable technologies. Most major companies rather pursue a regionalization strategy, though. After a first wave of acquisitions during privatizations in Latin America, Eastern Europe, and Central Asia and a second wave of acquisitions of renewable energy projects and companies, European utilities may now become the target for outside investors. China's Three Gorges outbid European majors to acquire an important share of Energias de Portugal. A number of smaller European players may turn into attractive targets, too.

Irrespective of their strategy, the biggest challenge for all European players is the decentralized energy revolution: the sun may become the world's largest source of electricity by 2050, ahead of fossil fuels, wind, hydro, and nuclear, according to the International Energy Agency (IEA, 2014a). Renewable energies, in particular solar, will redefine the role of electric utilities in the future. In the rural and suburban context, autonomous microgrids, as Marney points out in this book, or "Local Energy Networks," similar to Local Area Networks for data transmission in the Internet, and self-sufficient households will become the dominant form of energy supply. Companies that focus on the Utility 2.0 strategy will have to compete with one-stop-solution providers such as SunEdison. For their target customers, the value added will be the combination of management of assets and management of information, both in the residential, as well as in the commercial and industrial segments.

In the urban context, where self-supply is less likely, due to a high population density, utilities may be able to build upon their cumulated expertise along the energy value chain, from managing generation assets to delivering the electricity to the final customer, which enables them to transform themselves into safeguards of system stability, even if the costs for producing energy become negligible.

Energy utilities that pursue internationalization strategies will benefit from the rise of the global middle class, and increasing wealth of nations. According to the International Energy Agency, around 7200 GW of generation capacity will have to be built to satisfy worldwide electricity demand until 2040 (IEA, 2014b). Conventional generation technologies will remain an important source of electricity for the next 20–30 years, but then the more favorable cost structure of photovoltaics—and to a lesser extent wind parks, onshore and offshore—will crowd out other generation technologies. Utilities that have started early to gain expertise in international renewables will have an easier stance vis-à-vis energy giants from emerging markets, in particular China and India, than those utilities that focus on conventional generation.

New entrants will threaten the position of European energy companies in their regional markets, as well as on the international scale. Only if utilities succeed in convincing their workforce that a cultural change is necessary, they will not only survive, but prosper in the future.

REFERENCES

Borghese, F., 2010. Automated Meter Management Roll-Out—Enel's Experience. Enel, Rome.

Bugge, A., 2011. China Three Gorges Buys EDP Stake for 2.7 Billion Euros. Reuters, December 22.

Burger, C., Pandit, S., 2014. ESMT Consolidation Index 2012—Dataset. ESMT, Berlin.

Burger, C., Weinmann, J., 2012. ESMT Innovation Index 2010—Electricity Supply Industry. European School of Management and Technology, Berlin.

Burger, C., Weinmann, J., 2013. The Decentralized Energy Revolution—Business Strategies for a New Paradigm. Palgrave Macmillan, Basingstoke.

Burger, C., Weinmann, J., 2014. ESMT Innovation Index 2012—Electricity Supply Industry. European School of Management and Technology, Berlin.

Clercq, G.D., 2013. Iberdrola Bets on Foreign Grids as EU Utility Industry Struggles. Reuters, October 22.

Drozdiak, N., 2014. E.ON to Split Into Two Companies. Wall Street Journal, November 30.

E.ON, 2015. E.ON Connecting Energies. Available: http://www.eon.com/en/about-us/structure/company-finder/eon-connecting-energies.html

Enel, 2014. Sustainability Report 2013. Enel, Rome, Italy.

Ericsson, 2014. Transforming Industries: Energy and Utilities. Ericsson, Stockholm.

European Commission, 2015. Smart Grids and Meters. Brussels. Available: https://ec.europa.eu/energy/en/topics/markets-and-consumers/smart-grids-and-meters

Farrell, J., 2014. Beyond Utility 2.0 to Energy Democracy. Institute for Local Self-Reliance, Minneapolis.

Flauger, J., 2013. Abschied vom Gigantismus. Handelsblatt, November 29.

Gamaroff, L., 2015. Blockchain-Aware Smart Metering: Bitcoin's Killer App. www.mindthegapexpo. com. Available: https://www.youtube.com/watch?v=7f7hE0K8-OA

GDF-Suez, 2015. Positive Energy & Territories. Available: http://www.gdfsuez.com/en/innovation-energy-transition/sustainable-cities-regions-mobility/positive-energy-local-initiatives

GETEC, 2014. Annual Report 2013. Magdeburg.

Iberdrola, 2015. Renewable Energy Business. Available: http://www.iberdrola.es/about-us/lines-business/renewables-business/

IBM, 2013. Smart Grid Analytics: All That Remains to be Ready is You. White Paper. IBM.

IEA, 2014a. Technology Roadmap—Solar Photovoltaic Energy [Online]. OECD, Paris.

IEA, 2014b. World Energy Outlook 2014. OECD, Paris.

IEA, 2015. About Renewable Energy. Available: http://www.iea.org/topics/renewables/

Impax Asset Management, 2015. European Renewable Energy M&A Trends. Available: http://www.impaxam.com/sites/default/files/Impax_PROOF_Final(amended).pdf

Jacobs, C., Nair, A., 2012. GDF Suez Takes Full Control of International Power. Reuters, April 16.

Kolk, A., Lindeque, J., Van Den Buuse, D., 2014. Regionalization strategies of European Union electric utilities. Br. J. Manag. 25, S77–S99.

Meiners, J., 2015. Tesla wird scheitern. Handelsblatt, May 27.

Musiolik, J., Markard, J., 2011. Creating and shaping innovation systems: formal networks in the innovation system for stationary fuel cells in Germany. Energy Policy 39, 1909–1922.

Powell, W.W., Grodal, S., 2005. Networks of innovators. The Oxford Handbook of Innovation, pp. 56–85.

RWE, 2014. RWE Offers Charging Solutions for Sustainable Mobility. Available: www.rwe-mobility.com

RWE, 2015. Focus on Our Core Regions. Available: http://www.rwe.com/web/cms/en/1857096/rwe/investor-relations/

Schultz, S., 2015. Tesla und Lichtblick schmieden Stromspeicher-Allianz. Spiegel, May 1.

smh/Reuters, 2015. Norwegen: Ärger im Elektroauto-Paradies. Spiegel, April 21.

Stahl, L.-F., 2014. LichtBlick ZuhauseKraftwerke wechseln vom KWKG zum EEG. BHKW-Infothek. Available: http://www.bhkw-infothek.de/nachrichten/19380/2013-06-25-lichtblick-zuhausekraftwerke-wechseln-vom-kwkg-zum-eeg/

Vanamali, A., 2015. The Urgent Need to Shift to "Utility 2.0". Available: http://blogs.worldwatch.org/the-urgent-need-to-shift-to-utility-2-0/

Vattenfall, 2014. Markets. Available: http://corporate.vattenfall.com/about-vattenfall/operations/markets/

Vodafone, 2015. RWE ist ein Ready Business. Available: http://www.vodafone.de/business/firmen-kunden/loesungen/referenzkunden.html?deeplink=rwe

Wohlsen, M., 2014. What Google Really Gets Out of Buying Nest for $3.2 Billion. Wired, January 14.

Chapter 17

Thriving Despite Disruptive Technologies: A German Utilities' Case Study

Sabine Löbbe*, Gerhard Jochum[†]
*Reutlingen University, Reutlingen, Germany; [†]Büro Jochum, Berlin, Germany

1 INTRODUCTION

With sales of €590 billion, Germany's electricity market is the largest in Europe. In electricity production, market share of the former "Big 4" E.ON, RWE, Vattenfall, EnBW, amounted to about 74% in 2013, declining from 84% in 2010. Some 1.5 million individuals and corporations produce electricity. Regarding sales, some 900 incumbents and some hundred newcomers serve end customers, with the established companies holding regional market shares of 70–80%, steadily declining.

In fact, profitability of German utilities suffers from stagnation. This needs to be differentiated: while energy sales, networks and some niche providers in generation tend to continue generating positive EBIT, nearly all conventional generation, as well as certain cogeneration assets and some renewables projects, for example, offshore wind, are not profitable. According to a recent study, two in three companies assessed cannot finance investment from their own assets (PwC, 2015a). The difficult situation is often accompanied by a lack of acknowledgement concerning declining sales and market shares, whereas over-regulation is considered to be the major threat. Companies tend to underestimate competitors from other industries, and are not sufficiently customer orientated. Innovation expenditures are low and stagnate; the share of new products compared to existing products is declining (ZEW, 2014).

So, why are the existing players threatened? Section 2 provides answers to this question. An analysis of market development will differentiate between stable and disruptive, and predictable and unpredictable parts of German energy transition.

Utilities' inertia with regard to strategy development and implementation is based on bottom-up approaches focusing on today's often unpredictable and unstable regulatory environment. Entrepreneurial thinking and action is scarce,

Future of Utilities - Utilities of the Future. http://dx.doi.org/10.1016/B978-0-12-804249-6.00017-8

323

whereas knee-jerk and opportunistic reactions upon government's amendments are frequent. This, in turn, blocks resources, sometimes without adding value. Therefore, Section 3 proposes strategic approaches for these utilities. The vital importance of a clear positioning, and differentiated future strategies, is explained. This includes corporate as well as business unit strategies related to distributed and centralized production, networks, sales, and energy-related services. Section 4 lays out success factors for strategy development and realization, followed by the chapter's conclusions.

2 THE ENVIRONMENT

Since the 1990s, introduction of competition has opened up markets, whereas climate protection led to a variety of market oriented political instruments dominated by a "command economy." In 2011, this political and regulatory development was renamed "Energiewende" (energy transition). The following section presents a closer look at the framework that utilities will be facing. In this context, a development is called

- stable, if trend scenarios are likely to come true, based on experience, and in correlation with well-known parameters; it is disruptive, if there is a high chance of one or several game changers, that is, new developments that do not correlate with experience or indicators
- predictable, if chances are high that a prediction becomes reality, and unpredictable, if no one dares to bet what might happen in the future.

2.1 Societal Development

We live in a world of growing complexity that sometimes is confusing to people. In their role as consumers, they can choose from a lot more options than before. At the same time, traditional relationships within our society tend to be replaced by virtual links. There is one important remedy for this: reducing complexity and confusion helps to evoke transparency (Luhmann, 2000) in order to boost trust. The need for, and impact of, transparency and trust can be observed in the rising demand for credence goods. Credence goods are goods for which the quality cannot be assessed objectively and reliably. In order to create trust in ecological electricity providers, and show transparency, markets and government develop tools such as tests, quality standards, or obligations to declare the origin or use of electricity.

2.2 Energy Policy

In 2014, the German Federal Ministry for Economy and Energy explained: "Germany sets a good example with its energy transition" (BMWi, 2014, p. 144), meaning: Germany's zealous energy policy is supposed to inspire the rest of the world to follow.

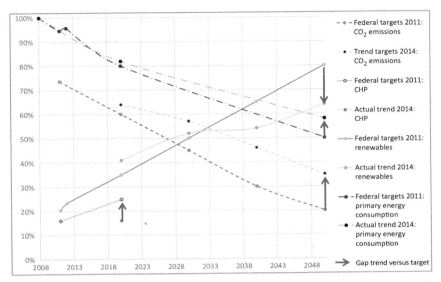

FIGURE 17.1   Review after 3 years of Federal targets as defined in 2011 versus actual trends in 2014. *(Source: Dyllong and Maaßen (2014).)*

By 2030, the European Council is striving for a reduction of EU greenhouse emission by 40%, Germany is aiming for a 54% reduction based on the level of 1990. The EU is planning to attain a renewable energy share of 27% by 2030, while Germany's goal is to reach more than a 50% renewables share by 2030. Overall EU energy use is to decrease by 27%, Germany is striving for a 50% reduction against the energy use level of 2008. However, Germany is not likely to achieve its ambitious goals, as illustrated in Fig. 17.1.

As a consequence, Germany's position and reliability of political environment is at stake. Unpredictability will prevail, and disruptions are likely to happen.

2.3   Regulatory Measures

Anticipation of future regulatory activity in electricity, gas and heating leads to the following results in the medium term, as conceptually illustrated in Fig. 17.2.

The support for newly built renewable energies is to be replaced by market integration. However, this will lead to more and more regulation: the first renewable energy law passed in 2011 had 11 pages, the current law dated 2014 runs to 300 pages. Evolving or newly emerging technologies might call forth new regulations—meaning that support schemes will survive.

The European Emissions regulation is susceptible to unpredictable and slow development. The key question is: who will take action to change the regulations? In fact, the authors expect continued minor short term interventions to improve

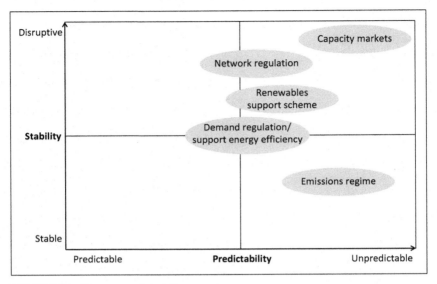

FIGURE 17.2 German regulatory framework.

efficiency and effectiveness. An example of this is the re-regulation of the European Emission Trading System regarding its introduction of the need to produce a certain level of reserves in order to guarantee market stability that will be a legal requirement from January 2019, and the reduction of trading in emission exemption rights, called "backloading." However, this development that looks stabilizing at first glance does not contribute to the long-term investment incentives needed. A systemic view would be necessary in order to achieve an optimization across the value chain, and regional, as well as timeline, optimization.

EU-wide developments in national reserve and capacity markets are incoherent and inefficient. In Germany, future market design is highly unpredictable, disruptive developments threaten the continuation of fossil power plants and energy security.

Demand-side regulation and support schemes for energy efficiency, meant to implement EU Directives, need to be enhanced, and the German government is adapting legislation while this book is being published. Otherwise, political objectives need to be reduced fundamentally. Higher financial subsidies and similar instruments are easily predictable. At the same time, disruptive development is unlikely, due to lack of financial resources for high impact financial instruments.

The framework for network regulation is set through unbundling, and incentive regulation. However, tariff structure might be adapted to the needs of a higher amount of distributed generation, and the "smartification" of the grid.[1]

---

1. See the chapters from Gellings, Biggar and Reeves, and Knieps in this volume.

In this context, the interdependency of dynamics of subsidies needs to be taken into account. The more renewable energies are subsidized, the more energy efficiency needs to be subsidized, due the energy price reduction achieved through subsidies. Requirements and incompatibilities will increase in the complex relations between the regulatory practice, the need for efficiency, the call for the better integration of renewables, and the development of customer behavior. This raises the question if there is any party that actually represents an integrated view of the energy system.

Overall, this promises a rather instable environment (Fig. 17.2):

- The energy production infrastructure, both the existing one and the one that is deemed necessary for future energy needs, is put under pressure: the differences and incompatibilities between the production that can be planned and the one that feeds in stochastically continue to increase. Just as well, gaps and incompatibilities between production and grid, and between production and consumption continue to increase. Regulation shifts cost and yield arbitrarily between the actors, and thus discourages investment.
- The spheres of interest collide: the energy economy versus the "rest" of the economy, energy politics against other political subjects, German action plans with the action plans of its neighbors or the international community.
- This leads to rising costs and reduced supply security. Thus, the unsuccessful pursuit of political goals incurs high costs, accentuating the inconsistencies and increased distributional conflict. Finally, this situation demands that the society realigns and reassesses its whole energy supply system.

For companies, this creates a question as to the long-term challenges for their strategy: what direction needs to be pursued?

2.4 Technology

In the short-term, the energy economy will need to find replacements for conventional power plants reaching the end of their lifespan, and, in the midterm, for renewable energy power plants running out of service. Furthermore, interfaces need to be created to cope with the combination of central and decentral energy production. New technologies shall contribute to solving these challenges.

Sustaining innovations maintain the existing performance structure, and improve an existing technology or system. Disruptive innovations change and disturb existing processes, and search for performance improvements outside the existing system and the established technology in the market (Chesbrough, 2007; Richter, 2013). Naturally, disruptive innovation cannot be foreseen. New technologies that might replace existing technologies offering the same value or creating new values might change the game, as conceptually illustrated in Fig. 17.3.

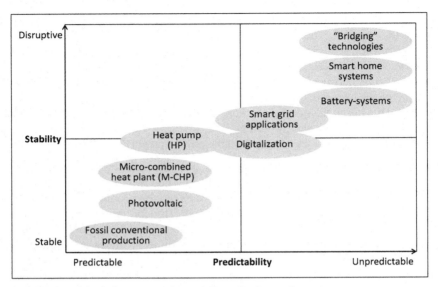

FIGURE 17.3 Technology: sustaining and disruptive innovation.

Innovation for gas and coal fueled production shall further ameliorate existing processes and performance. For heat pumps, technological change might lead to further improvements, as development of zeolite gas heat pumps demonstrate. The development of micro combined heat plants (micro CHP) is influenced by available material efficiency of large scale combined heat plants that has reached a high level. Most of the research results can be applied in micro CHP. Currently, the first combined heat plants with fuel cells are placed on the market (Dieckhöner and Hecking, 2014). In photovoltaics (PV), performance, resource efficiency, and module longevity will be further improved. Specific PV materials for building integration, road integration, and other specific supports, will lead to some disruptive innovation (IEA, 2014a).

As for smart home products, IT appliances and customer value is crucial, including energy and nonenergy related benefits (Balta-Okzana et al., 2014). Regarding the smart grid, the products' main functions and concepts are clear. Political action and regulation are key drivers of this market, so predictability is low. Concerning battery storage systems, it is not clear which technology will win the race. In the midterm, operation management of battery systems, system performance and safety, in the context of multiple applications, will improve and will be further developed (IEA, 2014b).

Technologies and product packages "bridging" centralized and local systems are already cocreating markets, such as virtual power plants, solar thermal/heat pumps, PV cells/heat pumps, PV cells/batteries. Finally, digitalization is

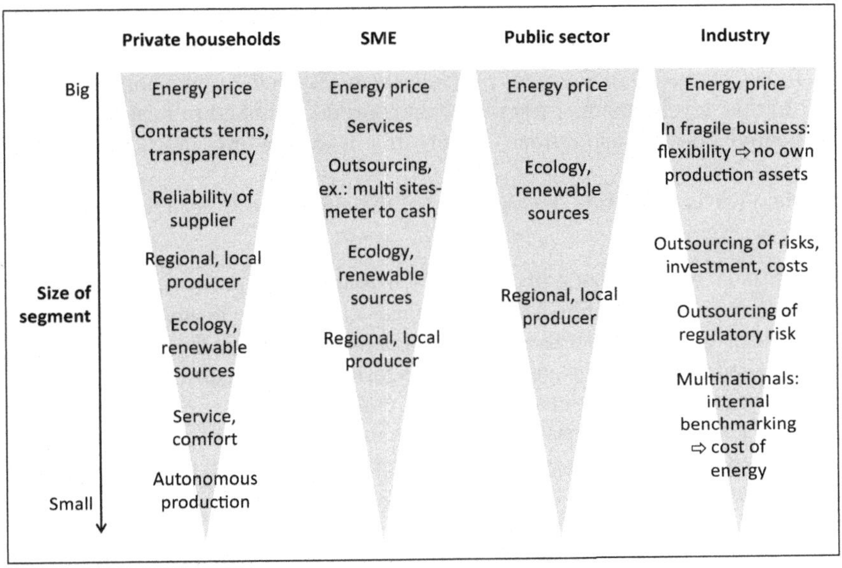

	Private households	SME	Public sector	Industry
Big	Energy price	Energy price	Energy price	Energy price
	Contracts terms, transparency	Services		In fragile business: flexibility ⇨ no own production assets
	Reliability of supplier	Outsourcing, ex.: multi sites-meter to cash	Ecology, renewable sources	
	Regional, local producer	Ecology, renewable sources		Outsourcing of risks, investment, costs
Size of segment	Ecology, renewable sources	Regional, local producer	Regional, local producer	Outsourcing of regulatory risk
	Service, comfort			Multinationals: internal benchmarking ⇨ cost of energy
Small	Autonomous production			

FIGURE 17.4   Typical factors influencing buying decisions of customer segments.

the basis for many of these new products, as described by Cooper (see chapter 5: Innovation Platform Enables the Internet of Things).

2.5   Customers' Needs

Energy demand is relatively easy to predict. Studying the analyses, a good guess might be a reduction in electricity consumption by –0.5 to –1% per annum,[2] and a reduction in heating/gas consumption by around –1 to –1.8% per annum.[3] This is not comfortable, but it is relatively predictable and stable.

It's the churn rates that arouse instability and unpredictability for decision makers. The German energy market is competitive—on average, every private household can choose from 80 electricity suppliers. So far, 65% changed either their supplier (20%), or their contract (45%).

To define future needs of energy customers, we exemplify motivational aspects of private households (see Fig. 17.4 for details): for German energy customers, price is the key, as every market survey indicates. Next is ecology: more than 8 million customers buy eco energy—in addition to energy taxes, that is, for feed-in-tariffs, accounting for about 50% of the energy bill. Regional

2. Target Energiekonzept Bundesregierung: net energy consumption to be reduced by 2020 by 0.87 %/annum, compared to 2008.
3. From 2008 to 2020 heating consumption is to be reduced by 20%.

provenance today is an important criterion, with an interesting potential for differentiation. Retailers in fact offer corresponding products and services.

Today, about 2% of Germans produce electricity in their own homes. In contrast to this development, energy efficiency products linked to heating systems lag behind political aspirations, as investment in building and heating appliances is slow.

Main drivers of future development of the autonomous production segment are:

- Attitudes and behavior of potential customers: as recent market research indicates (PwC, 2015b), 65% of Germans can imagine producing electricity themselves, with a preference for solar installation on the rooftop (54%), followed by CHP units and small wind turbines in the garden, depending on customer's access to financial and real estate capital.
- Only a small group acts to attain autonomy of energy supply. Customers are getting more energy literate, thereby appreciating the value of grid connection.
- The profitability of autonomous production in private households depends on the regulatory framework (Institut der deutschen Wirtschaft Köln, 2014), that is, the price autonomous producers receive for feeding energy back into the grid, and exoneration from grid fees.
- Future regulation: if this opaque and one-sided burden on the whole system, and on energy supply security, will be shifted toward a more cause- and cost-effective allocation of costs, this might mark growth limits of this specific motivation for distributed generation.

In the next chapter, Smith and MacGill (see chapter 16: The Future of Utility Customers and the Utility Customer of the Future) go further into detail of customers' preferences regarding electricity demand and distributed energy. As a consequence, retailers should continue developing products and services with segment-specific selling propositions such as price, ecology, and local production, simultaneously (Section 3.2). In general, customer segments need to be defined by their future attitudes and their behavior—which become more and more fragmented. In fact, product development needs to be adaptable, corresponding to changing customer's needs.

2.6 Dynamics via Partners and Competitors

In the previous chapter, Burger and Weinmann explain roles of industrial competitors and cooperation partners along the value chain. Energy companies, as well as other industries—IT and IT based service providers, energy service aggregators, or sales channel based service providers, social networks, etc.—develop new business. This will happen through interfaces, horizontally, vertically, or laterally. Customers might respond to innovative products and services with action plans ranging from consumption to cocreation of products.

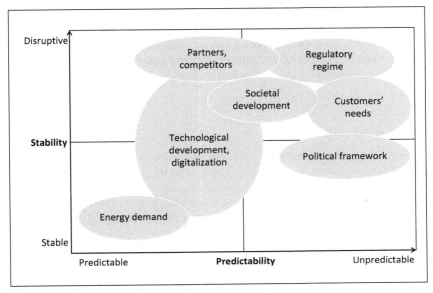

FIGURE 17.5   Stability and predictability of environmental development.

In the capital intensive energy business, influence of capital cost is crucial. With the share of renewable energies increasing, capital intensity will grow even further. Therefore, financial investors for project and company financing, for debt and equity financing, are important stakeholders. As the sector environment becomes more disruptive and less predictable, risk increases, and as capital cost increases proportionally, its optimization becomes more important. Companies need to professionalize their risk management, and their assessment and selection of investment projects.

2.7 Main Challenges

Disruptive and unpredictable developments need to be taken into account, as conceptually illustrated in Fig. 17.5.

Stability and predictability vary along the value chain, and for different customer segments, regions, and energy sources, with different intensity and timelines. This, in turn, increases complexity, volatility, and risk. The following section focuses on strategic answers to these challenges.

3 BUILDING BLOCKS FOR STRATEGIC CHOICES

The conflict between in-/stability and un-/predictability, on the one hand, and its crucial importance for the planning of utilities opens up the question: what are the strategic consequences of this situation? Of course, every company will have to find its own specific answers to this question, referring to the specific environment, specific shareholder expectancies, with specific objectives, specific time horizons, a specific policy towards risk, and specific capital, human resources and assets. Therefore, only a number of general questions and dynamics can be considered for every company's strategic answer, as further described later.

3.1 Positioning

Often, strategy development in the energy industry is focused on parts of the value chain, neglecting the overall objectives and the need for a consistent positioning of the company. Yet, sustainable development and responsible entrepreneurship need a good understanding of the entire context. Two concrete issues need to be considered in this context. First, in today's German energy transition, the main thrust of building the ultimate infrastructure for renewable energies has to be brought into compliance with the needs for supply security, profitability, and the fair allocation of the costs between different social strata. Second, the main thrust of decentralization still needs to be synchronized with the requirements of grid access, grid utilization, and overall profitability.

Utilities' positioning incorporates reliability, long term thinking, and a sense for good governance and good citizenship. In turn, these companies incorporate responsibility for the cited pitfalls and problems of the energy transition. For example, the following operational situations result in positive business cases, but with negative external effects on the energy system:

- Production of heat in peak load boilers, instead of cogeneration: this happens when the electricity trading market sends signals to stop electricity production, whereas heat demand is high. Example: times with high wind and PV production, low electricity demand and high heat demand—as it may happen on Sunday around 11 am in winter.
- Supply from local batteries in times when PV modules of a house are not producing energy, but at the same time low demand and high wind feed-in leads to high load in regional or local grids.

Promoting such solutions does offer positive margins, but it does not fit with positioning a company as a responsible actor for a sustainable development because external effects are negative. Therefore, utilities tend to refrain from offering such products or services, or do so without conviction. As a consequence, the market does not reward these half-heartedly developed products with no link to the company's positioning and strategy.

However, not offering such products or solutions means leaving the market to competitors. To become or stay convincing, this means: only offering solutions and products without a clear contribution to a functioning energy transition as long as is necessary to obtain a relevant position, in order to then be part of the "real" energy system of the future. However, this strategy requires not only a

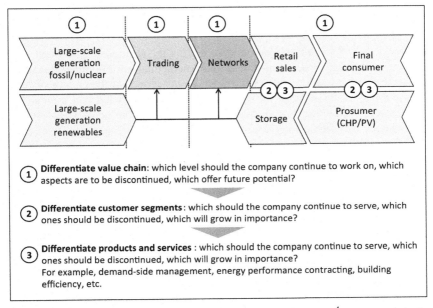

FIGURE 17.6   **Key questions for differentiated strategy development.**[4] For value chain description refer to Burger and Weinmann (2014).

high level of professional expertise, but also a substantial ethical standing, and a stable company structure in times of instability and unpredictability.

3.2   Developing a Differentiated Strategy

The more the environmental development becomes uncertain, the more management and employees themselves get confused and disoriented. Opinions about where to direct the company diverge, decisions are delayed and avoided. On the contrary, to be successful, companies definitely need a clear view of the future, with positive and distinct objectives and strategies. This means the certainty inside the company must be increased. To achieve this, a differentiated strategy, corresponding to the divergent and turbulent environment, is necessary.

The classic model of creating value through sales, grid supply, and distribution, as well as energy services, will need to be redeveloped, taking into account the instability and unpredictability of different value chain levels, different customer segments, different energy sources, and different regional contexts, as can be seen in Fig. 17.6. This new strategy will further need to include all the different factors that bear on a company's success, described in Section 2.

4. For value chain description, refer to Burger and Weinmann.

This differentiated strategy development requires the examination of specific parameters in the areas of

- Production. What are the factors determining the relationship between conventional energy production and production on the basis of renewables within the company? How much centralization is needed for the energy system, and what does that mean for the company today, and what will it mean for tomorrow? How much decentralization can the system bear? How can the company decide between Energy Only Markets 2.0 versus capacity or flexibility markets of different sizes and structures? What threats and opportunities arise for the company's production in its different markets?
- Trading. How will liquidity of different products evolve? How relevant are different trading places? How will the regulatory and the physical conditions of trade evolve in the future? What is the desired position and role of the company's energy trade? One of the highly debated niches is flexibility markets, as recent studies in Reutlingen Research Institute demonstrate. One of the actors in this market, Next Kraftwerke, a startup company with 100 employees (2015), operates one of the largest virtual power plants in Germany. Primarily, the company markets frequency control with the help of production and load shedding units. This is mainly effectuated via biogas units and—to a minor extent—load shedding of large customers, delivering negative minute and secondary reserve). In February 2015, Next Kraftwerke managed 2786 units with a capacity of 1539 MW.
- The energy grid. How will the regulatory framework for the energy grid develop? How will different political interventions shape the system parameters—that is, the priority of feed in for renewable energy—and how will the new regulations affect investments in the grid infrastructure and grid availability? What actions will (local) political expectations with regard to linkage between distribution networks and other regional infrastructure stir? What business model is the company to adopt with regard to its supply grid—for electricity, gas, or heat?
- As Eurelectric (2014) puts it, grid operators will not change their role—but they will need new tools to fulfill it: to continue exerting their role as system operator, they actively manage the reliability and the quality of service[5]; in order to fulfill their role as neutral market facilitators, grid operators will need to manage increasing amounts of data. Electrical and thermal systems shall need to be interconnected, in order to fulfill these roles. Multiple research studies within operating systems, integrating smart meters, demand side management devices, virtual power plants etc., are being implemented.
- Energy sales. How will the needs of existing customers and potential customer segments differ? How will relationships between vendor and buyer develop? How will the competition intensity evolve, what kind of stakeholders will develop? What specific threats and opportunities will arise from these factors? For example, most German utilities differentiate customers

---

5. For further details, see Hanser and Van Horn in this volume.

segment-wise, with offers and specific sales channels for out of area and in area customers; however, differentiation will need to be more skillful and unique in order to be competitive.

- Services in the area of energy and energy efficiency. What specific customer requirements will emerge, and what requirements for customer-specific tailoring of integration services and demand-side solutions will become relevant? What are the threats and opportunities that emerge from this in a specific region? Some concrete examples, such as GETEC and RWE Effizienz, are described by Burger and Weinmann in this volume.[6]
- How can the rising demand for credence goods be satisfied? Which products, what selling proposition, what credence tools and which underlying (digitalized) processes can utilities develop in order to open up this growing market? What is the value of established tools, such as expert surveys, tests, supplier warranties and branding? How can autonomous-generation be made available to customers in a simple, transparent, and locally specific way? How can related partnerships be created? Virtually, all German utilities offer eco-energy products. Meanwhile, for a certain (and growing) customer segment, renewables certificates as a basis for these eco-products are considered as untrue greenwashing. For this target group, local and regional production is more attractive. However, it is hard to offer competitive products in a highly regulated "market" dominated by feed-in-tariffs.

As there is no "blueprint strategy" for future utilities, these questions need specific answers that the management needs to find, based on the specific regional situation, its goals, and its strategy. This will also lead to answers regarding choice of business models—as described by Nillesen and Pollitt in this volume (chapter 15). Specifically, this requires decisions regarding:

- cutting existing asset-lock-ins, discontinuing existing products and services, leaving previously supported customer segments or regions. A prominent example of a differentiated strategy concerning the overall company strategy is E.ON splitting up into E.ON New Energies and Uniper, described by Burger and Weinmann in this volume (chapter 16); and
- committing to other business activities, such as offering new products and services along the value chain, and targeting new customer segments in or out of area.

3.3   Best Practice Strategies

Which modules are useful for making progress in the differentiated strategy development?

3.3.1   Efficiency and Effectiveness

Instability and unpredictability increase the pressure to be effective—to do the right thing, and to be efficient—to do this in the right way. It is therefore

6. Boscàan and Poudineh, as well as Rai and Zarnikau, provide further examples in this volume.

essential to create economies of scale and economies of scope, within the company or beyond, with stakeholders.

3.3.2 Flexibility and Options

In order to have attractive and timely offers in the market, in an instable and unpredictable environment, flexibility is needed at all levels, including:

- Flexible strategies. Following expectedly stable markets while at the same time taking up opportunities that emerge from a less certain environment. Burger and Weinmann in this volume present the example of LichtBlick, a company that defines flexibility as a basis of their strategy: the company succeeds in constantly adapting to a changing environment.
- Company structures. Actively developing different structures for different business units, for example, production, networks, sales, energy services. This may include founding (legally) separate entities to enhance the ability to correspond to customers' needs. One interesting example is BEEGY, a joint venture of the utility MVV Energie AG, BayWaRe as an investor in renewable energy, Glen-Dimplex, being a supplier for electrical heating and cooling systems, and GreenCom Networks, offering energy data management. BEEGY offers services in distributed energy systems. With this affiliate, MVV Energie is in good position to address future customers' needs within new and flexible structures.
- Company processes. Adapting to market processes and demand: all mass processes need automated digitalized processes, whereas customized B2B offers, such as contracting, have to be adaptable.
- Time-management. Taking decisions at the right moment, implementing them within a set time period: companies have to be able to react immediately, later, fast or slow, depending on the context and circumstances. Moreover, time management involves optimizing the pace of development. It is preferable to continue an existing business activity as long as its net present value is higher than the net present value of the alternative activity the company could engage in—but to discontinue this activity as soon as a the new activity will provide a higher net present value. This is the very key question in most conventional generation assets: is the specific production unit still in-the-money or not? What is the resulting net present value? One recent example is the shutdown of two Irsching plants ($\sim$1100 MW): the gas fired units, owned by E.ON, Mainova, and HSE belong to the most modern units in Europe, put into operation in 2010/11, and to be shut down in April 2016. The reason: influx of subsidized renewable electricity and cheap CO_2 emission rights reduce the wholesale price under the level of marginal cost for this plant.
- Balance sheet design. Actively shaping relevant ratios, margins, and limits for consolidation.
- Financial structure. Including the use of off-balance or nonrecourse financing.
- Human Resources. Enabling a company to vary capacities and qualifications. For example, Thüga, a company holding shares in, and actively

managing, 120 local and regional utilities all over Germany, contributes to value generation of their companies by offering HR development instruments, implementing job markets for their subsidiaries, and career management, in order to keep HR resources within the group.
- Flexible legal structure. Applying flexible supply contracts or structures, allowing new cooperation.

Only an organization that is flexible enough can follow the developing dynamics and offer the solution that the market demands. Given the often homogenous strategies, structures, and cultures of today's utilities, the development and realization of flexibility and options is one of the main challenges the companies have to cope with.

3.3.3 Diversification
In-/stability and un-/predictability are a potential threat to company's survival. A solution to minimize this risk is diversification.

Experience shows that companies tend to diversity when challenged by an opportunity, lacking a clear vision of what the company should look like in future. This often leads to badly steered or integrated activities, participations, and structures. Therefore, prior to diversifying, companies need to clarify, within their core energy business, questions like: "Should we develop new large scale (f.ex. renewable) production units and / or develop distributed generation?" This results in a concentration or diversification strategy based on the specific competences and resources in capital, capabilities, and cooperation partners.

To mitigate risk and achieve growth further, utilities may decide to diversify into related fields, such as telecom and cable TV, water supply, waste water and sewage, or waste disposal.

In fact, today's and tomorrow's challenge is: How can we implement both options? How can we stay competitive in existing markets—as for example in conventional production—and, at the same time, develop new business models, for example, in distributed generation or in infrastructural services? (Smith et al., 2010) The important challenge then is to deal with uncertainties in all fields of strategy, as well as with paradoxical situations, as one option tends to cannibalize the other one.

And, in most cases, this option might not fit into the available organizational structure—which in turn means that the organizational structure might need to change. This leads on to the next strategy:

3.3.4 Cooperation
Cooperation is a powerful option to combine effectiveness with efficiency. It becomes essential to optimize the scope and depths of the cooperation by developing cooperative business models for each specific situation, for example,

- in some activities, the company's strengths and development might best be served by "going it alone"

- in others, it might be good to integrate partners that will help achieve success, but be the leading partner
- in yet other activities, it might be good to keep this activity available to the company by becoming the junior partner of a company with more experience
- and for some activities, it might be good to de-invest, to develop an exit strategy.

The German energy sector consists of an abundance of horizontal, vertical and lateral cooperations in energy production, supply, and capital market (Cf. Section 2.5). "Big 4" companies, utilities, as well as service providers and cooperations have been developing competitive offers. For example, Trianel, founded in 1999, owned by more than 50 communal utilities, delivers trading and back office services, develops and operates joint production units and white label products for their partners. Trianel itself cooperates with partners like IT service providers or in delivery of services in order to ensure competitive and sound processes.

To decide which aspect or type of cooperation is most suitable, a detailed assessment of the company's strengths is needed. These strengths can be "hard skills" such as capital, or "soft skills," that is, the company culture or the performance record, the speed, the ability to cope with opportunities, threats and complementary structures, the innovation and cooperation potential.

3.3.5 Sustaining and Disruptive Innovation

Sustaining innovations (Section 2.4) consisting of improvements of an existing technology might include:

- Digitally enabled customers demand different points of contact for commanding energy and for information about energy related issues—internet, apps, etc. Utilities develop new processes and sales channels. To be efficient, companies need to cooperate with professional, reliable, and cost efficient partners with complementary skills and resources.
- Digital utilities: processes are to be digitalized—here again, utilities tend to find appropriate solutions, even though, to date, cost efficiency and time to market pose serious challenges.

Regarding disruptive innovations, "the problem of incumbents is not the technology itself but the inability to commercialize these new technologies respectively to perceive related opportunities" (Richter, 2013). This is where utilities so far do not excel. For example, developing a profitable business case from contracting for small residential customers has hardly been realized to date. In future, new offers combining product features from different sources and competitors need to be developed. Trianel (see earlier) is one of several actors implementing a white label contracting product for a PV/storage system. While utilities are attracted by this kind of cooperation, sales success usually still is moderate. The authors presume that, on the one hand, in many cases, sales processes still need to be professionalized; on the other hand, the market segment is limited, and still needs to be developed.

4 SUCCESS FACTORS FOR STRATEGY DEVELOPMENT AND REALIZATION

The development of specific business strategies requires:

- An analysis of strengths, weaknesses, opportunities, and threats: *what is the company's position, where does the company want to be in the future, what are the options and conditions the company operates in?*
- A concrete development process with clearly defined goals: *what does the company need, when and how can we achieve them?*
- Specific strategies for business units and functions as paths to reach the Goals: *how can the company get there?*
- The strategies have to be translated into processes: *How can this be driven forward, how can this process be set in motion?*
- These processes have to be given structures (*whose responsibility, whose competence, where are the capacities, who has the relevant qualifications etc.?*)
- Then, these structures have to "come alive" in the company culture: *how will processes proceed—open or hierarchical, risk-averse or risk-conscious, fast, slow, meticulous, or broad strokes etc.?*

Although this sounds like common sense, empirical evidence suggests that for many utilities systematic approaches in strategy development and execution are challenging tasks. One particular challenge is to be able to manage parallel and multilayered strategies, sometimes even contradictory strands of strategy development. One such example can be seen in the business model for decentral energy production:[7]

- Markets with disruptive innovation and explorative strategies, such as new combined products in the area of autonomous production, cannot be unlocked based on one reliable business plan, but rather with a trial-and-error approach and a start-up mentality.
- For exploitation strategies that are required in the conventional production business, a stringent optimization of the asset management and of the operative business, and a high performance focus, are needed. This requires a more structured and deterministic decision-making process in future than what can be experienced today.

Conflicts in resource allocations of these divergent and, in some cases, even cannibalizing business processes have to be assessed and resolved openly. The

---

7. The impact on the network and distribution strategy and planning is outlined by Hanser & Van Horn in this volume.

resulting tensions have to be managed—this requires excellent leadership skills, and the ability to work in the context of changing roles (Smith et al., 2010).

5 CONCLUSIONS

While some utilities of the future will be different than they are today, others will not. Depending on their specific environment, stakeholders, market position, strengths and weaknesses, they may evolve along one or more of the following paths:

- continue to be an integrated energy supplier, with different depths of internal value added
- develop towards a manager of complex energy systems integrating value generation from cooperation partners or external sources, on a long or short term basis
- develop toward a facilitator of central and distributed components and structures, reducing internal value generation and offering a platform to aggregators and other partners or service providers, based mainly on grid activities
- diversify from energy toward infrastructural services, from regional to national/international energy production or sales.

Indeed, instead of waiting for and constantly adapting to details of political interventions, utilities need to focus on their environment from a holistic perspective. The unique position of the company—be it a local utility, a bigger player, or an international utility specializing in specific segments—has to be the basis of goals and strategies. But without consistent translation of these goals and strategies into processes, structures, and company culture, a strategy remains pure theory. Companies need to engage in a continuing learning process. This means being willing to pass on strategies, to slow down or speed up, to work from a different angle etc. This must be based on a clear definition of:

- *What*: continue or close down production or projects, develop new parts of the value chain, engage in new regions, address new customer segments.
- *How*: alone, in cooperation with a partner, as junior partner, with business models that make use of more cooperative value creation, that is, with other service delivery partners, and with customers becoming integrated partners.

Making use of these dynamics of strategic development can be a response to instability and unpredictability, can reduce complexity, overcome or compensate volatility, reduce threats, and maximize opportunities.

REFERENCES

Balta-Ozkana, N., Botelerb, B., Amerighic, O., 2014. European smart home market development. Energy Research and Social Science 3, 65–77.

Bundesministerium für Wirtschaft und Energie, 2014. Die Energie der Zukunft, Erster Fortschrittsbericht zur Energiewende. BMWi (Federal Ministry for Economic Affairs and Energy), Berlin.

Burger, C., Weinmann, J., 2014. Germany's Decentralized Energy Revolution. In: Sioshansi, F.P. (Ed.), Distributed Generation and its Implications for the Utility Industry. Elsevier, Amsterdam, p. 69.

Chesbrough, H., 2007. Business model innovation: it's not just about technology anymore. Strat. Leadersh. 35 (6), 12–17.

Dieckhöner, C., Hecking, H., 2014. Developments of the German Heat Market of private households until 2030. Zeitschrift für Energiewirtschaft 38, 117–130, p. 126.

Dyllong, Y., Maaßen, U., 2014. Bewertung der Studie "Entwicklung der Energiemärkte—Energiereferenzprognose" aus Sicht der Braunkohle. Energiewirtschaftliche Tagesfragen 12.

Eurelectric, 2014. DSO Declaration: Power distribution: Contributing to the European Energy Transition. Eurelectric, Brussels.

International Energy Agency, 2014a. Technology Roadmap—Solar Photovoltaic Energy. OECD, Paris.

International Energy Agency, 2014b. Technology Roadmap—Energy Storage. OECD, Paris.

Institut der deutschen Wirtschaft Köln, 2014. Eigenerzeugung und Selbstverbrauch von Strom. Energiewirtschaftliches Institut an der Universität zu Köln. Cologne. "im Auftrag des BDEW." In German.

Luhmann, N., 2000. Vertrauen—Ein Mechanismus der Reduktion sozialer Komplexität, 4th Lucius and Lucius, Stuttgart.

PwC, 2015a. Finanzwirtschaftliche Herausforderungen der Energie—und Wasserversorgungsunternehmen. Frankfurt (German).

PwC, 2015b. Bevölkerungsbefragung Stromanbieter. Frankfurt (German).

Richter, M., 2013. Business Model innovation for sustainable energy. Energy Policy 62, 1226–1237.

Smith, W.K., Binns, A., Tushman, M.L., 2010. Complex business models: managing strategic paradoxes simultaneously. Long Range Plann. 43, 448–461.

Zentrum für Europäische Wirtschaftsforschung GmbH, 2014. Innovationsverhalten der deutschen Wirtschaft. ZEW, Mannheim.

Chapter 18

The Future of Utility Customers and the Utility Customer of the Future

Robert Smith*, Iain MacGill[†]
*East Economics, Sydney, Australia; [†]Centre for Energy and Environmental Markets (CEEM) and School of Electrical Engineering and Telecommunications, UNSW, Sydney, Australia

1 INTRODUCTION

Electricity utilities have a reputation for being cautious, conservative, and inward looking. This wasn't always so. Early industry players such as Edison were highly entrepreneurial in delivering customers tailored distributed energy services (mainly lighting, motive power for industry, and traction for rail) (Smith and MacGill, 2014). However, with maturity and continued growth, and as electricity became central to people's wellbeing, the industry transitioned to a more staid and stable business model: providing an essential public good through natural monopoly infrastructure. The grid's role was keeping the lights on and our role as appreciative energy consumers was to pay for it. The magic that happened on the utilities side of the meter was their business, what we as customers did on our side of the meter was our business.

This model has proved remarkably resilient over the past century accommodating not only new technologies, in both generation and end-use equipment, but also changes in public policy, in some cases including removing vertical and horizontal integration to create wholesale competition and retail customer choice. However, because distribution networks have continued as regulated monopolies, even many competitive market sectors still operate within the constraints of earlier arrangements, including simple metering and tariffs, constraints designed as much for public equity objectives as economic efficiency.

Now, however, real change appears to be unavoidable. Utilities that have enjoyed unrelenting growth and protection from competition now find usage declining, and load profiles shifting, as customers improve energy efficiency, deploy distributed generation, and start to manage their loads more actively. Distributed generation, most notably photovoltaic (PV) systems, is a major disrupter but other technologies, including distributed storage, "smart" end-user equipment,

Future of Utilities - Utilities of the Future. http://dx.doi.org/10.1016/B978-0-12-804249-6.00018-X

343

and electric vehicles, may prove as disquieting (as described in the chapters by Cooper and Grozev et al.). Network monopolies are being challenged and regulators are left to ponder what tariff, market, and regulatory arrangements might be required for this brave new world.

The role to be played by energy consumers, particularly residential customers, is the main focus of the chapter, structured as follows. Section 2 briefly considers competing futures of the grid from technology, business model, market reform, and policy perspectives, highlighting some of the limitations in the framing of these issues. Section 3 focuses on the consumer's perspective, and the underappreciated importance of considering end-users preferences—what customers want. Section 4 looks at some key energy technologies from the perspective of customer's effectual demand, then Section 5 covers grid parity, followed by the chapter's conclusions.

Providing a counterweight to the industry's focus on new technologies and business models, and policy maker's focus on competition and tariff reform, Section 6 concludes that customers' preferences and responses, not technologies alone, are fundamental to the shape of the utility of the future.

2 WHERE NEXT FOR THE GRID—FATE AND ITS DRIVERS

Where to now? As evident from this book and elsewhere, there is no one clear path ahead. Nillesen and Pollitt describe five megatrends, Buger and Weinmann describe incumbent strategies, and Nelson and McNeill discuss Australia's specific strategies. Yet, the grid seems to face one of three possible fates: *mostly business-as-usual*; a more *distributed grid*; and *grid defection*, where a growing number of energy users depart it entirely. Which fate triumphs will depend on the interplay of varied drivers, including technology progress, prices, market and regulatory arrangements, and broader societal objectives, including environmental challenges. However, often overlooked, it also depends on energy consumers themselves. Customers' needs are what drives the value chain, what underpins the business case of industry structure options, and what determines how surplus value is created and shared.

2.1 Mostly Business-as-Usual

There is no doubt about the transformative power of electrification, or the value people place on what it does. "We are creatures of the grid."[1] Although there are still more people without access to electricity than when Edison introduced his light bulb in 1888, most of the world can take reliable grid electricity for granted.[2]

---

1. The phase is Achenbach's (2010). Off-grid applications provide valuable but different services to those experienced from the grid; also see chapters by Cooper and Covino et al.
2. The important issues for developing countries of access to, and the quality of, electricity supply and of energy poverty in developed countries are not covered in the chapter. This paper's focus is on utilities in a future of relatively prosperous developed nations.

Centralized grids owned and operated by vertically integrated monopoly utilities have proven remarkably successful. Grids represent the largest machines on the planet, and electrification has been judged the greatest engineering achievement of the twentieth century, ahead of the automobile, the telephone, and the computer (Constable and Somerville, 2003). The success reflects the benefits of combining the centralized geography of traditional energy resources (eg, hydro systems and coal basins) and major economies of scale in conventional distribution and generation (eg, hydro, thermal, and nuclear), with historical load growth, and the load diversity. The most straightforward, and still the predominant, business model to deliver electricity has been to make power utilities, whether private or public, regulated monopolies. Indeed, only a relatively small proportion of energy consumers are served by utilities in restructured, market-oriented energy industries, and for these network monopoly charges still remain a major part of bills.

Until recently centralized monopoly grid electricity has been able to provide cheap and reliable power that outcompeted other energy options for the major share, but not all, stationary energy uses. This positioned electricity as an essential public good and supported subsidized access for less fortunate or especially favored customers. Hence, cost recovery for utilities could be managed with "postage stamp" pricing across customer classes, largely determined by equity concerns, rather than cost reflective tariffs.[3]

How might new technologies, business models, market arrangements and policy drivers change this? Scale and diversity advantages have allowed the vertically integrated grid model to assimilate successfully both new policies and new energy technologies—generation technologies (including efficient gas-fired plants, as well as highly variable and unpredictable wind and PV), and new loads like peaky reverse cycle air-conditioners. But while life has become more complex and riskier for many generators and retailers, only recently is the network's traditional business model being confronted by changing drivers of customers' usage.

The drivers of customers' energy usage can be grouped as growth increasers (population and prosperity), growth slowers (prices and policy), and growth muddiers (productivity and preferences).[4] Population and prosperity, the two traditional long-term growth drivers, will continue to play a role. If the past is any guide, there will be more of us, we will be richer, and we will spend more. Yet, while energy usage remains linked to economic growth, this relationship is weaker now than in the past and is being swamped by other counterbalancing factors, chiefly prices and policy.

---

3. Chapters in Part II, as well as those by Biggar and Reeves, Knieps, Gellings, and Nelson and McNeill explore aspects of pricing issues and the need for change.
4. These have parallels with the megatrends and disruptors in chapters by Nillesen and Pollitt, as well as Rowe et al. Also see Sioshansi (2013)."

Prices, volume-based electricity tariff increases, are clearly driving a slow-down in usage across many jurisdictions. But are continuing grid price rises as inexorable as some assume? The key role of relative prices and "grid parity" are discussed further in Section 4 but highly visible price developments may be being given more prominence than they merit.[5] Less visibly, policies underpinned by environmental concerns have been driving reductions in electricity usage through energy efficiency schemes and education, appliances and building standards, and the impacts of Feed-in-Tariffs and carbon prices. And if voters' preferences support cleaner energy futures, policy drivers will continue to reduce energy usage.

The two final drivers of energy usage, productivity and preferences have muddier, mixed, and less certain impacts. Productivity is the efficiency benefit technology delivers, getting more with less, and is most often reflected for generation technologies in falling product and output prices. Productivity is also seen in the growing availability and competitiveness of efficient end-use technologies, such as LED lighting and heat pumps in refrigeration and reverse cycle air-conditioning.

Most forecasts of the utility of the future hinge on increased productivity making new technologies cost effective.[6] Yet this risks missing something important—preferences drive customers' choices. This also applies in the policy space where voters' environmental views will be critical to support for lower energy use and renewable generation. Ultimately, policy consists of peoples' preferences as filtered through the political process.

Energy customers' responses to shifting drivers is leading to flat or falling demand, and increasing distributed generation. This is making utilities' traditional business models, tariff, and market arrangements appear outdated. As what utility business's do on their side of the meter is now being threatened by what customers' do on the other side, business-as-usual seems likely to become a nostalgic scenario.

2.2 The Distributed Grid

Users of the grid have always been distributed, in terms of numbers and location. Only the most energy hungry industry would chose to locate near power supplies or to develop onsite generation for more than backup supply. But now that the scale economies and the geographical distribution of traditional energy sources (thermal, hydro, and nuclear) that kept generation big and centralized are less compelling, it seems natural to assume that energy supply will become distributed to match energy demand.

This may or may not follow. Distributed technologies such as PV and battery energy storage may not require major economies of scale, but neither do

---

5. Recent Australian price rises represent huge step changes or price shocks, but a price driven utilities "death spiral" has been widely debunked, for example, the Chapter by Athawale and Felder, and Smith and MacGill (2014).
6. See the discussion in the chapter by Nelson and McNeill.

they have major diseconomies. Household sized PV systems can have similar costs to utility size plants, so there may no longer be a single best size. For such technologies, bigger is no longer necessarily better, but neither is smaller. A fully scalable generation technology can be effectively deployed in multiple locations and multiple sizes, both beyond and within a network. For example, Marnay's chapter makes the case for an integrated grid comprising multiple connected but semiautonomous microgrids.

The rapid uptake of household PV systems in a number of jurisdictions seems to point towards a fully distributed "small is beautiful" future. Australia's 1.4 million household PV systems, discussed in the chapter by Nelson and McNeill, are notable examples and cover the roofs of 15% of all houses, and 40% of owner occupied houses in some states. Most were installed in the past 5 years during a time of generous government subsidies and gross metered feed-in-tariffs and when PV system costs fell markedly. Now household's future financial benefits from PV's will come predominantly from offsetting their use at the equivalent of retail KWh tariffs, as policy support has been reduced, PV systems are net metered and exports are paid a much lower rate—perhaps a quarter of the retail tariff.

At some tariff level, PV systems make good financial sense although, as we discuss later, this point of "grid parity" for PV is both crucial and problematic. In particular, PV exposes the lack of cost reflectivity in volumetric network tariffs, as households' daytime PV generation does not necessarily reduce network costs driven by residential evening demand peaks; see chapter by Grozev et al.[7]

Without cost reflectivity, under the flat volumetric tariffs almost all households and small businesses currently pay, efficient behaviors and choices are not incentivized. Suggested industry responses vary from specific "solar tariffs" to structural tariff changes toward fixed and demand charges. But, as most of the authors in this volume conclude, it seems fair to say that policy makers and regulators are struggling to develop coherent market and regulatory frameworks for incorporating these new developments within the grid.

A more distributed grid seems the most likely electricity industry fate, leaving utilities somewhere betwixt the old business-as-usual world and a brave new world, where their old role is substantially diminished. For some however, progress in distributed battery storage technologies is seen as foreshadowing a radical shift to grid or load defection.

2.3 Grid Defection

A growing number of industry observers see solar + storage[8] "grid parity" priced household PV combined with low cost battery storage, making grid defection both technologically possible and potentially economically advantageous.

---

7. Cost and revenue mismatches are not restricted to distributed generation, for example, peaky reverse cycle air-conditioning can create similar problems, see Department of Industry and Science (2015).

8. The chapter is focused on households, but related issues are being faced from business customers.

The Rocky Mountain Institute (RMI) present a strong case for solar + storage to undermine the economics of the existing grid through "load defection,"[9] and reflects a growing consensus that the grid will be significantly stressed and downsized, if not made redundant. This view is mirrored in financial analysts' reactions, such as the following:

> In the 100+ year history of the electricity utility industry, there has never been a truly cost-competitive substitute available for grid power. We believe the solar + storage could reconfigure the organization and regulation of the electricity power business over the coming decade.[10]

> Over time, many U.S. customers could partially or completely eliminate their usage of the power grid.[11]

> Our view is that the "we have done it like this for a century" value chain in developed electricity markets will be turned upside down within the next 10–20 years, driven by solar and batteries.[12]

The RMI paper concludes that "it remains unlikely that large numbers of customers would leap directly from grid connection to grid defection," but the "economically optional generation mix" *shifts* toward solar + storage in 10–15 years until by 2050 the "grid takes a backup-only role."

The RMI's conclusions naturally depend on many assumptions, including zero demand and capacity charges, falling solar and storage costs, and rising grid electricity costs. In particular, assuming a 3% per annum rise in grid electricity cost effectively assures long-run success for solar + storage as by 2030 grid electricity prices are 60% higher and by 2050 close to three times the current levels.[13] And, expected cost reductions in stand-alone household scale solar + storage will surely also apply to small and large scale grid applications as well. Nevertheless, the message is clear, the grid is under threat from load defection if not grid defection and a very different electricity industry future now seems possible.

Yet, while Tesla's Powerwall home battery launch has ignited imaginations, and grid parity seems nigh, talk of distributed storage currently outstrips both its application and the industry's understanding of how customers will respond. The seemingly underappreciated role consumers and their preferences have in creating defection, a more distributed, or a mostly business-as-usual grid future is discussed next.

---

9. Mandel et al. (2015), which builds on RMI's earlier paper on grid defection.
10. Barclays, quoted in Mandel et al. (2015).
11. Morgan Stanley, quoted in Mandel et al. (2015).
12. UBS, analyst note in Mandel et al. (2015); also see Hasnie (2015) and Bayless (2014).
13. Major issues are also left unaddressed, such as "Grid-facing costs such as T&D maintenance and central generation, as well as costs for grid-dependant customers who can't or don't invest in solar-plus-battery systems, are important issues beyond the scope of this analysis." Mandel et al. (2015).

3 PREFERENCES—MASLOW'S BASEMENT, COOPETITION, AND ENERGY ECOSYSTEMS

At the end of the 20th century, just as Y2K was about to disable the planet and before the Dot-com bubble burst, Hal Varian—now the chief economist at Google—cautioned: "Technology changes. Economic laws do not" (Shapiro and Varian, 1999). Economic laws, however, have limits. Economics' strength is in its abstraction of reality to simplify and model a complex world but this is also a weakness. First year economics student are taught to "assume preferences are given," and then ignore them for the rest of their studies (and possibly their careers) (Dietrich and List, 2013). Yet, down the hall, first year marketing students are discovering that customers' preferences are elusive and fluid, as they are taught to gauge, shape, and fulfil consumer desires.

Understanding the future requires a closer look at customers' preferences, behavior and responses, within a complex household energy ecosystem where incumbent grid electricity still has valued characteristics: reliability; affordability (driven by economies of scale); invisibility, a homogenous commodity supporting mostly passive customers; and ubiquitousness, an essential service and "health and hygiene" factor available everywhere on the grid. But while electricity remains an essential service, how people value it can vary with their circumstances.

For people on very low incomes, grid electricity is liable to be a superior good, where an increasing share of extra income would be spent on meeting basic "heat and light" needs better. As incomes rise, grid electricity becomes a normal good, where use rises with rising incomes but not as quickly. Finally, at the higher income levels experienced by well off customers today, and increasingly electricity customers of the future, grid electricity may be an inferior good, where usage falls with rising income as more expensive new appliances and dwellings are inherently more energy efficient and solar + storage is a possible lifestyle choice, rather than a cost effectiveness decision.

Frei (2004) describes an "energy policy needs pyramid" based on Maslow's hierarchy of needs in order to explain different countries' policy approach to climate change. He contends that, once access to commercial energy is achieved as a policy goal, "it can be observed that the question of supply security prevails over cost-efficiency, environmental and social issues."[14] These basement needs in Maslow's hierarchy—physiological and security—also underpin household energy use.

The many shapes and forms of household energy use are so commonplace they tend to be overlooked. Smartphones, laptops, TV, and upmarket kitchen appliances are what engage customers' interest but the majority of household energy use is in heating and cooling loads: hot water; space heating and cooling; fridges and freezers; the kitchen/cooking; and even lighting, most of the energy for which is lost as heat. "Low tech," "old school," and "low involvement"

---

14. Frei, who is consistent with Kranzberg's fourth law of technology "Although technology might be a prime element in many public issues nontechnical factors take precedence in technology-policy decisions"

FIGURE 18.1 **Energy usage for an all-electric household.** *(Source: Photo from Ausgrid Energy Efficiency Centre.)*

appliances consume the most energy, not the stuff of YouTube videos, Pinterest, and tweets (Fig. 18.1).

Internet-ready fridges notwithstanding, most basic energy needs and many appliances have not fundamentally changed. What has changed are their prices. A "Flame shell" 2000 W electric heater, which in the late 1950s cost £5 17 shillings and sixpence (about $11.75 then, or $140 in today's money), does the same job as a 2000 W electric heater that sells for only $14 today. Put another way, it took a third of a week's pay to buy 2000 W of heating capacity on an Australian minimum wage in the mid-1960s. Today, the same can be bought for less than an hour's work (Fig. 18.2).

It is easy to overlook such dramatic changes when they occur in small steps over long periods. Yet dramatic changes have occurred, unremarked upon, in the relativities of appliances cost of purchase and cost of use.[15] Depending on local tariffs, leaving a 2000 W heater on for a day costs more in kilowatt-hour tariff charges than the heater's purchase cost. And two such heaters create as much load as charging an EV, or running a large air-conditioner in summer. The "flame shell" example of cost relativities is replicated in the broader appliance

---

15. See Nordhaus (1998) and Fouquet and Pearson (2006) on lighting, and "The people history" website for appliance's history. Moore's Law for computers is an exceptional case, but air-conditioners, for example, are one-third more efficient than a decade ago, and the least efficient air-conditioner (of 4 kW or less) on sale in 2015 is better than the most efficient available in 2002 (BREE), and can be more cost effective for heating than an electric bar radiator. The people history: 1940s appliances including price, http://www.thepeoplehistory.com/40selectrical.html

FIGURE 18.2   The 1950s Vulcan Conray "Flame Shell."

price index measures in Fig. 18.3. This shows how households' spending to meet basic energy needs is increasingly about kilowatt-hour operating costs and less about appliance purchase costs, thereby making additional features, including energy efficiency, increasingly affordable.

The premium customers pay for higher order needs over low level needs can be substantial. Like an electric heater, an electric kettle's function has not changed for almost a century. Modern examples have some improvements but a $7.50 Kmart model and a $199 SMEG version have similar features and do the same job. Rather than basic functionality, most of the $191.50 or 2550% price differences in the two kettles can be attributed to meeting consumers' higher order needs. And, while not

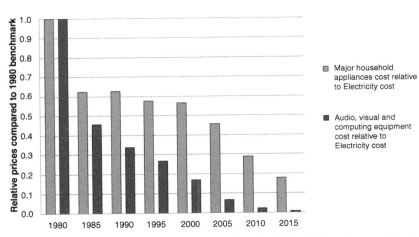

FIGURE 18.3   Falling appliance prices and rising electricity tariffs. (*Source: Australian Bureau of Statistics, 6401.0.*)

Kmart $7.50   **SMEG $199**

FIGURE 18.4   Electric kettle preferences.

everyone chooses to boil water in a SMEG designer statement,[16] it is this sort of higher order preference that affluent customers aspire to, is a driving factor in existing energy preferences and will shape customers' future energy choices (Fig. 18.4).

Some household spending, particularly on energy, will always be about Maslow's basement "physiological and security" needs. Yet after these needs are met the majority of spending, now and in the future, will likely be about the upper reaches of Maslow's pyramid—preferences, wants, and aspirations for social standing, positioning, and esteem. About who customers are, who they want to be, and how they fit in. Influences that are already evident in households' energy choices.

Grid electricity is the dominant form of energy supply for the majority of uses in the majority of households but it is not an energy monopoly. Grid electricity's major advantages, compared to traditional competitors, particularly local cleanliness, low maintenance, and low costs, has not wiped out all competition. Customers' preferences support diversity. Grid electricity often coexists and competes with gas, fuel oil and wood, in order to supply households with energy services, particularly cooking and heating loads.[17] Product characteristics and people's preferences, not only cost effectiveness, drives these choices.

Moreover, households' energy use encompasses a diverse and eclectic collection of sources and options as complements or specialist niche players, as well as substitutes. Esoteric energy sources lie hidden within households including: chemiluminescence in glow sticks; pyrophoricity in lighter flints; exothermic chemical reactions in heat pads and cold packs; piezoelectricity in a self-winding watch; pneumatic pressure (in aerosols, bottled gas and CO_2

---

16. See Webb and Suggitt (2000) on the Alessi kettle design classic.
17. Across Australia, only 6.5% of Victorian households are "electricity only," compared to more than half of Tasmanians, and 32% nationally, 13% still use wood as an energy source, and candles still light celebrations and romances (ABS 4670.0).

canisters); and hydro pressure in water pipes. Energy diversity, not grid monopoly fuels households.

Household energy storage is already here as more and better batteries play a bigger part in consumers' lives. Mobile devices make batteries ubiquitous in phones, tablets, laptops, toys, and rechargeable appliances. Almost unnoticed, these mobile devices have reduced the potential disruption to households from short term electricity outages as, in combination with thermal storage (in hot water systems, building design, and refrigerators), other energy sources and efficient uses (such as LED lights), they create a viable off-grid service buffer of 1–4 hours. So, electricity, the textbook natural monopoly (Kishtainy, 2014), already operates in coopetition with a growing number of energy supply, service, and storage alternatives.

In the midst of this energy coopetition, what the electricity utility industry does best is "keep the lights on"—maintain reliability. In modern cities, this reliability is a fundamental need; an essential health and hygiene factor that allows the ordinary business of life to continue.[18] But once lower order needs are met higher needs take over and customers are likely to seek a bundle of characteristics to fit their needs and preferences. While environmentalists may be queuing on the left for solar + storage, survivalists may be queuing on the right to buy the same thing for very different reasons.

To fit into customers' households, new energy technologies, like solar + storage, need to balance price and product characteristics against both customers' basic needs and higher order preferences.

4   GRID PARITY IN PRICES AND PRODUCT

A key trigger seen as pushing the grid to endangered species status is "Grid parity," the current or imminent equating of distributed small scale solar + storage with centralized grid electricity prices.

The potential benefits of solar + storage are substantial, notably: free energy from the sun; independence; protection from grid price rises; no grid outages; and reduced environmental impacts. But the indirect costs include "insourcing"[19] operation, and maintenance of panels, inverters, and battery storage. Most significantly, indirect costs include the reliability impacts of being off-grid, or else the additional cost of retaining a grid connection.

---

18. While the worlds' biggest blackout, in India in 2014, occurred with minimal disruption because people had adapted to regular localised outages, this does not imply this was acceptable to consumers. The World Bank concludes "Access to electricity is a necessary input to economic growth and poverty reduction. It is also a key ingredient for achieving welfare outcomes of the Millennium Development goals ... in addition up to a billion people... are subject to unreliable and low quality power supply ...thus the electricity access challenge goes beyond basic connectivity to ensuring that the supply is affordable, adequate, reliable and of acceptable quality" IEG (2013).

19. Insourcing runs against the trend for affluent consumers to outsource to others dealing with basic health and hygiene needs. Although more than half of all Australian surveyed "insource" some part of their food supply with a herb/veggie garden (Wise, 2014) this is more about higher order gourmet/freshness/hobby/leisure preferences than price competitiveness with supermarkets and becoming food prosumers.

	Redundancy	Faults and reliability	Basic need buffer	Network profile impact
The Grid	N-1	99.98% two faults needed for an outage	1-4 h from thermal storage and batteries in key devices	Peak demand hot summer evenings
Grid and PV	N-2 in sunshine	>99.98% three faults needed in daylight outage	Night time 1-4 h thermal storage, batteries in key devices	Daytime demand reduced, lesser impact on evening peak
Grid, PV, and battery	N-2, N-3 in sunshine	>99.98% four faults needed for a daytime outage	No outages for sufficient battery size and normal weather	Throughout day peak demand reduced
PV and battery	N, N-1 in sunshine	<99.98% two faults needed for a daytime outage	Based on battery size, outage length is repair time for PV or battery at night	Peak and usage reduced
PV, battery, and genset	N-2, N-3 in sunshine	>99.98% two faults needed for an outage	No outages if sufficient battery size and genset buffer for bad weather	Peak and usage reduced
Grid and battery	N-2	>99.98% three faults needed for an outage	No outages as a function of battery size	Peak reduced

FIGURE 18.5   Reliability with Grid DG and storage options.

Assumptions of "grid parity" in current retail electricity prices are not necessarily accompanied by a like-for-like grid parity in product characteristics or underlying costs. True grid parity for solar faces two issues: volumetric tariffs are not cost reflective, and product characteristics are not the same.

The cross subsidies in volumetric kWh tariffs for solar are well recognized.[20] So it is likely that utilities will move to increase capacity and connection charges, while lowering kWh tariffs and thus reducing the volumetric component of customers' bills. This is a substantial readjustment for incumbent electricity utilities, but is both technically achievable and capable of customer acceptance. In Australia, as in most OECD countries, capacity and demand tariffs only exist for larger commercial customers. Yet households cope with significant fixed charges, and declining block tariffs (lower usage pays a higher average per unit cost) in gas bills, water bills, and local government rates, so it seems entirely possible that households will eventually accept this in their electricity bills, just as business customers have.

Predictions of "grid parity" solar + storage leading to mass grid or load defection reflect a combative "all or nothing" approach—Schumpeter's "creative destruction" view of innovation meeting Schumacher's "small is beautiful" and "appropriate technology" philosophy. Reality is more nuanced. Even with significant price falls for solar + storage, a mixed model with load defection but a low possibility of grid defection appears most likely because of the reliability characteristics of household grid, PV, and storage options, shown in Fig. 18.5.

This is supported by the modelling by Khalilpour and Vassallo (2015) of real PV and household consumption data that uses combinations of tariffs, sizes, PV, and battery costs to estimate financial benefits and unserved energy:

A small PV-battery system has the highest NPV, but also the highest amount of unserved energy... The results of the study imply that leaving the grid is not a feasible option even at low PV and battery installation costs.[21]

---

20. See Nelson and McNeill amongst others.
21. Khalilpour and Vassallo (2015) conclusions have parallels with the electric boosting of solar hot water. Perversely overwhelming success of new energy technologies creates sunk grid costs leading to marginal-cost-only grid tariffs (Hartley and Medlock, 2014).

So, grid defection may be palatable only where people are willing to forego reliability, to buy uneconomically large combinations of solar + storage, and/or an additional supply like a small genset.

The off-grid reliability outcomes shown in Fig. 18.5 also assumes household solar + storage can be islanded from the network, in the event of an outage, currently an additional expense only justified for major commercial and critical services customers. Without an islanding option, the reliability experienced by customers is the same with or without PV and batteries but the costs are borne more by customers without generation and storage.

Technology-based forecasts seem to favor a heavily distributed grid or grid defection, over a mostly business-as-usual scenario. But, while little is really known about customers' preferences, it seems reasonable to assume customers are primarily concerned with outcomes, and should be largely indifferent to the generation structures and business models that deliver them. Indeed, the majority of the world's population now chooses to live not with greater self-sufficiency but in cities that create value from the opportunities they offer for shared infrastructure, endeavor, and exchange.

Once generation is freed from the economies of scale constraint placed on cost effective fossil fuel and nuclear plants, there is no longer a size barrier to decentralization. Generation and storage may now be cost effective at many sizes and many locations, not just the smallest and nearest to home. Scalable solar + storage may be as effective at an aggregated local level, or within and supporting a distribution network, as when distributed with individual end users. What can we expect from new energy technologies?

5   S-CURVES, S-BENDS, AND CAN EV'S SAVE THE GRID?

If the cost of solar + storage falls substantially, it will entail lower costs for household batteries and hence batteries everywhere, removing one of the major barriers to Electric Vehicle (EV) adoption. So, just as solar + storage appears to be threatening the grid, could EVs emerge alongside to be its savior? Could the EV's combination of significant load growth, inbuilt battery and charging flexibility counterbalance load defection from solar + storage?

EVs provide the prospect of a disruptive technological change, yet they have been the "car of the future" for over 100 years, so also provide cautionary lessons on technology challenges from the past, as in the quote below.

The electric vehicle is coming! It is an integral part of man's future and survival on this planet. Today we are observing the stepping stones that will bring technology and imagination together to create truly efficient vehicles and energy systems worthy of the 21st century... The threshold of a new age is upon us. It is ours to behold.

The Complete Book of Electric Vehicles, 2nd edition, 1981 (Shacket, 1981)

Toyota with the Prius, and now Tesla, have demonstrated the EV's mechanical reliability, desirability and commercial viability, at least in niche markets, so EV's do now seem poised to reshape the transport and energy industries. Yet

EV sales in Australia and internationally are still below a "take-off point" for widespread acceptance and mass market sales. Of more than 1.1 million cars sold in Australia in 2014, only 1,181 (around 1 in 1000) were EVs, and in only five countries are EVs more than 1% of new car sales. The focus for EVs is now on range anxiety, energy density, cost, and recharging infrastructure—barriers predominantly around battery development and customer preferences.

The typical forecast path of a new product involves applying S-curves after a trip thorough the "valley of death." This involves surviving a period of negative development, set up, and production cost cash flows in bringing a product to market before revenue from accelerating sales volumes are sufficient to offset costs.

S-curves models allow mathematical forecasting of an innovation's long-term future from small preliminary data sets that effectively overlay a predetermined S-curve success story on top of the data and make it fit.[22] Rather than progressing up an S-curve, most technologies falter before crossing the "valley of death," and disappear down the S-bend. One reason is the overestimation by innovators of their product's benefits, compared to end customer's views (Gourville, 2006).

Success of EVs will need to be part of the long and winding road taken by the modern passenger car.[23] Today, backward looking technological determinism would see the success for petrol engine vehicles as inevitable yet, like the EV today, the petrol driven car needed to find what Adam Smith called "effectual demand."

A very poor man may be said in some sense to have a demand for a coach and six; he might like to have it; but his demand is not an effectual demand, as the commodity can never be brought to market in order to satisfy it.[24]

Effectual demand can be interpreted as the combination of price and product characteristics, compared to other alternatives, that matches customers' needs, preferences, and budgets. Historically, EVs have failed to satisfy effectual demand and the path to commercial success is strewn with the wreckage of carmakers' unmet ambitions and customers' unmet preferences, like the Segway and Sinclair C5.

The Segway, middle right in Fig. 18.6, was the secret US$ 100 million "project Ginger" launched to great publicity in 2001 by its inventor, Dean Kamen, who predicted it "will be to the car what the car was to the horse and buggy'.

---

22. Debeacker and Modis find S-curve fitting requires data for a minimum of 20% of the curves full range, well below where EVs and battery storage are at present, see AEMO forecasts, Cooper covers innovation cycles and the role of Grief and Mastery in the last chapter.

23. Cars are as good example because, unlike electricity, they are a high profile, high emotion, high status and high involvement purchase for customers. For more on EV's, see the chapter by Cordani et al.

24. Smith (1776): The world's richest woman, the Queen of England, has two "coach and six" horse carriages, the Australian and Diamond Jubilee State Coaches, but even a fairly poor man in the developed world now has the "effective demand" for a car. Meanwhile, India's Tata nano "one-lakh" car—four seats and 37 hp for less than $US 2000—has struggled to find effectual demand partly because "no one wants to be seen driving the world's cheapest car."

FIGURE 18.6 Electric vehicle innovation gallery.

The Sinclair C5, computer pioneer Sir Clive Sinclair's three-wheeled vehicle, in the bottom right hand corner of Fig. 18.6, was supposed to revolutionize personal transportation for Britons in the mid-1980s, but bankrupted its inventor, and has been voted "the worst gadget in history."[25]

Effectual demand for EVs will come when prices fall (which requires battery costs to plummet) and customers are satisfied that EV's characteristics can meet their preferences, particularly around range anxiety, but also for status and acceptance.

When we do a reality check, imagined technology futures have a tendency to turn out more mundane than we expected. The robot we looked forward to owning in the 1920s, Maria, star of Fritz Lang's Metropolis, and again in the 1960s, Rosie, star of the Jetsons, is here—although not as we imagined it as Roomba, star of the shopping channel (Fig. 18.7).

---

25. Shai Agassi of Better Place continued the computer industry mogul jinx on EV projects, but Tesla's Elon Musk is bucking the trend. Notwithstanding the ambitions of IT entrepreneurs, broadly defined, electric transport already stretches beyond niche applications in milk vans, golf buggies, moon buggies, forklifts and wheelchairs to encompass not only trams and trains but also lifts, escalators, travelators and inclinators, and approaching 500 million electric bicycles, motorbikes and scooters.

| 1927 | 1962 | This morning |

FIGURE 18.7   Imagined futures in robotics.

Even adoption of simple, proven, and cost-effective products can lag expectations. Proponents had championed the whole-of-life cost effectiveness advantage of fluorescent lights over incandescent from the 1950s onward, yet their eventual success took multiple advances in product characteristics, such as size, shape, light output, color temperature, color rendering, mercury content, dimability, distribution channels, and social acceptability. CFL's "price parity" with incandescent lights was a necessary but not sufficient condition for widespread adoption. Ultimately, government policy, through efficiency standards effectively banning incandescents from most uses, combined with price and product improvements, was necessary to put CFL into households. Price parity and payback periods in energy markets don't guarantee "effectual demand" and customer acceptance.

Currently, "Smart" technologies are being championed with smart meters, smart appliances and the Internet of Everything promising unprecedented flexibility and control (including economic efficiency through prices to devices). Unfortunately, this seems to rely on a value proposition driven by technology push, rather than customer preference pull. For example, recent trials of energy usage in-home-displays failed to engage customers, and only delivered high MTKD—mean time to kitchen drawer—performance.[26]

Google-owned multibillion dollar home automation company Nest Lab have plans to change energy management, "one unloved appliance at a time,"[27] starting with thermostats and smoke detectors. But Nest's initial products have a significant price premium—like Smeg kettles—and the combination of low price, ubiquity, and functionality to deliver an Internet of Everything may not

---

26. Ausgrid http://www.smartgridsmartcity.com.au/
27. https://nest.com/

Nous	Knowledge of the issue and what to do
Squint	Financial myopia to investing in long term savings
Din	Competing messages and priorities
Clout	Split incentives and ability to contract for savings
Grunt	Low energy cost and low on customers priorities
Nimby	Unrecognised externalities and local opposition
Sloth	Inertia even among the willing

FIGURE 18.8 Barriers to new energy technology adoption. In behavioral economic terms: Nous is imperfect information; Squint is hyperbolic discounting; Din is cognitive overload and attention constraints; Clout is principle agent problems and split incentives; Grunt is satisficing behaviour for basic needs; Nimby is unpriced externalities; and Sloth is endowment effects, status quo effects and non-optimising behavior. See Gillingham and Palmer (2013) on this and the energy efficiency gap.

have arrived just yet. If and when it does, utilities might not have a central role.[28] Utilities' view that smart utility meters will be a gateway, portal, or hub to the home could prove to be wishful introspection. Ubiquitous low cost computing power and internet conductively could make an electricity meter no more special than a kettle—a thing or at best a node amongs many within a complete home automation system. When all household equipment becomes smart, the main advantage of a smart meter are likely to be its old school virtues—a hard-wired connection for mains power and outside communications. Creating attractive cost-effective solutions to the unrecognized problems of customers' unloved appliances will not be an easy task.

Forecasts of energy technologies, including household solar + storage, are being made principally in terms of price parity, rather than how they meet customer's needs, compared to viable alternatives. Yet, there is a long history of substantial consumer surplus in grid supply supported by the low consumer involvement, satisfying behavior, and bounded rationality to be expected of a basic need. So, even where prices and preferences are favorable, new technologies may still face practical barriers to customers' acceptance of new technology (Fig. 18.8).

The immediate future holds out the prospect of exciting technological change, within the last year Tesla's PowerWall and Model S, Google's NEST, and the Oculus Rift have hit the market and are pushing against the barriers to effectual demand. Sometimes, the future arrives on time, often it is too early or too late. Effectual demand for new energy technologies may be hard to achieve while the understanding of customers' needs remains elusive, and engaging them problematic.[29] Consumer research and literature review by the CSIRO (2015) supports this as:

28. See chapter by Cooper for a description of the IoT and the potential role for utilities in supplying Personal Energy as a service.
29. Industry discussions on energy customer futures contrast with the views of outsiders like Coates (2002), Underhill (1999) and Norman (2004) of what drives customer choice.

Consumer engagement with their electricity supply has recently increased, but it is uncertain how much consumers will want to engage in the future... up until recent price events and solar uptake, electricity use was invisible to the residential consumer.

Despite saying they are willing to change their behaviour to reduce their energy bills, many residential consumers continue to behave in ways that are contradictory to their intent (for example, they increase their use of energy-intensive appliances). Research suggests that motivations (for example, to help the environment) do not necessarily translate into behaviour (such as turning off lights or installing solar panels) and other factors also come into play (for example, social norms, ingrained habits, and the extent to which the person believes it is easy or difficult to take action).

(CSIRO, 2015)

So, while completely business-as-usual is unlikely to remain the fate of the utility of the future—given the history of customers' responses to technology—the anticipatable benefits of current technologies don't yet appear sufficient to push consumers' effectual demand through the valley of death, past the S-bend, and up the S-curve, and warrant a switch to a major grid substitute in the near term.

6 CONCLUSIONS

Technologies alone will not shape the utility of the future—customers also will. Both supply and demand will. Technology and policy shape customer's options but their budgets, needs, and preferences drive effectual demand. The customer of the future will still be rational and value cost effective solutions, but will do so on their own terms. Terms which, once basic needs are met, may include higher order fashion, status, esteem and social cachet preferences, and that could favor renewables, distributed generation, and storage. Or they may not.

Customers in the future will be different than today. On reasonable expectations and existing trends, customers will be: older; more educated; more urban; higher tech and more connected; richer; healthier; greener; and there will be a lot more of them. How will this affect electricity utilities and distributed generation's future? Part of the answer will be in what will not change. Customers are unlikely to become mechanistically rational in their choices, making decisions based on cost benefit analysis and spreadsheets. Neither will Maslow's higher order self-actualizations goals overwhelmingly drive their energy use. Despite the turbulence in the industry, electricity is likely to remain a "low involvement product," stuck in Maslow's basement, while customers' attention is elsewhere, focused where their higher order interests and preferences are met.

A mass exodus off-grid or a death spiral from load defection remains unlikely in the reasonably foreseeable future. Ultimately, customers' "basic need" for grid-like reliability and preference for trading DG output should prevent grid defection.

Grid parity prices of solar + storage will need to be accompanied by "grid parity" service for customers. Product differences, particularly reliability levels, still provide a damper on grid and load defection rates. At high levels of grid defection, tariffs will adjust, or stranded assets and sunk costs will come into play and this will extend the life of the grid beyond the true price and product parity point of solar + storage.

Technological determinism may suggest an imminent triumph of cheap PV and batteries, compared to the current grid. Yet it is people, Kant's "the crooked timber of humanity, out of which no straight thing was ever made," that are the hardest part to predict, not technology. Actual market outcomes involve mixes of price and product where, once their basic needs are met, customer's decisions and effectual demand are defined by perceptions and preferences, as much as simple cost comparison. The utility of future will be shaped not only by electricity supply technologies and their cost but what they do for us as the customers of the future, including, surprisingly, how they make us feel.

REFERENCES

Achenbach, J., 2010. The 21st century grid, can we fix the infrastructure that powers our lives? National Geographic Magazine, July, 2010 accessed at, http://ngm.nationalgeographic.com/print/2010/07/power-grid/achenbach-text

AEMO, 2015. Emerging Technologies Information Paper, National Electricity Forecasting Report, AEMO, June 2015, accessed at, http://www.aemo.com.au/

Bayless, C.E., 2014. The Death of the Grid? Public Utilities Fortnightly.

Bureau of Resources and Energy Economics (BREE), 2014. Energy in Australia 2014, Canberra, November, Commonwealth of Australia.

Coates, D., 2002. Watches Tell More Than Time: Product Design, Information, and the Quest for Elegance, third ed. New York, McGraw-Hill.

Constable, G., Somerville, B., 2003. A Century of Innovation: Twenty Engineering Achievements That Transformed Our Lives. National Academies Press, Washington, DC, Chapter 1, see http://www.greatachievements.org/.

CSIRO, 2015. Change and choice: The Future Grid Forum's analysis of Australia's potential electricity pathways to 2050, CSIRO, Clayton.

Department of Industry and Science, 2015. Energy White Paper. http://ewp.industry.gov.au

Dietrich, F., List, C., 2013. Where do preferences come from? Int. J. Game Theor. 42 (3), 613–637.

Fouquet, R., Pearson, P.J.G., 2006. Seven centuries of energy services: the price and use of light in the United Kingdom (1300-2000). Energy J. 27 (1), 139–177.

Frei, C.W., 2004. The Kyoto protocol—a victim of supply security? Or: if Maslow were in energy politics. Energy Policy 32, 1253–1256.

Gillingham, K., Palmer, K., 2013. Bridging the energy efficiency gap: insight for policy from economic theory and empirical analysis. Resources For the Future Discussion Papers 13-02.

Gourville, J.T., 2006. Eager Sellers and Stony Buyers: Understanding the Psychology of New-Product Adoption. Harvard Business Review 84 (6), 98–106.

Hartley, P., Medlock, K., 2014. The Valley of Death for New Energy Technologies 2014. University of Western Australia, Perth, RISE Working Paper 14-021, RISE initiative for the study of Economics.

Hasnie, S., 2015. Are you killing your Electricity Utility? LinkedIn Pulse, May 1, available on https://www.linkedin.com/pulse/you-killing-your-electricity-utility-sohail-hasnie.

IEG, Independent Evaluation Group, 2013: Evaluation of the World Bank Group's Support for Electricity Access: Approach Paper.

Khalilpour, R., Vassallo, A., 2015. Leaving the grid: an ambition or a real choice? Energy Policy 82, 207–221.

Kishtainy, N., 2014. Economics in Minutes: 200 Key Concepts Explained in an Instant. Paperback, Quercus, London.

Mandel, J., Guccione, L., et al., 2015. The Economics of Load Defection: How Grid-Connected Solar-Plus Battery Systems Will Compete With Traditional Electric Service, Why it Matters, and Possible Paths Forward. The Rocky Mountain Institute, Boulder, CO, www.rmi.org.

Nordhaus, W., 1998. Do Real-Output and Real-Wage Measures Capture Reality? The History of Lighting Suggests Not. Yale University, New Haven, CT, Crowles Foundation paper No. 975.

Norman, D.A., 2004. Emotional design: why we love (or hate) everyday things. Basic Books, New York, NY.

Shacket, S.R., 1981. The Complete Book of Electric Vehicles, second ed. Domus Books, Northbrook, IL.

Shapiro, C., Varian, H.R., 1999. Information Rules: a Strategic Guide to the Network Economy. Harvard Business School Press, Boston, MA.

Sioshansi, F.P., 2013. Energy Efficiency: Towards the End of Demand Growth. Elsevier Science/AP, Amsterdam.

Smith, A., 1776. The Wealth of Nations.

Smith, R., MacGill, I., 2014. Revolution, revolution or back to the future. In: Sioshansi, F.P. (Ed.), Distribute Generation and Its Implication for the Utility Industry. Elsevier Academic Press, Amsterdam, Chapter 24.

Underhill, P., 1999. Why we Buy: The Science of Shopping. Simon & Schuster, New York, NY.

Webb, P., Suggitt, M., 2000. Gadgets and Necessities: An Encyclopaedia of Household Innovations. ABC-CLIO, CA.

Wise, P., 2014. Grow Your Own. The Potential Value and Impacts of Residential and Community Food. The Australia Institute, Canberra, Policy brief 59.

Chapter 19

Business Models for Power System Flexibility: New Actors, New Roles, New Rules

Luis Boscán*, Rahmatallah Poudineh†
*Department of Economics, Copenhagen Business School, Frederiksberg, Denmark;
†Oxford Institute for Energy Studies, Oxford, United Kingdom

1 INTRODUCTION

The significant increase in the share of renewables in the generation mix poses a number of planning and operational challenges to power systems, raising the need for flexibility more than ever. At the same time, the emergence of innovative solutions is catalyzing the development of new, flexibility-enabling business models; adding activities to the existing supply chain. New actors, sparking innovation in software, hardware, and market design, are defining new roles. For example, aggregators are linking small-scale suppliers of flexibility to electricity markets. Likewise, consumers are not passive anymore, but instead are evolving into active participants: *prosumers*, with an active role in the supply side.

The key element in the emergence of new business models for power system flexibility is, unequivocally, technological change. The context of this evolution is, in most cases, a post-liberalization power system, characterized by unbundling of activities, with transmission and distribution operating as regulated monopolies, and competition being promoted in generation and retail. After several years of experience with reforms throughout the world, market power has been mitigated, efficiency has increased, but many firms still retain a dominant position. On the other hand, market mechanisms are well established now and relied upon. Wholesale and intraday markets are generally used to allocate and price electric energy. Ancillary services and capacity are also competitively procured.

Sioshansi (in chapter: What future for electric power sector?) explored current trends in power systems, including the rapid uptake of distributed generation and renewables, microgrids, storage, and so on. With the increase in the cost efficiency and the competitiveness of renewable resources, they become a

Future of Utilities - Utilities of the Future. http://dx.doi.org/10.1016/B978-0-12-804249-6.00019-1
363

more serious alternative to traditional power plants. However, the operational challenges derived from power system operation with intermittent resources require planners to actively incentivize the adaptability of systems to the challenge posed by stochastic variability.[1] The IEA (2014), for example, claims that integrating a significant share of renewables is dependent on an overall transformation that increases system flexibility and advocates for further development of market-based, short-term balancing mechanisms that create reliable price signals for it.

In addition, the rapid progress of information systems, the declining cost of computing, and the swift evolution of software are creating the conditions for smart grid solutions to become feasible. Coupled with progress achieved in areas like electricity storage, home automation, and electric vehicle development, synergies among energy sectors, such as transportation and heating, are also becoming viable.

In light of recent developments, this chapter reviews the evolution of operational flexibility issues and its associated business models, with a particular focus on short-term flexibility services and the role of emerging players. Long-term issues of market based capacity arrangements have been discussed in the chapter by Woodhouse.

Section 2 discusses the concept of flexibility and reviews the resources that can enable flexible operation of the power system. Section 3 reviews the issue of trading flexibility as a commodity and describes some of the challenges associated with contracting for flexibility services. Section 4 is about the emerging business models for flexibility services and the role of new players, followed by the chapter's conclusions.

2   FLEXIBILITY IN THE POWER SYSTEM

In recent years, the technical literature has coined the term "flexibility" in relation to the requirements of power systems to integrate intermittent resources. However, its definition remains vague and implies different meanings depending on the context. In this chapter, flexibility refers to the ability of power systems to utilize its resources to manage net load variation and generation outage, over various time horizons. *Net load* is defined as load minus supply from intermittent resources, such as wind and solar. As a commodity, flexibility has several dimensions, including capacity, duration, and ramp rate or lead time, for demand-side resources. Boscán and Poudineh (2015) distinguish between short-term flexibility, associated to real-time balancing of the grid, and long-term flexibility, which relates to the adequacy of generation capacity and investment.

---

1. The technically oriented reader is referred to Morales et al. (2014), who devote an entire book to the analysis of operational problems associated to the integration of renewables into electricity markets. Chapter five of their book studies flexibility, originating from different sources in the power system, as an alternative to deal with the stochastic nature of renewable sources of generation.

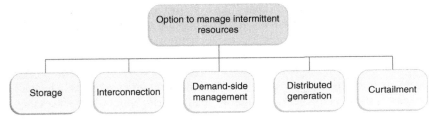

FIGURE 19.1   Options to manage variability of renewables.

It is also helpful to distinguish between *resource flexibility*, which refers to the built-in flexibility of a particular resource, such as demand response; and *system flexibility*, which comprehends transmission, network flexibility, and market design. The transmission network is not an additional source of flexibility per se, but the lack of an adequate transmission network severely affects power system flexibility.

2.1 Flexibility-Enabling Resources

There are various options available to manage the variability of intermittent resources. As shown in Fig. 19.1, these range from storage technologies, interconnections, demand-side management to distributed generation, and curtailment.

Electrical energy storage technologies are among the most effective ways of absorbing net load variability, and although there are various options available, not all of them are commercially viable. Fig. 19.2 (IEC, 2011) classifies existing technologies into five main categories: mechanical, electrochemical, chemical, electrical, and thermal. Of these, the most widely used form is mechanical: specifically, pumped hydro, which accounts for 99% of global energy

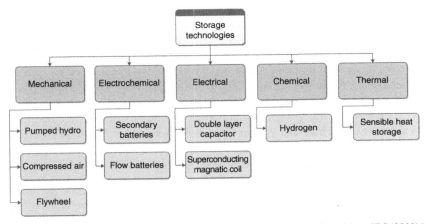

FIGURE 19.2   Storage technologies classification. *(Source: Authors, adapted from IEC (2011).)*

storage (127 GW of installed capacity). Given its unparalleled startup and ramp rate capability, it is a particularly attractive option to address variability from renewables. The second largest electrical energy storage in operation is compressed air but, compared to pumped hydro, it has a negligible global capacity (440 MW). Other means of storage, such as batteries, capacitors, or heat storage have very low penetration levels currently, but recent improvements in technology and cost of electrochemical batteries (particularly, lithium-ion) makes them a promising source of electrical storage, with various benefits to the power system, including flexibility. However, the key to the success of storage technologies is the viability of business models that allow the industry to move forward, beyond demonstration cases and toward massive penetration (Section 4.1.4).

The interconnectivity of power systems is a determinant factor in the extent to which power systems are flexible. In fact, not only interconnections have the potential to facilitate integration of variable generation, but also can contribute to energy security, decarbonization and affordability. In Europe, for example, where there is a strong interest to create an integrated, sustainable, and competitive energy market, there is a specific target to achieve 10% of interconnection (as a share of the installed production capacity) for each member state. Although the European interconnection capacity has increased considerably during the last decade, there remain member states that have less than the 10% goal, and are thus isolated from the internal electricity market (EC, 2015). Fig. 19.3 shows the countries with interconnection that is higher and lower than 10%. Countries such as the United Kingdom, Spain, Italy, and Ireland need to invest in their interconnection capacity. In contrast, Denmark, with a high penetration of wind power, has benefited significantly from the interconnection with countries such as Germany, besides the existing interconnections with Nord-Pool countries.[2] The EU third energy package clearly states the need for cross border interconnections, but for this to become a reality, it is required to design an efficient regulatory framework that incentivizes investment. The existing legal framework seems to favor a regulated business model for interconnection expansion, but it also allows for private merchant transmission initiatives.

Because of its suitability for relieving network congestion and providing ancillary services, such as fast and long-term reserve requirements, distributed generation, such as combined heat and power, is well positioned to increase power system flexibility (IEA, 2005). Traditionally, large conventional power plants served this purpose, and depending on their types, have been an effective source of flexibility. The most important requirements of flexible operation for conventional plants are startup time, ramp rate, and partial load efficiency (Boscán and Poudineh, 2015), but these are not fully available in all types of

---

2. Interestingly and widely cited by various media outlets, on Jul. 9, 2015, Denmark generated 140% of its electricity demand with wind power. However, Denmark managed the excess production by exporting to neighboring Norway, Sweden, and Germany. In relation to the relevance of interconnections, Green and Vasilakos (2012) perform an econometric analysis of Denmark's electricity exports and find that exporting on windy days is a cost-effective way to deal with intermittency.

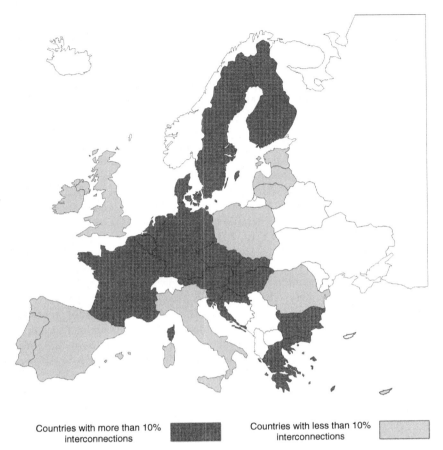

Countries with more than 10% interconnections

Countries with less than 10% interconnections

FIGURE 19.3 **The European electricity interconnection as a share of total installed capacity in 2014.** *(Source: Authors, based on the information from EC (2015).)*

conventional generation. For instance, cycling capability of most current coal power plants is limited and their ramp rate is generally low.[3] The same applies to nuclear power plants, with even more degrees of inflexibility. The most flexible types of thermal generation are gas fired power plants. However, cycling and ramping increase the wear and tear of plants, as well as their heat rate.

In recent years, the need for an efficient portfolio of flexibility resources has drawn attention to demand-side flexibility. In fact, with the advancement in information and communication technologies (ICT), many of the generation services can also be provided through demand response. In the United Kingdom, some forms of demand-side flexibility are currently being traded in the balancing market. For example, through National Grid's Frequency Control

3. Coal power plants, however, can be designed to operate flexibility.

Demand Management scheme, frequency response is provided through automatic interruption of contracted consumers, when the system frequency transgresses the low frequency relay setting on site. Furthermore, National Grid is utilizing slower responding demand response for load following services. Similar arrangements exist in other countries, as demand-side schemes gradually find their way into balancing markets.

Curtailment, a form of negative dispatch in which the system operator reduces the output of wind and solar generation to maintain stability, happens more frequently in the absence of sufficient flexibility. The issues that trigger curtailment are related to system balancing, system dynamics or grid constraints and, therefore the level of curtailment can be used as a negative metric for measuring power system flexibility. Although many countries with increasing shares of renewables have attempted to improve the flexibility of their systems, there remain some with high levels of curtailment. For example, China had an average curtailment rate of 18% in 2012 (Li, 2015), whereas this figure was 4% for the United States during the same period (NREL, 2014). As more renewables are integrated, these figures will rise, unless more flexibility is enabled. For example, the risk of overgeneration in the afternoon (low demand periods) is high in California, and this is likely to become even worse when the renewable portfolio requirement increases from 33% by 2020 to 50% by 2030, as currently proposed.

The use of flexibility services is not limited to addressing net load variation. Indeed, flexibility has three different functions in the power system, and three final users of flexibility services. An important role of flexibility is to ease the integration of intermittent resources. The transmission system operator (TSO), which is responsible for balancing the grid, is thus one of the main procurers of flexibility services. Another function of flexibility is to manage congestion in the electricity distribution network for which the distribution system operator (DSO) is the buyer of flexibility. The third usage of flexibility is for portfolio optimization. The market players (eg, aggregators, suppliers, balancing responsible parties) can obtain flexibility services to fulfill their energy obligations in a cost-efficient way by, for example, arbitraging between generation and demand response. Table 19.1 presents the parties involved in the procurement side of flexibility services in liberalized electricity markets. It is worth mentioning that although TSOs or DSOs procure flexibility services in a competitive manner, these companies recover their costs in a regulated fashion.

3 TRADING FLEXIBILITY SERVICES

The ability to trade flexibility services is important for the reliable operation of power systems. In the currently liberalized electricity sector, flexibility services are traded in intraday and day-ahead markets as an energy product, or in ancillary service markets, as control reserve products (Boscán and Poudineh, 2015). Market design has important implications for procuring flexibility in an efficient and reliable manner: even when there are sufficient resources available for managing

TABLE 19.1 Flexibility Service and Their Final Users

Party	Activity	Business model	Commodity	Use	Final objectives
TSO	Balancing the grid	Regulated business	System flexibility service	System-wide	Grid planning and operational efficiency maximization
DSO	Managing distribution grid	Regulated business	System flexibility service	Local, regional, or national	Grid planning and operational efficiency maximization
Market player	Trading electricity	Price set by market rules	Resource flexibility (portfolio optimization)	System-wide	Profit maximization

Source: Adapted from EDSO (2014).

variable generation, the market may not have been designed to incentivize efficient use of them. For example, in some US regions where there are no subhourly electricity markets, variations in the net load need to be met by regulation services, which have a high ramping rate, and thus are among the costliest flexibility services. This unnecessarily high cost results from the market design because it has been shown that variable generation requires does not require a faster ramping rate than the contingency reserves (Boscán and Poudineh, 2015).[4]

As the current electricity markets in many countries were not originally designed to manage a large share of intermittent resources, further penetration of variable generation might lead to increased market power, reduced competition, and reliability degradation (Ela et al., 2014). Additionally, it is not clear whether the current market design can provide a sufficient level of flexibility when the need for it increases in the system. In the US electricity market, several mechanisms are in place to incentivize flexibility, for example, centralized scheduling and pricing, 5-min settlements, ancillary service markets, make-whole payments, and day-ahead profit guarantees (Ela et al., 2014). However, a different design might be required to incentivize the right amount of flexibility resources both in the short run and the long run. Nontraditional resources such as demand response, storage, and even variable generation itself can contribute to system flexibility when the incentives are provided. Evidence from Great Britain's electricity market shows that, with more uptake of variable generation, the real-time price volatility increases

---

4. On Nov. 1, 2014, California ISO has introduced an energy imbalance market (EIM), mostly to allow grid operators of adjacent areas to share and economically dispatch a broad array of resources for efficient renewable integration.

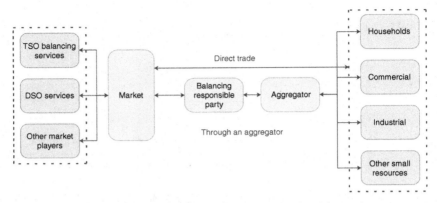

FIGURE 19.4 Trading demand-side flexibility services in the electricity market.

much faster than the day-ahead price volatility and flexible resources can take the advantages of this volatility (Pöyry, 2014).

Flexibility enabling contracts can be traded either directly between the final user and the resource provider or through an aggregator. The capacity of supply is an important factor for the way that trade can happen: transaction cost is an impediment for the small capacity resources to participate directly in the market. The small resource providers, such as households, can thus be aggregated and offered to the market through an intermediary. Fig. 19.4 presents the way that demand-side flexibility-enabling contracts can be traded in an electricity market.

3.1 Designing Contracts for Flexibility Services

Flexibility is a multidimensional commodity, and the marginal cost at each dimension is the private information of the resource provider. Therefore the procurer should design the contract in such a manner that informational rents are minimized, and the cost of integrating renewables is efficient. Designing optimal contracts for flexibility services under multidimensional information asymmetry is challenging, and becomes even more important when the cost of balancing services increases with an increased uptake of intermittent resources.

In bilateral contracts (between the resource providers and the final users or an aggregator), when the sellers differentiate themselves by concentrating on different dimensions, the procurer can design the contract in a way to extract all informational rents (Li et al., 2015). For example, consider a system operator who aims to control thermostats in two households' premises and for this she offers a contract based on two parameters of lead time and duration of load control.[5] Under the condition that the households are very similar, in terms of the disutility they experience at each dimension (lead time and duration of load control), there is no way for the system operator to design a truth-telling contract, which extracts

---

5. So the contract is in the form of a payment for specific lead time and duration.

all informational rents. In this case, the system operator needs to give up some rents by distorting downward the contract specifications (lead time and duration of load control) from the optimal level for one of the households. However, if the two households differ significantly at each dimension, the system operator can extract all the rents. This happens when, for example, the flexibility procurer knows that one household incurs a high disutility for the short lead time and the other for the long duration of load control. Naturally, the former household prefers a contract with higher lead time, but can sacrifice on load control duration, whereas the latter value more a contract which has a shorter load control duration. In this case, both households select a contract, which is optimal for them.

The previous results also hold when there are multiple flexibility resource providers. Therefore more differentiation across the dimensions of flexibility, by the resource providers, benefits the buyer, and vice versa. If the contract is designed (and offered) by flexibility resource provider rather than the system operator, the results are not necessarily symmetric to the previous case. For example, double marginalization[6] can happen although the supplier can change the specification of contract to avoid this. Additionally, when the resource provider enters into a contract with an aggregator who faces an uncertain demand for flexibility in the market, the optimal mechanism requires reducing the specification of contract at each dimensions, that is, it is optimal for the aggregator to buy less compared to the case of a deterministic demand.

An intermediary (for example, an aggregator) might face a demand for multiple flexibility products, with various specifications in terms of capacity, duration, response time, and ramp rate. This is because the impact of intermittent resources on the power system can be considered in four time frames: frequency regulation, load following, scheduling, and unit commitment (Boscán and Poudineh, 2015). Frequency regulation requires very speedy response and ramp rates, and thus is costly. The requirement for speed of response decreases, as the time frame moves toward load following and beyond. Therefore for each time frame a different flexibility service and, consequently, flexibility contracts are needed. In this case, the intermediary needs to make a decision between supplying all range of flexibility products, or only some of them. Theoretically, there is a fundamental trade-off in the intermediary's product selection decision in this case. This trade-off results from the market share of slower responding flexibility resources, versus the revenue obtained from more expensive flexibility services (eg, regulation services).

3.2 Next Generation Utilities and System Flexibility

As the traditional utility model is evolving, next generation utility concepts emerge as a result of rapid advancement in ICT. Demand response, electric vehicles, energy efficiency, and intelligent grid management, will have an evolved

---

6. Double marginalization happens when the two actors across the supply chain apply their own markups over the price, which results in higher deadweight losses.

TABLE 19.2 Next Generation Utility Concepts

	Traditional approach	Conventional wisdom now	Next generation concepts
Demand response	Emergency curtailment	Peak shaving	Resource for capacity and balancing service
Plug-in electric vehicles	R&D only	Flexible load	Vehicle-to-grid storage resource
Intermittent resources	Marginal fuel saving, no capacity value	Some capacity value with gas fired firming	Resource for capacity and balancing service
Grid automation and intelligence	Unidirectional from source to load	Some intelligence to automate loads	Omnidirectional web of sources and loads
Energy efficiency	Up to the customer	Component-based utility programs	Breakthrough-level system efficiencies

Source: Adapted from Hansen and Levine (2008).

function, as described in Table 19.2 (Hansen and Levine, 2008). For example, demand response, traditionally used for emergency curtailment to protect grid frequency under an emergency condition (load shedding), gradually enters the electricity markets as a capacity resource, as well as balancing service at all time frames. In the current electricity markets, the need for system flexibility may not be critical yet, but it is not clear that this will remain the case in the future, as long as renewables gain a greater share in the generation mix. Access to various sources of flexibility services, both on the supply and demand side, along with appropriate market design, provide an opportunity to profit from short-term spikes in spot prices, balancing markets and specific contracts with grid operators.

It is likely that next generation utilities will be more reliant on ICT, and they are already storming the industry with "smart," programmable, communicable gadgets, and "Internet of Things" (see the chapter by Cooper). Although ICT has always been important in the power sector, especially for system protection, with the need for more flexibility and a reliable real-time operation, the role of ICT becomes even more critical. Smart grid, smart meters, intelligent home management systems, and various forms of advanced technologies, will enable utilities to profit from trading flexibility services.

4 NEW BUSINESS MODELS

The electricity sector landscape is changing rapidly with the integration of renewables, technological advancement in ICT, and the emergence of various new players. Amidst this environment, entrants are coming to participate in electricity markets, but in completely novel ways. Decentralized generation units are beginning to compete with traditional generators. Aggregators, acting as intermediaries, acquire the right to modify energy consumption from end users, and sell it in the form of available capacity. Software and technology developers offer energy management solutions, intelligent devices, and storage capability. Ventures among these new players, teaming up to offer new services, are becoming more frequent, and the sum of it all depicts a creatively chaotic picture. Yet, these entrants share some common features: relative to incumbents that rely on traditional, large scale industrial assets, entrants have considerably lower fixed costs, and depend on nontraditional, knowledge-based assets.[7]

Taken together, they constitute a *layer of innovation* that is being added to the existing structure of power systems and challenges the traditional business model in which utilities enjoy a relatively undisputed position, and consumers act as the passive end of the supply chain. All of this contests the status quo, motivates incumbents to reconsider their roles and, potentially, adopt new ones in accordance with the changing environment. Regulators, in consequence, are being led to consider new, previously unforeseen sources of involvement and potential dispute among entrants and incumbents.

4.1   Partial Taxonomy of New Actors, New Roles, and New Business Models

The rapidly evolving nature of innovation and frequent function overlap prevents an exhaustive enumeration, and mutually exclusive categorization of agents involved.[8] To contribute in the understanding of business models leading to increased levels of power system flexibility, a simple, yet partial, categorization is proposed as follows:

1. *New actors* are the constituents of the innovation layer, composed of entrants sparking innovation through new software, technology, and market design proposals. Firms, researchers and, to a lesser extent, regulators can also be identified here.
2. *New roles* are defined by new actors, and are assumed by existing market participants and new actors alike. Aggregators and prosumers are two good examples of this category.
3. *Business models* are the commercial outcome of innovation brought about by the new actors. In a well-defined business model, the sources of revenue,

---

7. Rodgers (2003) identifies three categories of knowledge-based assets, namely *human assets*: attitudes, perceptions, and abilities of employees; *organizational*: intellectual property such as brands, copyrights, patents, and trademarks; *relational*: knowledge of and acquaintance with communities, competitors, customers, governments, and suppliers in which the company operates.

8. In a recent discussion paper by Ofgem (2015) on the topic of nontraditional, business models, they find the same difficulty.

cost and, therefore profitability, are unambiguously defined. Furthermore, business models are subject to evolution and depend on the overall economic environment: some will appear, consolidate, and evolve into new areas of action, while others will disappear, given their lack of viability (for more on this, see the chapter by Nillesen and Pollitt).

4.1.1 Aggregation for Demand-Side Management

Aggregation for demand-side management is one of the most consolidated existing business models for power system flexibility.[9] The role of aggregator, fulfilled by energy management software developers and other traditional retailers with real-time metering, is to bundle "negawatts" (unused capacity)[10] offered by commercial and industrial (C&I), and residential consumers of electricity. In exchange for capacity and energy usage payments or rebates in their electricity bill, consumers adjust consumption at times of peak demand, or when required by grid operators. Aggregators sell negawatts in different outlets, including capacity, balancing, and ancillary services markets, or as part of demand response programs carried out by utilities.

This business model has grown in several countries, including Europe and Asia, but has shown particular strength in the United States. As an example of its relevance, consider the 2014 capacity auction results for PJM, the largest wholesale electricity market in the United States: 10.9 GW of demand response capacity were procured, the equivalent to more than 6% of the total. Nevertheless, in the latest episode of a legal battle between power companies and aggregators, demand response in the United States has recently received a regulatory setback. The federal order that set demand response and generation on equal footing regarding payment received by grid operators was vacated last year, on the grounds that demand response is being overcompensated, inducing inefficient prices that discriminate against generators.[11] The final say, though, has not been declared yet: at the time of writing, the US Supreme Court of Justice has decided to reconsider the case.

From a more general perspective, though, the role of aggregation for demand-side management goes beyond conventional demand response. Aggregators typically rely on software solutions and other hardware to realize efficiency gains and, therefore they are shifting to developing integrated energy management solutions. As a result, some of them are rebranding themselves

---

9. The term "demand-side management" is used to encompass both participation of demand as a resource in markets, such in capacity markets; and in the conventional demand response sense, which includes interruptible loads, load management, peak shaving, and so on.

10. The term "negawatts" has been attributed to Amory Lovins, cofounder and Chief Scientist of the Rocky Mountain Institute, by a number of publications, including The Economist's special report on energy and technology (2015) and Maurer and Barroso (2011).

11. Order 745 issued by FERC (Mar. 15, 2011) on the topic of "Demand Response Compensation in Organized Wholesale Electricity Markets."

as software developers, while others are emphasizing the role of hardware as a tool for demand-side management, while retaining their role as aggregators. They are also entering agreements with utilities and grid operators to manage intermittency from renewables with demand-side resources, an element that emphasizes their growing role as a supplier of flexibility. For example, Ener-NOC—a US-based aggregator known for its demand response operations who is refocusing its business toward software development—ran a pilot project with the Bonneville Power Administration to show the capabilities of demand response to deliver short-term balancing. Such new services from demand response are especially valuable, as the likelihood of overgeneration increases in places such as California. Also, as the role of distributed assets increases, aggregators will not only manage demand, but will make a transition into virtual power plant managers.

4.1.2 Thermostats as a Demand-Side Management Tool

Although thermostats are key for controlling energy consumption in residential and C&I buildings, they have rarely been a particularly interesting object of attention for retail consumers. With the majority of sales channeled through dealers offering service contracts, well-established products developed by long-standing incumbents have taken the lead.

Nevertheless, the usually undisrupted retail market for thermostats became invigorated once Nest Labs transformed this typically uninteresting device into an appealing gadget for tech-savvy consumers, through the development of user-adaptive technology, and a well-designed marketing strategy.

While the argument for significant product differentiation and technological breakthrough by the Nest thermostat is not easily argued for,[12] more significantly, their contribution has been to introduce innovative business models for flexibility, in which smart thermostats are the key element to enable demand-side management.

According to these business models, smart thermostat users are given the choice to surrender control of their load at peak demand hours or when there are seasonal weather variations, and allow the utility to adjust consumption, following user-defined comfort levels. In exchange, utilities compensate consumers with rebates on their final bill, or through direct payments. Not only have other smart thermostat developers followed suit, but utilities have also created *bring-your-own-thermostat* demand-side management programs. Moreover, nontraditional demand response services (increasing consumption as opposed to reduce demand) are also becoming increasingly important. These forms of demand-side management can be particularly relevant for places with overgeneration, like California, Texas, Denmark, and Germany.

---

12. Ecobee, a Canadian competitor to Google-owned Nest Labs, introduced the first WiFi connected thermostat at least 2 years before the Nest thermostat hit the market.

Looking forward, there is ample room for these business models to develop further, as the penetration of programmable thermostats still remains very low.[13] The upfront cost of deployment, though, is a barrier for many customers. As a solution, and in a similar vein to business models in the telecommunications industry where service carriers and hardware providers team up, utility companies are subsidizing the deployment of smart thermostats. In summary, thermostat-based demand-side management is setting new standards in the adoption of new technology and in the development of demand-side management models that could easily extend to other devices.

4.1.3 Software Developers

The emerging business models for power system flexibility are closely intertwined to smart grid development. Coupled with hardware, software is pervasive across processes and solutions. Remotely controlling devices, smart metering, and identifying consumption patterns to reduce demand charges, are just some of the many examples that highlight the role of software (for more details, see the chapter by Cooper).

In some models, software is bundled with hardware as part of the complete solution. For example, on-site energy storage vendor Stem describes its system as composed of three elements: software, batteries and a real-time meter. Others focus on software development and work with any kind of hardware. Such is the case of software vendor BuildingIQ, specializing on demand-side management for heating, ventilation, and air conditioning systems in C&I buildings.

More generally, and as part of an emerging trend, many agents currently developing new business models for power system flexibility are expanding their role into *software-as-a-service* suppliers. This licensing and delivery model has consolidated in recent years among software developers, because it allows end users to reduce hardware, upfront, and maintenance costs, and has enabled scalable usage and payment. On the other hand, vendors obtain a recurring revenue stream from subscription payments. Firms like EnerNOC, a leading aggregator, are following this trend as a growth strategy, and are also creating interactions with other existing business models. In summary, software is already playing a central role in the new business models for power system flexibility, and its relevance will only continue to grow.

4.1.4 Storage Providers

Location within the supply chain largely defines the scale, response time, size and, therefore suitability of different storage solutions to increase flexibility

---

13. Consider, for example, the American market, where (according to the US Energy Information), 85% of American homes with central heating own thermostats, but less than half of these are programmable. Similarly, 60% of those with central cooling own them, but approximately a half of these are programmable.

in the power system. Although not fully consolidated,[14] recent years have witnessed a considerable expansion of electricity storage business models, with greater emphasis on *behind-the-meter* (distributed) than in *front-of-the-meter* (grid-level) solutions. A report by the firm GTM Research (2015), sponsored by the US Energy Storage Association, reveals that distributed storage deployments increased more than threefold between 2013 and 2014 in the United States, and the nonresidential sector accounted for the lion's share of this amount. They expect the distributed storage segment to continue growing in years to come, outpacing grid-level storage, until it reaches 45% of the total market share by 2019.

Distributed storage targeted at C&I, large residential, and institutional consumers is one of these models. In most markets, these clients pay for the energy they consume, plus a share of their peak demand within a billing period. Coupled with software analytics and real-time metering to analyze *peak-shaving* opportunities, suppliers offer on-site storage systems to go off-grid when demand is high. Typical agreements between storage suppliers and their customers are based on revenue sharing, but initial investments, operational and price risk are assumed by suppliers.

The economic case for residential energy storage is different and, given current conditions in most retail markets, difficult to make. To begin with, most residential customers have fixed price retail contracts and, therefore price arbitrage and peak shaving become mostly irrelevant. Furthermore, in markets where residential solar PV systems are becoming widely adopted, it is sensible to acquire storage if customers wish to become entirely independent of the grid. Yet not only are such green energy oriented customers a well-off minority, but net metering—an incentive that is particularly relevant in many US states—is at odds with it: being completely off grid would imply cutting off a source of revenue that helps to pay the cost of the solar facility investment. Unless the cost of residential storage is competitive enough, or there is an economic incentive to install it, this business model is not viable.

However, distributed energy storage is one of the most effective resources to enable power system flexibility, as it can balance power supply and demand instantaneously. The aggregated deployment of storage capability creates virtual power plants that, depending on market design innovations at the distribution level, may create sources of revenue for owners of this kind of resources. Grid operators interested in the procurement of capacity, reactive power, and voltage management might well provide the necessary source of revenue to further boost the adoption of distributed storage, including residential applications. In fact, following the recent introduction of Tesla Motor's batteries for C&I and residential applications, at an approximate price of $500/kWh, its partner company Solar City clarified that the 10-year lease agreements for solar and storage systems, with which they typically operate, contemplated revenue sharing of grid service income.

---

14. Energy storage, mostly pumped hydro, accounts for 2% of total US generation capacity.

4.1.5 Market Design Innovation

Driven by technology, the new business models are already transforming the way in which power systems operate. However, given the crucial role of incentives, market design and regulation can either hinder or help their consolidation and evolution as a tool to increase flexibility.

Regulators and system operators in areas where renewables are on the way to playing a more relevant role are considering different market design innovations. Many of these, though, still appear to have a piecemeal and tentative approach. Some of them prioritize the role of short-term balancing, whereas others emphasize the role of demand-side management, and long-term resource adequacy. While there is no one-size-fits-all solution, restructuring existing electricity market designs to enable flexibility requires a holistic approach. Hogan (2014) argues that adapting existing markets to renewables requires, first, recognizing the value of energy efficiency, including demand-side management; second, upgrading grid operations to increase short-term flexibility; and third, incentivizing long-term flexibility investments, that is, adopting flexible resources.

An interesting example comes from the California Independent System Operator (CAISO) which is currently developing a *flexible ramping product,* aimed at minimizing short-term (5-min to 5-min) load variations. In contrast to conventional ancillary services, this product focuses on addressing net load changes between time intervals, and not on standby capacity aimed at meeting demand deviations within a time period. In addition, an innovative feature of this proposal is that it is continuously procured and dispatched.

Another interesting experience is Southern California Edison's recent capacity procurement of 2.2 GW of behind-the-meter solar PV generation, storage, and demand-side management to alleviate congestion in particular zones of the grid. Besides being a complex process because of the necessary cross-comparisons between technologies, location of assets, and the diverse nature of contracts with suppliers, it reveals emerging business models in which generation and distributed energy resources are treated on a par with conventional generation. Of particular interest is the agreement with distributed solar generation company Sun-Power which assumes and enhances the role of aggregator. Upon requirement of the utility, the aggregator commits to achieving savings through solar power, which it procures at specific sites from generation facilities scattered throughout different grid locations—a Virtual Power Plant—without exporting it to the grid.

Also, in the context of a comprehensive review of their power system, the single electricity market for Ireland has decided on a number of measures aimed at adapting it to the 2020 goal of 40% of renewables in Irish electricity demand. On the market design front, relying on a hybrid regulated tariff/auction mechanism to procure contracts with maturities from 1 to 15 years, it has been agreed to increase the number of ancillary services procured from 7 to 14, including specific ramping products with horizons of up to 8 h.

To sum up, market design innovation is already playing a key role and it will have a substantial impact on the consolidation of emerging business models.

4.2   New Business Models and the Future of Utilities

The absence of large-scale economically viable storage, and an entirely passive demand-side have justified the existence of the traditional power system business model, but technological breakthrough has begun to challenge this approach. From this follows a central question for the future, namely: *what is the impact of new business models on existing utilities?*

The immediate consequence is that the business-as-usual operation of utilities is challenged, but the extent of the impact depends on the strategic decisions that both incumbents and entrants make. Incumbents can choose a confrontational approach to deter consolidation of the emerging business models, or can accommodate to entry (see the chapter by Burger and Weinmann).

Evidence shows that confrontation is already happening. The extended legal battle between power producers and aggregators in the US over Federal Order 745 mandating equal treatment between demand response and conventional generators in wholesale markets is an example of this. In France, a similar conflict over imbalance mechanisms arose between retailers and aggregator Voltalis.

Nonetheless, the line between confrontation and adaptation is not clearly delineated, because several incumbents are extending their activities into new business models. Big players, including large vertically integrated energy holdings, are entering the aggregation business, and are acquiring stakes in energy management developments, effectively extending their scope. For example, NRG, which owns 50 GW of fossil-fuel dominated generation assets in the United States, acquired Energy Curtailment Specialists in 2013, a leading US aggregator with a portfolio of 2 GW. In France, Schneider Electric acquired leading European aggregator Energy Pool in 2010, which controls more than 1.5 GW in demand response assets. Also, Swiss generator Alpiq, which owns a generation portfolio of 6 GW including hydro, fossil, and nuclear, acquired British aggregator Flexitricity in 2014.

Entrants, on the other hand, are partnering in their offers, bundling products in markets that show potential first-mover advantages. For example, Tesla and Advanced Microgrid Solutions (AMS) have recently announced a sales deal to install up to 500 MWh in battery capacity, as part of a grid-scale storage project. EnerNOC and Tesla have also announced a partnership to bundle batteries with software solutions to enable demand-side management. Google's acquisition of Nest Labs in 2014 for US$3.2 billion is yet another indication of the rapidly changing face of the new business models.

The future will depend, mostly, on these strategic interactions, and while it is impossible to predict the future, one thing is certain: utilities as we know them today will definitely change. Table 19.3 summarizes the emerging players and associated new business models for the flexible power systems of future.

TABLE 19.3 Summary of New Business Models for Power System Flexibility

Business model	Characteristics of the agreements
Aggregation for demand-side management	• Consumers obtain capacity and/or energy usage payments • Negawatts are sold in organized markets or as part of bilateral agreements with utilities
Thermostats as a demand-side management tool	• Consumers acquire the device with a subsidy from utility • Consumers enter direct-load control agreements, allowing load to be adjusted to predefined comfort settings • Consumers are given rebates or paid for energy not consumed • Utility manages peak load with higher cost efficiency • Hardware sales increase
Storage (C&I clients)	• Consumers pay no upfront cost for software or hardware deployment. Alternatively, supplier delays deployment costs until first revenue streams are realized • Revenue from demand charge reduction is shared between consumer and supplier
Storage (residential)	• Upfront deployment cost is borne by households • Consumers benefit from going off-grid, price arbitrage, or grid service payments
Market design innovation	• Utilities procuring services through a number of bilateral contracts with suppliers of flexibility • New ancillary services • New short-term services focused on short-term balancing
Software	• Software-as-a-service • Vendors collect subscription fees • End users reduce hardware, upfront and maintenance costs

5 CONCLUSIONS

Integrating renewables efficiently requires increasing flexibility and technological progress is facilitating this process. Over the last few decades, technology has pushed the operational boundary of utilities away from a traditional paradigm, but the changes happening now are paving the way for a next generation of utilities. According to this emerging paradigm, new actors sparking innovation are defining new roles and, as a result, interconnectors, distributed generators, storage providers, and suppliers of demand response are competing with incumbents, while relying on the novelty of their business models to provide flexibility services. Meanwhile, the interaction of regulation, technological innovation, and business model evolution are shaping the strategic interaction among players, who have several pieces of private information. Markets, contracts, and regulatory frameworks will have to change to become more compatible with the requirements of the new environment.

These trends provide a sense of the forces shaping the emergence of a completely new state of affairs in which existing utilities will have evolved, and will coexist with new players in the provision of flexibility. The final shape of power systems will not be unique, as it is path-dependent due to the effects of technological, financial, and institutional legacies. What is certain is that the change is inevitable, and utilities as are known today will definitely change. In the new environment, the opportunities for utilizing competition among the suppliers of flexibility increase. In conclusion, as flexibility becomes scarce in the system, innovative flexibility-enabling business models initiated by new actors will be highly valuable and critical for the efficient provision of flexibility services.

REFERENCES

Boscán, L., Poudineh, R., 2015. Flexibility-Enabling Contracts in Electricity Markets. Unpublished mimeo, Oxford Institute for Energy Studies.

EC, 2015. Achieving the 10% Electricity Interconnection Target: Making Europe's Electricity Grid fit for 2020. European Commission, Brussels, 25.2.2015, COM(2015) 82 final. Available online: http://ec.europa.eu/priorities/energy-union/docs/interconnectors_en.pdf

EDSO, 2014. Flexibility: The role of DSOs in tomorrow's electricity market. European Distribution System Operators for Smart Grids. http://www.edsoforsmartgrids.eu/wp-content/uploads/public/EDSO-views-on-Flexibility-FINAL-May-5th-2014.pdf

Ela, E., Milligan, M., Bloom, A., Botterud, A., Townsend, A., Levin, T., 2014. Evolution of Wholesale Electricity Market Design with Increasing Levels of Renewable Generation. Technical Report NREL/TP-5D00-61765, September 2014 National Renewable Energy Laboratory (NREL). www.nrel.gov/publications

Green, R., Vasilakos, N., 2012. Storing wind for a rainy day: what kind of electricity does Denmark export? Energy J. 33 (3), 1–22.

GTM Research, 2015. US Energy Storage Monitor (Q1 2015). http://www.greentechmedia.com/research/us-energy-storage-monitor

Hansen, L., Levine, J., 2008. Intermittent Renewables in the Next Generation Utility. Rocky Mountain Institute, Boulder, CO. http://www.rmi.org/Knowledge-Center/Library/2008-22_IntermittentRenewablesInNGU

Hogan, M., 2014. Power Markets: Aligning Power Markets to deliver value. The Regulatory Assistance Project. Available online at: http://americaspowerplan.com/wp-content/uploads/2014/01/APP-Markets-Paper.pdf

IEA, 2005. Variability of Wind Power and Other Renewables: Management Options and Strategies. International Energy Agency. http://www.uwig.org/iea_report_on_variability.pdf

IEA, 2014. The Power of Transformation. Wind, Sun and the Economics of Flexible Power Systems.

IEC, 2011. Electrical Energy Storage. International, Electrotechnical Commission, White paper. Geneva, Switzerland. http://www.iec.ch/whitepaper/pdf/iecWP-energystorage-LR-en.pdf

Li, X., 2015. Decarbonizing China's Power System With Wind Power: the Past and the Future. Oxford Institute for Energy Studies. http://www.oxfordenergy.org/wpcms/wp-content/uploads/2015/01/EL-11.pdf

Li, Z., Ryan, J.K., Sun, D., 2015. Multi-attribute procurement contracts. Int. J. Prod. Econ. 159, 137–146.

Maurer, L., Barroso, L., 2011. Electricity Auctions: An Overview of Efficient Practices. The World Bank, Washington, DC.

Morales, J., Conejo, A., Madsen, H., Pinson, P., Zugno, M., 2014. Integrating Renewables in Electricity Markets. Operational Problems. Springer, New York, NY.

NREL, 2014. Wind and Solar Energy Curtailment: Experience and Practices in the United States. National Renewable Energy Laboratory, Technical Report NREL/TP-6A20-60983, http://www.nrel.gov/docs/fy14osti/60983.pdf

Ofgem, 2015. Non-Traditional Business Models: Supporting Transformative Change in the Energy Market. Discussion Paper. Available online at: https://www.ofgem.gov.uk/ofgem-publications/93586/non-traditionalbusinessmodelsdiscussionpaper-pdf

Pöyry, 2014. The Value of Within-Day Flexibility in the GB Electricity Market. Pöyry Management Consulting. Available online at: http://www.poyry.co.uk/news/poyry-point-view-value-within-day-flexibility-gb-electricity-market

Rodgers, W., 2003. Measurement and reporting of knowledge-based assets. J. Intell. Capital 4 (2), 181–190.

The Economist, 2015. Special Report on Energy and Technology. Available online at: http://www.economist.com/printedition/specialreports?year[value][year]=2015&category=76986

Chapter 20

The Repurposed Distribution Utility: Roadmaps to Getting There

Philip Q. Hanser, Kai E. Van Horn
The Brattle Group, Cambridge, MA, United States of America

1 INTRODUCTION

The era of the distributed energy resource (DER) has begun for the electric utility industry.[1] Steeply declining costs for DERs, such as solar photovoltaic (PV) generation, continued development of localized storage, and regulatory incentives promoting DER expansion all have resulted in substantial investments in distribution-level DER installations over the previous decade (Barbose et al., 2014). Indeed, nearly 8 GW of PV capacity was installed on distribution systems in the United States between 2008 and 2014, representing 45% of total installed PV capacity over the same period.[2]

The dawning of this new era brings with it unprecedented challenges to the distribution utility's conventional role to provide on-demand access to electricity for customers in their service territory, at a level of reliability consistent with existing standards.[3,4,5] Moreover, existing utility infrastructure and operating practices reflect the expectation of unidirectional electricity flows from generators on the transmission system to end-use customers on the distribution system.[6]

---

1. The evolution of the utility industry in recent decades is discussed extensively in Fox-Penner (2014).
2. Distribution-level PV is defined here as PV installations with capacities less than 1 MW (Schmalensee and Bulovic, 2015).
3. For a detailed overview of such challenges see, for example, Kind (2013).
4. Utilities have traditionally operated as regulated monopolies trading the obligation to serve their customers for an administratively set rate of return capital investments recovered throw volumetric charges to their customers.
5. Distribution reliability standards are typically set on a state-by-state basis by state public utility commissions. For a discussion of distribution reliability standards in the United States and abroad see, for example, Hesmondhalgh et al. (2012).
6. The dividing line between what constitutes the transmission versus distribution system is typically drawn by voltage level. For the purposes of discussion here, all infrastructure operating below 35 kV will be considered as part of the distribution system.

Future of Utilities - Utilities of the Future. http://dx.doi.org/10.1016/B978-0-12-804249-6.00020-8

The distribution utility structure of today will not be adequate to address the DER integration challenges faced by the industry. To address these challenges effectively, and maximize the benefits of DER integration, today's distribution utilities must be repurposed—transitioning to become what we will call the repurposed distribution utility (RDU).[7]

This chapter explores the RDU transition from the frontline perspective of a utility planner faced with the daunting task of planning system upgrades and operational changes to accommodate the large-scale expansion of DERs on their distribution system—from conceptualizing the problem, to planning the RDU transition. Furthermore, using examples from California and Hawaii, this chapter delves into the approaches utilities are taking to integrate DERs that not only minimizes their potential negative impacts, but also maximizes their benefits.

Although the concentration of this chapter is on the decisions that must be made by distribution utilities to determine and plan their RDU transition needs, this assessment may also be an opportunity for distribution utilities to consider how to increase their resilience in the face of natural and cyber threats. Furthermore, though beyond the scope of this chapter, it must be noted that utility business model and rate design are important components of DER value maximization.[8]

The remaining sections of this chapter are organized as follows. Section 2 presents a brief review of the current situation for utilities. Section 3 provides an overview of the issues utilities face when making the RDU transition. Section 4 describes the changes to system planning required to meet RDU objectives and integrate DERs. Section 5 discusses in depth needed RDU operations reforms. The final section presents the chapter's conclusions.

2 THE CURRENT SITUATION

Existing utility infrastructure and operating practices reflect the expectation of unidirectional electricity flows from generators on the transmission system to end-use customers on the distribution system. The expansion of DERs in systems with high adoption rates, such as those of utilities in California and Hawaii, is beginning to undermine this expectation and cause, or exacerbate, a number of reliability and quality of service issues ranging from: (1) somewhat minor issues, such as voltage flicker and excessive switching of conventional discrete voltage regulation devices, which may lead to shortened equipment lifespan; to (2) major issues that may compromise reliability and safety, such as the failure of protection schemes to operate as designed, steady-state voltage

---

7. We have chosen the term repurposed distribution utility to avoid confusion with the State of New York's Reforming the Energy Vision and the State of California's Distribution Resource Planning initiatives, which are in the process of development and not yet fully described. Some of the motivations for these initiatives appear to be similar to that of this chapter.

8. Discussions of potential future utility business models can be found in chapters by Nillesen and Pollitt, Burger and Weinmann, Löbbe and Jochum, and Boscán and Poudineh, all in this book or see, for example, Hanser and Van Horn (2014). A comprehensive treatment of rate design issues related to DER integration can be found in chapter by Nelson and McNeil, in this book.

violations, reverse electricity flows, and thermal overloads, and compromised system stability.[9]

On the other hand, DERs present opportunities for greater local control and benefits such as reduced peak loading and distribution losses, as well increased resiliency due to locally available supply.[10,11] However, the magnitude of the benefits realized is contingent on the manner in which utilities integrate DERs into their systems—existing system structures and conventional approaches to planning and operations will not harvest the majority of the benefits.

The distribution utility structure of today will not be adequate to address the DER integration challenges faced by the industry. To address these challenges effectively, and maximize the benefits of DER integration, today's distribution utilities must be repurposed—transitioning to become what we will call the repurposed distribution utility (RDU).

The primary objective of the RDU is to maximize the value of distribution-level resources for ratepayers, while meeting reliability mandates. The RDU is similar to, but distinct from, the much talked-about distribution services platform provider (DSPP).[12] While the DSPP demands a complete divergence from the conventional role of the distribution utility, the RDU is the next step for utilities along the path toward the DSPP.

The RDU is built upon the infrastructure and operating practices of the present electric distribution utility. However, the RDU harnesses, to a much greater extent than current distribution utilities, the additional flexibility and adaptability that are facilitated by system-wide data collection and data-driven control algorithms, extensive two-way communication, and automatic control.[13] More importantly, in the context of this chapter, the RDU planners and operators view DERs as a resource whose benefits can be harnessed, rather than threat to be minimized and adopt planning and operational practices to support this view.

While steps are currently being taken by many distribution utilities toward becoming RDUs, such steps have not typically been part of a comprehensive push to make the RDU transition based on a detailed evaluation of the investment alternatives that will enable utilities to meet their mandate moving forward.[14] To make the RDU transition, utilities will be required, in many cases, to undertake substantial investments in new infrastructure, and significant

---

9. For an extensive discussion of the impacts of DER on distribution systems, see, for example Short (2014). See chapter by Rowe et al. in this volume.

10. See Gellings in this volume.

11. The use of PV inverters to provide voltage support is one major source of DER benefits. For a discussion of developments in this area, see CPUC and CEC (2015).

12. The DSPP concept was described extensively in NYSDPS (2014).

13. The key attributes utilities must incorporate to meet the challenges the technological, economic, and regulatory challenges they face moving forward have been the subject of much discussion. See, for example, Sioshansi (2014), Domínguez-García et al. (2011), Forsten (2015), Costello (2015).

14. One such step is the widespread deployment of advanced metering infrastructure (AMI), described in Lee et al. (2014).

operational reforms, which may result in significant transition costs. These investments can be made more efficiently if their economic and reliability impacts are considered, to the extent possible, in a common framework, rather than as piecemeal upgrades to address immediate problems.

Thus, there loom key questions that utilities continue to grapple with: how can a utility planner assess the individual needs of their system regarding investments in infrastructure and updated operational practices to make the transition from conventional distribution utility to RDU?

3 FRAMING THE RDU TRANSITION

RDU transition needs are specific to each utility and depend on the physical configuration of a utility's distribution feeders and the status of their equipment, as well as their operational practices. Thus, conceptualizing the requirement of an individual utility to make the RDU transition begins with a planner assessing the status of existing utility infrastructure and current operational practices—a system overview. The goal of such an assessment is to identify the unique characteristics of a utility's system and operational practices that impact the system's ability to integrate DERs, the so-called "hosting capacity,"[15] (eg, length of feeders, vintage of existing equipment).

A system overview helps the planner identify shortcomings of system infrastructure and operational practices, with respect to DER integration and the RDU's value maximization objective that must be addressed in the planning phase. Such an assessment may take the form of a survey, or series of questions, the answers to which can then be used to guide the planners when selecting infrastructure investments and operational reforms for detailed analysis. The information required for the system overview will come primarily from utility planning data and operating manuals.

An example system overview survey is given in Fig. 20.1. Key system characteristics relevant to the RDU transition are given on the left, and the extent of their presence can be assessed with a rating of one to three, on the right. The shading in the background indicates the relative impact each characteristic tends to have on system holding capacity, all else being equal (Black & Veatch, 2013; Navigant Consulting, 2013; Broderick et al., 2013). While this list is by no means exhaustive, it captures many of the key considerations relevant to the RDU transition.

An effective system overview will identify the distribution utility characteristics likely to have the highest impact on DER hosting capacity, and thus should receive more focus in subsequent planning activities. The system overview also points out where the current system infrastructure and operating practices may

---

15. Hosting capacity is defined as "the amount of DER that can be accommodated on a feeder without adversely impacting operations, power quality, or reliability" (Forsten, 2015).

Relative impact on DER hosting capacity	Lower		Higher
Utility system characteristics	**Classification**		
Mains and lateral length	(Long) 1 —— 2 —— 3 (Short)		
Remaining thermal capacity at peak load	(Low) 1 —— 2 —— 3 (High)		
Topology	(Radial) 1 —— 2 —— 3 (Mesh)		
Voltage level	(Low) 1 —— 2 —— 3 (High)		
Load density	(Low) 1 —— 2 —— 3 (High)		
Voltage drop under peak load conditions	(High) 1 —— 2 —— 3 (Low)		
Type of metering infrastructure	(Legacy) 1 —— 2 —— 3 (AMI)		
Prevailing type of protection equipment	(Analog) 1 —— 2 —— 3 (Digital)		
Prevelance of voltage control devices	(None) 1 —— 2 —— 3 (Many)		
Prevelance of two-way communication	(None) 1 —— 2 —— 3 (All)		
Prevelance of data collection devices	(None) 1 —— 2 —— 3 (Many)		
Extent of load control	(None) 1 —— 2 —— 3 (Full)		
Extent of DER control	(None) 1 —— 2 —— 3 (Full)		
Extent of control automation	(None) 1 —— 2 —— 3 (Full)		
Measurement-based EMS application use	(None) 1 —— 2 —— 3 (Many)		

Row categories (left margin): Physical, Communication and control, Operations.

FIGURE 20.1 Key RDU transition system characteristics. A "short" feeder is typically one that is less than 5 miles in length; load density refers to the total load on the feeder divided by the geographic area encompassed by the feeder; and the voltage drop under peak load conditions measures the voltage drop across the length of the feeder and is an estimate of the proximity of the system to its voltage limits. The thermal capacity should be considered with respect to peak loading conditions. A classification of "high" indicates that substantial additional thermal capacity remains under peak load conditions, for example, 50% of total capacity.

fail to align with the requirements of the RDU. These aspects can be divided into three primary and interrelated areas:

1. Inadequate physical infrastructure; for example, radial feeders with protection schemes geared toward electricity flows from substation to customer, but not vice versa;
2. Insufficient communication, measurement, and control infrastructure; for example, a lack of high-granularity measurements devices, such as microphasor measurement units, two-way communication, and control systems; and
3. Insufficient operational procedures; for example, little, if any, active control of loads and DERs, insufficient protection schemes for two-way flows, and few data-driven, automated operational procedures, such as online line health monitoring and dynamic rating of distribution lines/transformers.

Once the planner has broadly assessed the current status of system infrastructure and operational practices to achieve the RDU objective, the next step is to assess in a more detailed fashion the additional infrastructure and operational practices that will be required to overcome any barriers to meeting that objective. Doing so requires an expanded approach to conventional planning and operations.

4 PLANNING THE RDU TRANSITION

As described earlier, the goal of the RDU planner is to maximize the value of system resources for ratepayers while ensuring safe and reliable access to the distribution grid. Thus, when facing the expansion of their system in the presence of rapid DER deployment, the primary goals of RDU planners include the foundational goals of conventional distribution system planning: to ensure reliable, safe, and efficient access to electricity for consumers in the utility's service territory. However, the scope of these goals must be broadened, such that they apply not only to electricity consumption by end-use consumers, but also to production and other services provided by DERs.[16]

Moreover, the variability and uncertainty associated with DERs requires planners to expand the scope of conventional planning considerations, and demands updated planning approaches and tools. While planning with DERs is undeniably a more complicated undertaking than conventional planning, there are commonalities between conventional and DER-inclusive planning that enable the systematic assessment of DER-driven needs, and that reduce the burden on planners to invent an entirely new approach. Furthermore, as discussed later, the experience of "first adopters," such as the utilities in Hawaii and California, can act as a road map for other utilities seeking to plan their RDU transition.

Conventional utility distribution system planning is rooted in the steady-state analysis of the distribution system under forecast future peak load conditions.[17] Utility planners, using conventional planning tools such as power flow analysis, have typically assessed: (1) the adequacy of physical infrastructure to reliably meet future energy delivery and system security needs[18]; and (2) candidate upgrades to existing infrastructure in the event existing infrastructure is found to be inadequate.[19] A number of such candidate investments are evaluated and those deemed to

---

16. A discussion of the shifting roles of consumers on the distribution network can be found in chapters by Gimon, and Smith and MacGill in this volume.
17. For a representative summary of current planning practice, see, for example, PG&E (2015).
18. A typical goal of conventional planning is to maintain an Average Service Availability Index greater than or equal to 99.977% (Glover et al., 2012).
19. Conventional upgrades include new substations, thermal capacity upgrades for lines and transformers, and conventional voltage control devices, for example, tap-changing under load transformers (LTCs), static VAR compensators, and switched capacitors.

yield the lowest costs for ratepayers, while meeting the reliability needs of the future system are selected for investment.

While planning the RDU transition requires the assessment of (1) and (2) above, DERs complicate the process due to the uncertainty associated with their capacity, location, and the amount of power and energy they will produce, and the resulting impacts on power quality and voltage. These complications require additional planning processes, discussed next, that complement and expand the conventional approach. Furthermore, they require the assessment of current operational practices to determine where changes in operational practice may substitute for—or complement—infrastructure investment, in order to maximize the benefits of DER integration.

4.1   DER Expansion Forecast

A new and demanding task associated with planning the system for widespread DERs is DER expansion forecasting. A DER expansion forecast consists of forecasting, at a desired future point in time, the type and quantity, a single value or range, of DERs the utility anticipates coming onto the system, and the locations of those DERs. For many systems, such as those with long or thermally/voltage-constrained feeders, DER location is particularly important in determining the overall transition cost.

The DER expansion forecast may consist of a sophisticated forecast of future DER installations, if the data required for such an undertaking is available.[20] If the construction of such a forecast is not possible or not desired, the DER forecast may also consist of a set of potential DER growth scenarios against which the performance of the system can be assessed to "book end" the range of potential RDU needs. For example, a forecast may include two DER growth scenarios: a worst-case scenario in which DERs are highly concentrated far from the head of feeders, and a best-case scenario in which DERs are distributed more evenly along feeders, or at their strongest points of interconnection.

The use of scenario-based analysis for a multitude of forecast future DER growth scenarios is emerging as the primary methodology for assessing the diversity of system conditions distribution equipment will face with deep DER penetrations. For example, the California investor owned utilities (IOUs) were recently tasked with the development of Distribution Resource Plans, a key component of which is the production of a set of DER growth scenarios.[21] These

---

20. Such data would include, for example, demand growth and demographic data, current installation rates for DER systems, future end-use electricity rates, expectations for future DER incentive policies, and future DER system costs.

21. The development of the Distributed Resource Plans was a statutory requirement resulting from the passage of California Assembly Bill (AB) 327. For the first round of California IOU Distribution Resource Plans, see SCE (2015), SDG&E (2015), PG&E (2015).

scenarios define the potential future realizations of DERs against which the IOUs will assess the costs and benefits of DER integration, and plan their systems.

4.2   Planning Infrastructure to Maximize DER Value

Achieving the RDU value maximization objective requires planning tools that can assess the physical limitations of the system with respect to DER integration, as well as those capable of determining the various benefits of DER integration, and the system upgrades that maximize those benefits. These planning requirements are an extension of those used in conventional planning processes, which focused primarily on assessing cost and reliability of service. Table 20.1 compares conventional and RDU planning.

Physical infrastructure limitations to DER integration come in the form of: (1) inadequate thermal capacity of lines and transformers; (2) failure of the protection system to operate as designed; and (3) the emergence of voltage issues, such as flicker and steady-state voltage violations. Determining the extent to which each of these limitations exist in a particular system can be done with conventional power flow based tools. However, it requires planners to diverge from the conventional planning practice of assessing a few scenarios, or snapshots, of system operation, such as peak loading conditions.

Instead, the planner must expand the number of scenarios they consider to account for the DER expansion forecast scenarios. In addition, the focus in conventional planning on a few points in time, for example, the peak load hour, must give way to time series analysis to capture the impacts on the thermal capacity, protection systems, and voltage of period-to-period DER variability. The most challenging operating conditions with high levels of DER may not coincide with the peak load period, and this will be revealed through such time series analysis. Table 20.1 compares conventional and RDU planning.

Additionally, voltage regulation and protection equipment must be analyzed with respect to the higher variability in distribution line flow magnitude and direction. Present voltage control equipment is switched on time scales, on the order of minutes to hours, whereas DERs, such as solar, PV may cause voltage fluctuations on much shorter time scales (eg, seconds to minutes). Therefore, it may be necessary to perform power flow simulations for select periods on these shorter time scales, so as to assess the impacts of voltage control investments on smoothing out DER-driven voltage variability. Such variability will also necessitate the analysis of many additional fault scenarios in system protection studies, and require new models for DERs to capture their fault response adequately.

The existence of controllable equipment, such as generators and fast-regulating voltage equipment, on the distribution system will also require additional dynamic stability simulation in the planning process. However, in many cases, new dynamic models will be required for distribution-level devices. Such models will emerge from collaborations between device manufacturers, utilities, and the academic community.

TABLE 20.1 Comparison of Conventional and RDU Planning

Aspect	Conventional planning	RDU planning
Time frame	Few snapshots	Time series analysis
Time granularity	Single hour periods associated with snapshots	Multiple simulation periods used to assess different levels of variability, for example, hourly for net load ramping, minute-by-minute for short-term voltage fluctuations driven by PV
Number of planning scenarios	Limited to peak load and a few high impact contingency scenarios	Deployment of numerous scenarios for forecast DERs and net load
System protection assessment	Assess for one-way flows	Assess for variable, two-way flows
Voltage equipment assessment	Assess for one-way flows	Assess for variable, two-way flows
Dynamic stability analysis	Limited use of dynamic simulations	Feeder-level dynamic simulations required for feeders with widespread DERs
DER hosting capacity analysis	None	Foundation of RDU planning
Location-specific resource costs/ benefits assessment	None	Key for assessing value maximization objective
Assessment of operations solutions to infrastructure deficiencies	Little to none	Extensive

The extent of the system limitations to DER integration and the benefits of DER integration are highly dependent on the location of the DERs in the system. For example, DERs located near the substation may be less effective at reducing distribution-level losses than DERs located further down the feeder or closer to the loads, or the DER hosting capacity of individual buses may differ along the feeder. Therefore, to harness DERs efficiently, it is necessary for planners to assess their locational costs and benefits. The deployment of time series-based power flow simulation of distribution feeders enables such analysis.

Value maximizing planning also requires planners to not only expand the tools they use to plan the system, but also the set of potential candidate investments to address the infrastructure needs identified during the planning process.

For example, the most prudent approach to alleviating a voltage limitation that may restrict severely the DER hosting capacity of a feeder may be the installation of fast-acting voltage regulators, along with measurement and control infrastructure to enable advanced control. Furthermore, to address identified limitations efficiently, it may be necessary for planners to consider additional substations installations, as well as unconventional approaches such as more meshed topologies, distribution system interconnections, and neighborhood-level energy storage technologies.

4.3 Recent Developments in Advanced Distribution Planning

The rapid growth of DERs in Hawaii and California has prompted regulators and utilities to reevaluate distribution planning processes. The experience from the utilities in these states sheds light on some potential approaches to addressing the needs of the RDU transition, and, in particular, the integration of DERs into planning, so as maximize their value to the ratepayers.

As described earlier, evaluating the hosting capacity of a distribution feeder is one of the major challenges brought by DERs to distribution planning. To address this challenge, the California IOUs[22] have proposed a process termed Integrated Capacity Analysis (ICA), deployed in parallel with conventional planning processes (SDG&E, 2015; SCE, 2015; PG&E, 2015). The goal of ICA is to identify the hosting capacity of a distribution circuit, with respect to four limitation categories: (1) thermal ratings; (2) protection system limits; (3) power quality standards; and (4) safety standards. Feeder circuits are analyzed by breaking them into zones or line sections, typically four for a single distribution circuit. DER capacity is added to each zone or section individually, until one of the limits is reached.

Furthermore, the California IOUs have proposed an Optimal Location Net Benefits Methodology (LNBM) intended to assess the locational aspects of DER placement, and identify locations at which DER deployment is a viable substitute for conventional distribution investments (SDG&E, 2015; SCE, 2015; PG&E, 2015).

The Hawaiian utilities[23] have also been grappling with a flood of DER development, which has pushed them to develop new planning approaches. To this end, the Hawaiian utilities have proposed their own approach to integrating DER considerations into their distribution planning framework. The primary tool they proposed is the Distributed Generation Interconnection Capacity Analysis (DGICA), a "process for proactively identifying distribution system upgrades needed to safely and reliably interconnect DG resources and increase

---

22. The California IOUs include Pacific Gas and Electric Company, Southern California Edison Company, and San Diego Gas and Electric Company.
23. The Hawaiian utilities include Hawaiian Electric Company, Inc., Maui Electric Company, and Hawaii Electric Light Company.

circuit interconnection capability in capacity increments" (Hawaiian Electric, 2014). In a similar fashion to the ICA proposed in California, the DGICA outlines an approach to assessing the hosting capacity of distribution circuits, so that distribution planners prevent inadequate consideration of DER integration in planning from becoming a barrier to DER interconnection. Furthermore, Hawaii's utilities propose to modify the distribution circuit design criteria to facilitate increases in hosting capacity (Hawaiian Electric, 2014). The following additional criteria were proposed: (1) lowering impedance; (2) optimizing reverse flow on voltage regulation equipment; and (3) mitigating circuit-level transient overvoltage.

In both the California and Hawaiian cases, the need for time series-based power flow analysis, and more extensive scenarios of future system conditions, was explicitly acknowledged. Moreover, the utilities in these states, such as Pacific Gas and Electric Company in northern California, have made strides in incorporating such analysis into their existing planning framework, and provide examples for other utilities looking to follow a similar path (PG&E, 2015).

5   RDU TRANSITION OPERATIONAL PRACTICE REFORMS

Operations is perhaps the largest area in which utility practices will change as a result of the RDU transition. Due to the historical structure of the distribution system as a conduit for one-way flows from the substation to end-use consumers, the operation of the system has largely been passive in nature. What little controllable equipment has been present on the system, for example, switched capacitors or tap-changing transformers, was operated according to schedules or forecast system conditions, rather than according to real-time measurements and needs. Such an approach was reasonable and effective when the operator's goal was to meet a relatively predictable load reliably, and maintain voltages within specification. With DERs in the mix, however, such a passive approach is no longer tenable.

The RDU takes an active approach to operations in which it aims to maximize the value of DERs by fully utilizing advanced measurement, communication, and control infrastructure, as well as harnessing the control of DERs to shape the load and control voltage, among other objectives. Making this transition from passive to active operations will require additional measurement collection and control equipment, as well as changes in operating practices, all the way down to the energy management system (EMS) functionality. Table 20.2 gives a comparison of conventional and RDU operations.

The collection and use of high bandwidth, time-stamped measurements using equipment such as AMI, microphasor measurement units (PMUs) (Micro-Synchrophasors Project, 2015), or line-mounted sensors (Tollgrade Communications, 2015), will be the bedrock of this transition. The RDU will require a measurement system that has sufficient: (1) measurement granularity, for example, are measurements collected every hour, minute, or even more frequently, and do these align with the operational needs; (2) measurement coverage, for

TABLE 20.2 Comparison of Conventional and RDU Operations

Aspect	Conventional operations	RDU operations
Control objective	Serve end-use consumer, maintain voltage within specifications	Maximize DER value to rate payers while meeting reliability criteria
Use of measurements (system visibility)	Limited SCADA data collected at substations and advanced metering infrastructure (AMI)	Extensive measurement collection throughout system
Level of automation	Some automated breaker operation	Widespread distribution system automation and use of automatic DER control
Use of data-driven algorithms	Little to none	Algorithms become the cornerstone of operations
Two-way communication with resources	Modest, largely communication with loads via AMI	Ubiquitous two-way communications between operator and resources

example, are the key distribution buses equipped with voltage measurement devices; and (3) measurement type, for example, are sufficient voltage measurements being collected to enable advanced control. Furthermore, with extensive measurement collection, the capability will exist for data-driven analytics, such as online infrastructure health monitoring and DER dispatch, which will become a cornerstone of RDU operations.[24]

The deployment of measurement equipment throughout the system will present additional control opportunities for the system operator. However, to implement advanced control successfully, the RDU will require two-way communication with resources in the system to enable utility operators to collect data from devices in the system, and to send control signals to achieve desired operational objectives. While some two-way communication equipment, mainly AMI, has been installed in distribution systems, the rollout of such communication equipment must be broadened from AMI, and those resources participating in demand response programs, to a larger set of resources, including all new and existing DERs.

5.1 Recent Developments in Advanced Distribution Operations

Operational reforms have been a large part of the DER integration discussion taking place in California and Hawaii. In both states, utilities have proposed

---

24. For examples of such analytics using micro-PMU data, see, for example, Stewart et al. (2014).

operational reforms to facilitate higher DER penetrations, as well as pilot projects to test the effectiveness of the proposed reforms.

Southern California Edison, for example, has proposed a pilot project to "test its current operational capabilities and those capabilities that are needed to coordinate third-party DER and potentially utility-owned DER." Furthermore, "the technology infrastructure (eg, telecommunications, monitoring devices, and control systems) to be deployed in the area are aimed at providing additional capabilities (eg, monitoring, controls) that may enable coordination of higher levels of penetration throughout the SCE system."[25] Such pilot projects will reveal the extent to which utility operations must shift in order to enable the effective integration of DERs.

The Hawaiian utilities have been more explicit in defining operational reforms to facilitate DER integration. In their Distributed Generation Interconnection Plan (Hawaiian Electric, 2014), the utilities propose the following operational practice changes:

- Operating within voltage regulation bands
- Maintaining distribution circuit flexibility
- Lengthening reclosing time of feeder breakers and reclosers for islanding protection
- Monitoring voltage regulator tap operations
- Implementing SCADA at distribution substations

The Hawaiian proposal makes clear the need for a fundamental reassessment of distribution operations, with respect to increasing DER deployment. The proposal of basic reforms, such as the implementation of SCADA at substations, also sheds light on the long distance left for utilities to travel in order to complete the RDU transition.

6 CONCLUSIONS

All utilities will undertake the RDU transition, to some degree, the extent of their transition depending largely on the level of DER integration that will be required in their service area. While some steps have been taken by utilities in California and Hawaii toward making the RDU transition, there has been little experience so far with advanced planning and operations processes, so there is not yet a consensus on the "best practices."

This chapter presents an approach to conceptualizing and planning the RDU transition that enables planners to address DER integration, in a way that aims to maximize value for ratepayers. However, in any foreseeable future, the conditions under which utilities operate are rapidly evolving. Ultimately, it is up to individual distribution utilities to assess how they can adapt to their unique conditions best. The uniqueness of their conditions, however, does not preclude

---

25. For a complete description of the pilot project, see SCE's DRP (SCE, 2015).

them from taking a systemized approach to determining their needs. In fact, such an approach is both achievable and necessary to build highly complex distribution systems whose operation ensures, economically and reliably, access to electricity for consumers and producers alike.

REFERENCES

Barbose, G., Weaver, S., Darghouth, N., 2014. Tracking the Sun VII: An Historical Summary of the Installed Price of Photovoltaics in the United States from 1998–2013. Lawrence Berkeley National Laboratory.

Black & Veatch, 2013. Biennial Report on Impacts of Distributed Generation.

Broderick, R., Quiroz, J., Reno, M., Ellis, A., Smith, J., Dugan, R., 2013. Time Series Power Flow Analysis for Distribution Connected PV Generation. Sandia National Laboratory.

Costello, K., 2015. Utility Involvement in Distributed Generation: Regulatory Considerations. National Regulatory Research Institute.

CPUC, CEC, 2015. Recommendations for Utility Communications with Distributed Energy Resources (DER) Systems with Smart Inverters: Smart Inverter Working Group Phase 2 Recommendations. California Public Utilities Commission and California Energy Commission.

Domínguez-García, A., Heydt, G., Suryanarayanan, S., 2011. Implications of the Smart Grid Initiative on Distribution Engineering: Part 1—Characteristics of a Smart Distribution System and Design of Islanded Distributed Resources. Power Systems Engineering Research Center.

Forsten, K., 2015. The Integrated Grid: A Benefit-Cost Framework. Electric Power Research Institute.

Fox-Penner, P., 2014. Smart Power: Climate Change, the Smart Grid, and the Future of Electric Utilities. Island Press, Washington, DC.

Glover, J., Sarma, M., Overbye, T., 2012. Power System Analysis and Design, fifth ed. Cengage Learning, London.

Hanser, P., Van Horn, K., 2014. The evolution of the electric distribution utility. In: Sioshansi, F. (Ed.), Distributed Generation and its Implications for the Utility Industry. Elsevier, Waltham, (Chapter 11).

Hawaiian Electric Companies, 2014. Distributed Generation Interconnection Plan. Hawaiian Electric Companies.

Hesmondhalgh, S., Zarakas, W., Brown, T., 2012. Approaches to Setting Electric Distribution Reliability Standards and Outcomes. The Brattle Group.

Kind, P., 2013. Disruptive Challenges: Financial Implications and Strategic Responses to a Changing Retail Electric Business. Edison Electric Institute. http://www.eei.org/ourissues/finance/documents/disruptivechallenges.pdf

Lee, M., Aslam, O., Foster, B., Katham, D., Young, C., 2014. 2014 Assessment of Demand Response and Advanced Metering. Federal Energy Regulatory Commission.

Micro-Synchrophasors for Distribution Systems, 2015. About the ARPA-E Micro-Synchrophasor Project. http://pqubepmu.com/about.php

Navigant Consulting, Inc., 2013. Distributed Generation Integration Cost Study. California Energy Commission.

NYSDPS, 2014. Reforming the Energy Vision. NYS Department of Public Service.

PG&E, 2015. Pacific Gas and Electric Company Electric Distribution Resources Plan. Pacific Gas and Electric Company.

SCE, 2015. Distribution Resources Plan. Southern California Edison.

Schmalensee, R., Bulovic, V., 2015. The Future of Solar: An Interdisciplinary Study. MIT Energy Initiative. https://mitei.mit.edu/futureofsolar

SDG&E, 2015. Distribution Resources Plan. San Diego Gas and Electric Company.

Short, T., 2014. Electric Power Distribution Handbook, vol. 2. CRC Press, New York.

Sioshansi, F., 2014. Distributed Generation and its Implications for the Utility Industry. Elsevier, Waltham.

Stewart, E., Kiliccote, S., McParland, C., Roberts, C., 2014. Using Micro-Synchrophasor Data for Advanced Distribution Grid Planning and Operations Analysis. Ernest Orlando Lawrence Berkely National Laboratory.

Tollgrade Communications, Inc., 2015. Predictive Grid Quarterly Report: Building a Predictive Grid for the Motor City, vol. 1. DTE Energy.

Chapter 21

Distributed Utility: Conflicts and Opportunities Between Incumbent Utilities, Suppliers, and Emerging New Entrants

Kevin B. Jones, Taylor L. Curtis, Marc de Konkoly Thege,
Daniel Sauer, Matthew Roche
*Institute for Energy and the Environment, Vermont Law School, South Royalton, VT,
United States of America*

1 INTRODUCTION

A major communications company estimated that, by the end of the decade, 50 billion devices will be connected to the Internet. Approximately 80% of those devices will be talking almost entirely to other devices, not to people. In other words, 40 billion devices will be working behind the scenes, on our behalf. However, some experts argue that a paradigm shift in system integration, including the adoption of new business models, is required to revolutionize industry to take full advantage of this distributed communications network (Heck and Rogers, 2014, p. 124). Until recently, the electric utility industry has been an outlier to this digital communications revolution, since it has not fully leveraged information technology the way the other sectors of the economy have (Jones and Zoppo, 2014, p. 1). Recognizing the benefits of advanced information technology, both industry members and government regulators have begun to advocate for a smarter electric grid and other have considered the organization and structure of the "Utility of the Future." In New York State, the Reforming the Energy Vision (REV) proceeding has garnered national attention for its focus on a holistic utility business model reform and system integration. This chapter will explore the overall vision of the REV process, discuss the conflicts among the major players, and compare the reforms in New York to policy reform in California, Massachusetts, and Hawaii in order to better understand the key policy differences among leading states. The research and analysis of this chapter will provide a perspective of the utility of the future debate, including the challenges and opportunities that this new vision will pose for the utility,

Future of Utilities - Utilities of the Future. http://dx.doi.org/10.1016/B978-0-12-804249-6.00021-X

399

supplier and distributed sectors, and ultimately inform other states and countries about these reform efforts.

2 OVERVIEW OF NEW YORK'S REV

The New York State Public Service Commission's (NYPSC or Commission) landmark initiative, REV,[1] outlines how the state plans to reform the electric distribution grid, in much the same way that FERC Order 888 set in motion the restructuring of the wholesale transmission and generation sectors.[2] The NYPSC designed the initiative to "reorient both the electric industry and the ratemaking paradigm toward a more consumer oriented approach that harnesses technology and markets." The goal is to integrate distributed energy resources (DER) into the planning and operation of the system to achieve a more efficient, resilient, and climate-friendly energy system (NYPSC, 2015, p. 3). New York has leapt to the front of the policy debate, in terms of integrating DER. Yet it has one of the lowest rates of advanced metering infrastructure (AMI) deployment in the country, in contrast to California, which has already installed AMI for all customers.

REV seeks to reform the utilities' business models and services offered to encourage system security and resiliency needs in an increasingly decentralized, complex grid. REV encourages regulatory changes that promote energy efficiency, demand response, increased storage capacity, and renewable energy resources. These changes seek to empower end-users by providing more choice in how they manage and consume electric energy.

In order to provide incentives for utilities and markets for the development of a cleaner and more efficient electric system, REV also recognizes the need to reform the ratemaking process. REV calls for a fundamental shift from the existing ratemaking structure to meet the needs of a more distributed, consumer-focused energy system. Implementing the REV process will take years, not months. To account for this comprehensive reform process the NYPSC designed a two-track regulatory reform process. The first track involves a collaborative process examining the role of distribution utilities in enabling market-based deployment of distributed energy resources to promote load management and greater system efficiency, including peak load reductions. The second track will change regulatory, tariff, and market designs to better align utility interests with the Commission's policy objectives (NYDPS, 2014).

---

1. NYPSC (2013, Order Approving EEPS Program Changes) announced a fundamental reconsideration of NY regulatory paradigm and markets, examining how policy objectives are served both by clean energy programs and by the regulation of distribution utilities.
2. The report calls for state distribution utilities to achieve five policy objectives: (1) increasing customer knowledge and providing tools that support effective management of their total energy bill; (2) market animation and leverage of ratepayer contributions; (3) system-wide efficiency; (4) fuel and resource diversity; and (5) system reliability and resiliency.

A diverse group of stakeholders generated more than 600 comments in response to the REV proposal.[3] The interests of generators, utilities, distributed energy, customers, and environmental groups represent a wide array of issues with the REV proposal. Although all stakeholders generally supported the vision REV has created for transforming the electric grid in New York, there are divergent proposals and ideas concerning fundamental components of the initiative. Specifically, stakeholders differ on who should own DER, the benefit–cost analysis (BCA) for DER planning by the utility, and the structure and control of the distributed system platform. This chapter discusses each of these key issues in more detail to understand the differing views shared by the utilities, generation sector, and the emerging distributed energy industry.

2.1   Distributed System Platform and Provider

REV defines the distributed system platform as a new intelligent network platform that has three primary functions: integrated distribution planning, grid operations, and market operations. Central to REV is the identification of the entity that will manage the platform by taking on the role of the Distributed System Platform Provider (DSP). Dependence on a wide range of distributed resources for system reliability will increase the complexity and importance of distribution planning and operation. Today, the traditional distribution system is a one-way system with fairly predictable but nonresponsive demand. The introduction of DER, as well as intelligence on the grid, supports a much more dynamic, multidirectional, and efficient system. The DSP will be the interface between the wholesale bulk power system and increasingly diverse DER markets, consisting of customer load, new distributed sources of supply, and other services. The DSP will centralize sales and provision of a set products and services to customers, and service providers. Those might include transaction or usage fees, analytic services, interconnection services, pricing and billing, and metering. During the initial stages, the DSP market will likely consist primarily of open access tariffs, as opposed to market-based auctions. In order to operate this new system securely, the DSP will need to integrate distributed generation and other DERs into the grid in an efficient manner that does not compromise service reliability (NYPSC, 2015, pp. 31–33; NYDPS, 2014, p. 4).

Ultimately the NYPSC order, established existing distribution utilities as the DSP for their service areas. The NYPSC concluded that because the DSP core functions "would be highly integrated with utility planning and system operations, assigning them to an independent party would be redundant, inefficient, and unnecessarily costly." Largely duplicative of a distribution utility's functions in terms of system planning and operations, the NYPSC held that

---

3. The various comments and reports in the REV proceeding (New York State Public Service Commission, Case Number 14-M-0101) can be found at http://documents.dps.ny.gov/public/Matter Management/CaseMaster.aspx?MatterCaseNo=14-m-0101&submit=Search+by+Case+Number

utilities can most efficiently use their assets and existing functions to make the most efficient and economic decisions for all customers. Because the utilities have systems in place to maintain customer privacy and engagement and the ability to verify and measure performance for DER services, they are in the best position to be the DSP (NYPSC, 2015).

Other stakeholders had opposing viewpoints whether the utilities would be best suited as the DSP. Some environmental groups, energy service companies (ESCOs) and DER providers believe the DSP should be an entity unaffiliated with the distribution utility. Proponents of an independent DSP argue an independent body would take over the most complicated tasks (maintaining system reliability, opening the grid to third-party competition, and dispatching distributed resources), allowing the utility companies to focus on tasks already mandated. Former FERC Chairman, Jon Wellinghoff, believes that an independent DSP would create a competitive distribution energy market that would enable the full deployment of DER, while improving the utilization of existing grid resources (NYPSC, 2014a, pp. 3–5). This structure would also allow for greater consumer choice and participation in the electric grid of the future. Many incumbent generators also question whether the utility is best suited to be the DSP. They raise concerns about the utility exercising market power in their own interest, and then suppressing innovation, at the expense of customers and market participants. Despite these concerns, existing utilities were selected as the DSP. Interested parties encourage transparent performance standards and a well-defined succession plan for the utility-as-DSP if it fails to meet REV objectives. Many parties also requested effective separation between the DSP's market functions, and planning and system operations. As a result, the NYPSC determined that the utilities' performance as the DSP will be monitored, and evaluated through outcome based metrics (NYPSC, 2015, pp. 46–51).

2.2 Ownership of DERs on the DSP

DER ownership is a highly contentious issue. Many stakeholders would like to prohibit utilities from participating as owners of DER, where a market participant can and will provide DER services. A main objective of REV is to create a marketplace for DER based upon consumer information and choice. Opponents fear that utility ownership of DER will have anticompetitive effects. The NYPSC generally agreed with this position, and determined that utility ownership of DER would only be allowed if: (1) procurement of DER was solicited to meet a system need and the competitive alternatives are clearly inadequate or more costly than a traditional utility infrastructure alternative; (2) a project consists of energy storage integrated into distribution system architecture; (3) a project will facilitate low or moderate income residential customers to benefit where a market is not likely to satisfy the need; or (4) a project is being sponsored for demonstration purposes. The Commission has allowed utility affiliate ownership of DER outside of the utility's territory. Utility affiliates can own DER within their service territory, but only under stringent regulations, including third party evaluation of proposals when a utility affiliate is participating (NYPSC, 2015, pp. 70–71).

Numerous suppliers, DER providers and environmental groups oppose utility DER ownership. Concerned about utility market power, stakeholders argue that utility ownership would create unfair advantages because they have a natural inclination to limit competition in the market, when they benefit from doing so. Generators argue that the proposed NYPSC mitigation measures will fail to curb utility vertical market power, and will have a chilling effect on private investment. Many DER product and service providers, including ESCOs, believe nonutility companies should deploy and own DER. These industry players think that the nonutility companies owning DER should act either on behalf of the utility/DSP, or by participating in the competitive marketplace.

Those parties supporting the NYPSC approach argue that limited utility ownership could accelerate DER deployment (NYPSC, 2014b). Utility ownership would allow customers to avoid additional costs involved in getting services from both a utility and from a third-party DER provider. Additionally, utility involvement could assist DER deployment in low-income markets, which would not otherwise be served (NYPSC, 2015, pp. 63–64).

The utilities believe that their existing assets, particularly relationships with their customers, will help to catalyze DER markets. They argue that customers should be able to choose their DER provider, and that their ownership of DER services will increase reliability of service and customer convenience. By expanding utility DER participation to the customer side of the meter, utilities can collaborate with third parties to provide a pathway to a competitive market. Additionally, customers will be able to manage their energy services in one place (NYPSC, 2014c).

2.3 Benefit-Cost Considerations in the Implementation of the REV Policy

While a future goal of REV is market pricing of DER services, early phases require the need to determine both which services should be procured by the DSP, and at what price. BCA is the prospective tool that utilities and regulators use today to evaluate DER in distribution grid resource planning. A major point of contention in the development of REV stems from the Commission's proposal for utilities to overhaul the evaluation of costs and benefits, particularly for DER technology and services.

Environmental groups and DER providers think the BCA needs to extend beyond individual measures, and encompass DER programs and portfolios for deployment. They argue for a "societal cost" test that looks at DER deployment holistically, rather than a more narrow "rate impact" test. The utilities share the view that the BCA should evaluate their actions holistically and evaluate their portfolio of investments, rather than individual measures. Specifically, the utilities suggest the BCA framework should focus on the costs and benefits that are directly linked to aggregate customer bills, which include (1) The net impacts on T&D operations and costs; (2) The net impacts on wholesale energy and capacity costs; and (3) consideration of certain directly relevant and quantifiable externalities (NYPSC, 2014c).

Incumbent generators, on the other hand, are concerned the reformed BCA will favor DER systems in distribution grid planning and tariffs, rather than the services they currently provide. Generators fear that favoring DER products could affect wholesale market reliability negatively. Generators caution that prompting heavy investment in DER infrastructure and deployment would increase electricity capacity in the distribution grid and lower prices, and this would decrease their sale of power in the retail market. The effects of low electricity prices in the distribution system would spill over into the wholesale market, and suppress electricity prices there. Suppression of wholesale electricity prices would result in decreased revenues and potential death spirals of generators, effectively forcing them out of business. Their demise would likely be a disincentive to investment in the wholesale market for new generation entrants, subsequently straining system reliability. They argue that the summation of a poorly designed BCA could result in not only poor planning from a reliability standpoint, but that it could contradict its own overall environmental criteria by harming the financial stability of "clean" base load generation like nuclear generators in New York State (NYPSC, 2014d).

The NYPSC staff states that BCA standards will change at every significant change in DER deployment levels, and that improvement in information and communication technologies will improve the granularity and precision of the BCA on an ongoing basis. DER deployment will be evaluated on a portfolio wide basis and measuring environmental impacts will be an important aspect of the reformed BCA. It will quantify social costs of alternatives to DER when possible, and when not it will consider them qualitatively (NYDPS, 2015).

2.4   Other Policy Issues—Advanced Metering, Efficiency Programs, Renewables, Microgrids, and Demonstration Projects

While the NYPSC (2015) concluded that deployment of some form of advanced metering infrastructure (AMI) is necessary to implement REV, the Commission stopped short of supporting broad ratepayer funded installation of AMI. The NYPSC acknowledged that "dynamic pricing will require signals both to and from end-use equipment" and that "settlement data will often require time-stamped usage data." The NYPSC did not go further than acknowledging that each utility DSP will need to include a plan for dealing with advanced metering needs and noted that "third party investment may be preferred over sweeping ratepayer funded investments."

The NYPSC's REV proceeding also addresses existing utility energy efficiency programs, as well as the provision of large-scale renewables from New York's renewable portfolio standard. New York currently has a mix of utility managed efficiency programs as well as efficiency programs funded through NYSERDA. Additionally, New York has a unique renewable portfolio standard implemented through the centralized procurement of renewable energy credits by NYSERDA, rather than the individual procurement of renewables by the utilities. REV does not seek to modify the current NYPSC goals immediately, it wants to expand availability of these resources. Principal to the reformed vision is the eventual transformation of utility-directed efficiency rebates, and NYSERDA's centrally-procured large scale renewable requirements, into more market-based approaches. How this transition will proceed is unclear, but the NYPSC appears intent on facilitating forward progress on the Cuomo Administration's goals to replace subsidies and mandates with market responses. The REV proposal also supports the concept of microgrids, and includes provisions for encouraging demonstration programs to help inform which products and services should be procured through the DSP process.

2.5   REV Implementation

Implementation of REV will take years. While it is under development, traditional distribution rate cases will continue for the foreseeable future. However, major steps will begin now, as each utility is required to file a Distributed System Implementation Plan (DSIP). Each utility already files with the Commission, annually, a 5-year capital plan detailing system needs and the utility's plans to meet them. At a minimum, the DSIP will include:

1. Actual and forecast system loads and capital spending projections, at a level of specificity sufficient to inform market planning and participation by third parties
2. Actual and forecast levels of DER, including detailed analysis of system needs amenable to being met by DER
3. Plans for encouraging market development of DER; plans for increasing DER deployment in underserved markets
4. Specific plans including cost estimates for building DSP capabilities; and a description of internal organization of DSP and traditional utility functions

The assumptions and methodologies of the DSIP must be transparent and the results will be public, subject to any protections needed for purposes of system security (NYPSC, 2015).

3   CALIFORNIA'S PLANS FOR REFORMING THE DISTRIBUTED UTILITY

Similar to New York, California is exploring how investor owned utilities (IOUs) participate in creation of mechanisms and markets for DERs to offer their various values to the grid (De Martini, 2014). The utilities are facing significant changes in how they interact with customers and third-party owners of DERs, and how those aggregations of distributed resources are integrated within the state's broader grid operations.

As the national leader in installed capacity of solar PV, California faces unique challenges related to an existing scaled deployment of DERs that has not yet materialized in New York (Crosby et al., 2015). California's three largest IOUs, (San Diego Gas and Electric, Southern California Edison, and Pacific Gas and Electric) spend roughly $6 billion annually on distribution grid investments. Over the next decade, it's projected that California could see 15 GW of DERs come online, of which 12 GW of distributed solar, 1 GW of grid-scale energy storage, and 1 GW of demand response (De Martini, 2014). The utilities are clearly at a critical point and must understand how to utilize their resources most cost effectively (Wesoff, 2014). If these DERs are not effectively integrated into the grid, utilities will not be able to optimize this distributed resource to ensure that their $6 billion yearly investments will support a robust DER grid.

3.1   Current DER Reform in California

Following Assembly Bill 327 (California State Assembly, 2013), California set in motion a rulemaking proceeding in Jul. 2014 that requires California's IOUs to develop distribution resources plans (DRPs) in order to integrate DERs onto the grid better. The focus of DRPs, as laid out in AB 327, is on the technicalities related to DER integration, but does not explore explicitly changes to utility business models to integrate DERs more optimally. The California initiative does not propose to create a new distributed resource market, like the one envisioned in New York. However, like New York, the California process intends "to mov(e) the IOUs towards a more full integration of DERs into their distribution system planning, operations and investment" (Crosby et al., 2015). CPUC Commissioner Michael Pickering summed up California's focus in one word: "convergence" (Pickering, 2015).

In regard to California and New York, the fundamental distinction between their respective processes is twofold. While REV in NY looks to catalyze market forces to influence customer and third-party DER deployment to obtain more accurate valuation of resources, the policy development in California is technically focused and largely utility driven, with IOUs considering third party provided DERs in their planning process (Crosby et al., 2015). In addition, while there is a lot of public attention on a far-reaching regulatory proceeding in NY, focused on business model reform, California's state policy leadership has already resulted in near universal roll-out of AMI technology by its utilities, and a leading position in DER deployment.

4   CURRENT DER REFORM IN MASSACHUSETTS

Another state working on reforming their distribution grid is the Commonwealth of Massachusetts. To begin the reforms, Massachusetts passed two important pieces of legislation in 2012, in response to severe storms that caused power outages throughout the state. The Competitive Priced Electricity Act (2012) and the Emergency Service Response of the Public Utilities Commissions Act served as the beginning stages of grid modernization within the state. These laws increased the Department of Public Utilities (DPU) role in meeting electricity goals statewide. In response to this legislation, the process of DER reform in Massachusetts began in Oct. 2012, with a notice of investigation into the modernization of the grid, and the subsequent creation of a stakeholder working group to create a plan moving forward. In response to those legislative directives, a final order was issued on Jun. 12, 2014 by the MA Department of Public Utilities that shapes the requirements of grid modernization. It states that "grid modernization is an important means for advancing the statutory requirements and policy goals of further development of energy efficiency, renewable energy resources, demand response, electricity storage, microgrids, and EVs" (MA DPU, 2014, 12-76-B). The order emphasized that the end goal of a modern electric system builds upon the goals of clean energy through the maximization of integration of technologies such as solar, wind, and other local and renewable sources of power.

The DPU set out four benefits of grid modernization: (1) empowering customers to manage and reduce electricity costs better; (2) enhancing reliability and resiliency of electricity services; (3) encouraging innovation and investment in new technologies and infrastructure while strengthening market competition; and (4) addressing climate change and clean energy goals through the integration of clean and renewable power, demand response, storage, microgrids, and efficiency (MA DPU, 2014, 12-76-B).

In addition to benefits, the Department also laid out specific objectives of the grid modernization process. Its objectives are to reduce the effects of outages; optimize demand, including reducing system and customer costs; integrate distributed resources; and improve the workforce and asset management (MA DPU, 2014, 12-76-B).

The order requires every utility to develop and implement a 10-year grid modernization plan to carry out these objectives, and update it every 5 years.

In contrast with the NY REV proceeding, the Massachusetts order places a strong focus on advanced metering, and the infrastructure required to meet the goals proposed by the department. At the heart of the reform process in Massachusetts are distribution utilities implementation of AMI because it furthers all four of the grid modernization objectives. In contrast with the NY REV proposal, the MA proposal has a strong focus on advanced metering and the infrastructure required to meet the goals proposed by the department. Based on advanced metering infrastructure furthering all four of the grid modernization objectives, the department found it imperative that advanced metering serve as a priority for electric distribution companies. Additionally, AMI is viewed as necessary for all of the other benefits of grid modernization to be fully realized. The proposal included a requirement for a 5-year implementation plan that allows for preferential regulatory treatment for grid modernization investments.

Each utility is required to include its proposed technology and implementation plan, a BCA, and a request for preauthorization of investments. Since the customer has a large role in making the grid modernization work, each utility is required to incorporate a marketing, outreach, and education plan to educate customers of grid changes. One last key aspect of Massachusetts' grid modernization is the creation of targeted performance metrics to evaluate distribution utilities' progress (MA DPU, 2014, 12-76-B).

4.1 Other Massachusetts Actions Related to the Electricity Vision

The DPU undertook two other investigations related to grid modernization, focused on time of use rates and electric vehicle (EV) integration. The department hopes to create comprehensive metering policies with an approach that maximizes the benefits of time varying rates, and as a result, reduces costs for consumers (MA DPU 13-182). It also seeks to clarify whether electric distribution utilities should own and operate EV charging infrastructure, how companies are accommodating such infrastructure and the DPU's role in establishing and maintaining standards for EV charging.

5 DER BOOM IN HAWAII

Hawaii's geographic isolation, its reliance on fuel oil and the resulting economic challenges have led to rapid and sudden proliferation of distributed generation. As an island energy system importing expensive fuel oil, "Hawaii has the most expensive electricity of any state in the U.S. with an average rate of nearly 37 cents per kilowatt-hour" (Bade, 2015). High electricity rates and abundant solar radiation at the island's latitude has spurred a fast transition to distributed generation in Hawaii, testing the limits of rooftop solar penetration. As state regulatory Commissioner Lorraine Akibo (2014) of Hawaii stated, "the integrated grid of the future is one that requires strategic actions to realize in full value central power and distributed energy resources."

The Hawaii Public Utility Commission (referred to as PUC or Commission) has highlighted that Hawaii has "entered a new paradigm" where "the best path forward to lower electricity costs includes an aggressive pursuit of new clean energy sources" (HPUC, 2014b). "With Hawaii's high retail rates, rooftop PV has grown exponentially in recent years," according to Commission Chair Hermina Morita (Savenije and Cameron, 2014). Michael Champley, commissioner at the Hawaii PUC recently noted "12 percent of residential customers in Hawaii have rooftop solar—the highest percentage of any state in the nation" (Pyper, 2015). Further, solar-plus-storage has already reached grid parity for the commercial class and is nearing such a point for residential customers (Bronski, 2014). This reality presents a unique challenge for Hawaiian

Electric Company (HECO),[4] the largest utility in Hawaii. HECO reports that the company is seeing an increasing number of distribution circuits with rooftop solar exceeding "100 percent of daytime minimum load" (Wesoff, 2015). In Feb. 2014, HECO said that on the Big Island, 10 % of its circuits hit unstable levels (Trabish, 2014).

Rapid demand for solar PV has led to a serious backlog of solar PV approvals that HECO has been struggling to overcome. While many states are looking at the opportunity to integrate more renewables and the resulting challenge of operating the transmission and distribution grids, Hawaii is already facing this future. As the result of a fast pace transition, Hawaii does not have one regulatory comprehensive approach, but has been working at a more piecemeal pace.

5.1   Brief History of DER Reform in Hawaii

In 1992 the Hawaii Utility Commission established by order an Integrated Resource Planning (IRP) requirement for Hawaii Electric Companies (referred to as HECO Companies or Companies).[5] HECO is the largest supplier of electricity, supplying 95% of Hawaii's population (HSEO). In 2011 the Commission reviewed the background and history of the IRP and reestablished its importance (HPUC, 2011). However, in 2014 the Commission rejected the Companies' proposed plan as "clearly non-complaint and inconsistent" with the Commission's mandated IRP Framework (HPUC, 2014c).

Compelled by HECO's failure to articulate "a long-term, customer focused business strategy," the Commission issued a guidance document "Inclinations on the Future of Hawaii's Electric Utilities" which outlines a visionary map that aims to move Hawaii to a grid of the future. The guidance document outlines the HUC's "perspectives on the vision, business strategies and regulatory policy changes required to align the HECO Companies' business model with customers' interest and the state's public policy goal." The Commission made clear "[i]t is now incumbent on the HECO Companies to utilize this guidance in developing sustainable business model that explicitly governs the Companies' capital expenditure plans, major programs, and projects submitted for regulatory review and approval" (HPUC, 2014b).

Consistent with the guidelines, the HUC ordered each of the HECO Companies to file "Power Supply Improvement Plans" (PSIPs) to identify strategies, action plans and schedules to expeditiously achieve the results contemplated in

---

4. HECO serves approximately 450,000 customers or 95% of Hawaii's population through its electric utilities, Hawaiian Electric Company, Inc., Hawaii Electric Light Company, Inc., and Maui Electric Company, Limited and provides banking and other financial services to consumers and businesses through American Savings Bank, one of Hawaii's largest financial institutions.

5. See *In re Framework for Integrated Res. Planning,* Docket No. 2009-0108, Decision and Order, filed Mar. 14, 2011, at 3 (reviewing the background and history of the IRP requirement, which the Commission originally established by order in 1992).

the guidelines.[6] After the filing of the proposed PSIPs, stakeholders submitted petitions to intervene in the docket to review the plans. The public comment period resulted in gross opposition and criticism from a number of stakeholders including the state Energy Office (HPUC, 2014a).

Further complicating the regulatory landscape in Hawaii is the proposed merger of HECO with NextEra Energy, Inc. NextEra Energy is the parent of the utility Florida Power and Light Company, and NextEra Energy Resources LLC, the world's largest generator of solar and wind energy. Approval of the merger has been received by the Federal Energy Regulatory Commission, but is pending before the HUC.

In addition to the numerous orders by the Commission, the Legislature has recently enacted several laws providing express findings and mandates that parallel and support the Commission's work. Most recently, the Hawaii legislature passed a bill that would set a 100% renewable portfolio standard (RPS) by 2045 (Savenije, 2015). By far the most progressive RPS, the bill requires that the goal be met within the state's closed island energy system.

In addition, Act 37 and Act 109 passed in 2014 supplement efforts by the Commission. Act 37 sets forth numerous express findings regarding utility incentives. It is "imperative that Hawaii's electric utilities accelerate their efforts to acquire lower cost clean energy resources and reduce existing energy and other utility operating expenses." The Legislature authorized the Commission to establish a policy to implement incentives and cost recovery mechanism to "induce and accelerate electric utilities' cost reduction efforts, encourage greater utilization of renewable energy, accelerate the retirement of utility fossil generation, and increase investments to modernize the State's electrical grids." Act 109 also sets forth express findings, regarding the modernization of the grid, specifically associated with customer renewable generation. Hawaii solar industry has produced benefits including energy independence, job creation, and customer options, and "a long term, sustainable solar industry in the State's interest."[7] The Act pointed out, however, that the industry "is significantly impaired by the current interconnection process," and "the state needs a more transparent and timely process for electricity customers to exercise their options to manage their energy use."

5.2 Integration of Renewables in Hawaii

The Commission has specifically acknowledged the need to aggressively integrate renewable resources and reduce their restriction (HPUC, 2014b). The Commission's policy document states "Hawaii has already entered a new

---

6. Requirements to prepare Power Supply Improvement Plans are found in Order No. 32055, filed Apr. 28, 2014 in Docket No. 2011-0092; Decision and Order. 31758, filed Dec. 20, 2013 in Docket No. 2012-0212; and Decision and Order No. 32053, filed Apr. 28, 2014 in Docket No. 2011-0206. The Commission also issued a Policy Statement related to demand response programs.
7. Haw. Rev. Stat. Section 269-145.5(b) (Supp. 2014) (Act 109).

paradigm where the best path to lower electricity costs includes an aggressive pursuit of new clean energy source."[8] The case *In re Maui Elec. Co.* addressed the concerns of barriers to distributed energy resource implementation and cost (HPUC, 2014b). In this proceeding, the Commission required Maui Electric Company (MECO) to provide a System Improvement and Curtailment Reduction Plan to improve operational efficiency and reduce renewable curtailment. In reviewing the plan submitted by MECO, the Commission ruled that MECO had not set forth "a clearly defined path forward that addresses integration and curtailment of additional renewables, and that optimizes system operations through all of the tools that are available to MECO." The Commission specified "what was lacking is the vision of MECO as a utility of the future" (HPUC, 2014b, pp. 4–5).

5.3   Demand Response and DERs in Hawaii

The Commission is also increasing its focus on demand response. It noted that developing a unified plan is important so that "Hawaii would lead the nation in the use of advanced demand response for power system reliability service." The Commission has left it up to the utilities to create a unified plan, ordering HECO to file an Integrated Demand Response Portfolio Plan (IDRPP) to consolidate its programs into a "single integrated portfolio" (HPUC, 2014b).

Customer demand for DERs such as solar PV has seen rapid growth but with some resistance from the utilities. HECO notified applicants for grid interconnection of rooftop solar arrays in neighborhoods with high PV penetration that it will stop processing applications received after Oct. 22, 2014. In Order No. 32053, filed in the RSWG docket, the Commission found that HECO failed to be proactive and "have been quick to identify interconnection technical challenges, but slow to offer solutions to these problems" (HPUC, 2014b, p. 33). The Commission went on to say that "the lack of transparency and slow response to provide support technical information on reliability concerns fosters public distrust about utility management of the distributed generation interconnection challenges" (HPUC, 2014b, p. 34).

The Commission ordered HECO to provide a Distributed Generation Interconnection Plan (DGIP) to address "prioritization of proposed mitigation actions to focus on the immediate binding constraints for interconnection of additional distributed generation" and "focus on formulating well-reasoned technical strategies and resulting action plans that can be implemented expeditiously" (HPUC, 2014b, pp. 55–56). The docket review of the proposed DGIP, No. 2014-0192 remains pending.

---

8. Commission's Inclinations at 2; See *Application for Approval of Additional Waivers form the Framework for Competitive Bidding*, filed Nov. 4, 2013 in Docket No. 2013-0381, at 18 [According to HECO, the average levelized price for utility scale solar PV projects included in the Application is 15.576 cents/kWh (calculated without state tax credit), which is significantly lower than HECO's avoided cost of generation (22.697 cents/kWh at the time of filing)].

However, HECO and state regulators reached an agreement directing the utility to continue connecting customers with rooftop solar systems to the grid after a proposed moratorium. The agreement, signed by both "head of the PUC Randy Iwase and HECO CEO Alan Oshima, said: [s]tated simply, the policy is that the HECO Companies have an affirmative duty to interconnect a potential customer" (Walton, 2015).

As part of HECO's response to the PUC pressure to improve its interconnection process and eliminate the application backlog, it launched the web-based Integrated Interconnection Queues (IIQs) system (Trabish, 2015). The system shows applicants from all distributed generation programs, including those for net energy metering and feed-in tariffs. Although California's investor owned utilities have been publishing queue information about wholesale distribution Hawaii's IIQs are the first public interconnection queues that cover all types of DG installations (Trabish, 2015).

6 CONCLUSIONS

Growing concerns about climate change, the development of digital technologies for the grid, and continually declining cost of resources has led to an evolution of increasingly economically efficient distributed energy resources, and demand response technologies. As these technologies become more popular and policies supporting more efficient and clean energy resources advance, the emergence of a modernized grid is inevitable. This next generation of innovation is poised to change the way we think about and use the electric grid. Many of these innovations are more than just information flow—they are real energy resources like solar and energy storage, demand response, and energy efficiency.

Emerging technologies, shifting customer expectations, and new energy economics harken major changes for incumbent utilities, incumbent generators, and the new DER sector alike. In order to address this paradigm shift, industry leaders, policymakers, and regulators are rethinking the traditional regulated monopoly model. Stakeholders recognize that moving from a centralized grid to a more decentralized grid will require new business and regulatory models, as well as rate modernization.

Policy observers are accustomed to hearing about leadership on progressive energy reforms in California, including leading edge renewables mandates, clean transportation policy, smart grid investments, and now storage technologies. More recently, we have become aware of the practical realities and conflict between customers, policymakers, and the incumbent utility in Hawaii, where the DER rubber has really hit the road in terms of rapid growth of solar PV. Furthermore, states like Massachusetts continue their forward leadership on clean energy policy implementation.

Yet, it has been the REV process in New York, which has really captured both the national attention and conversation on what reforms to the distribution utility business model might actually look like. In New York, the major policy choices generating both regulatory interest and contentious debate are who should control the basic distribution system platform, and if the utilities should own DERs. The debate created significant divergences of opinion between utilities, on one hand, and the incumbent generators and DER providers, on the other. The NYPSC decision that the utilities will remain the DSP essentially aligns all four states reviewed here with the distribution utilities retaining ownership and control of the DSP. The most significant differentiation among the states, on this issue, is that New York has seemed to define more clearly that long-term choice of DERs will be guided by market forces, independent of utility control, while the other states will continue to have utilities responsible for access to the distribution grid. While New York's long-term intentions are clear, it seems, in the short-term, that traditional BCA, and utility administered tariffs will be essential in order to determining the role of DERs in the future.

While the utilities will retain control of the DSP in each of the states, it seems that New York is more determined to limit utility ownership of DERs, with the exception of a few predefined cases. In New York, the DER industry and the incumbent generators have been more successful in gaining NYPSC support for restricting utility DER ownership options, although utility affiliates will roam more freely in the DER market outside of the utilities franchise service area.

The conflict in Hawaii between the incumbent utility, customers, the emerging DER industry, and state policy leaders has revealed some of the potential challenges between a resistant distribution system provider and market forces driving DER growth. The roadblocks put in place by HECO were destined to result in regulatory reform and, in this case, a corporate restructuring through a proposed merger with NextEra. Depending on which side of the debate you choose to acknowledge, this proposed merger looks like a takeover of a recalcitrant utility by a renewable energy leader, or alternatively just a change of control, with another Florida monopoly utility in control. In stark contrast, in California we continue to see ongoing convergence of DER market forces, renewed but consistent state policy leadership, and effective utility implementation and system integration. This leadership continues to propel California to the forefront of both clean energy policy leadership, and DER adoption. The sustained and relatively smooth integration of renewables, electric vehicles, and now storage technologies, into the California energy market should continue to be admired for its successful execution, if not for radical business model reform.

In a quieter way, the Commonwealth of Massachusetts has taken significant strides toward grid modernization, in the last 3 years. In contrast with New York, the focus in Massachusetts is on the electric distribution companies and advanced metering. The interactions between customers and distribution companies appear to be the driving force behind their policy reform and the modernization in the state's grid. This approach is not as revolutionary in regards to business model reform, but nonetheless is a step toward reforming the utility DER platform.

The policy leadership in California and Massachusetts that uses AMI deployment to ensure the distribution platform is technically capable of integrating ubiquitous DERs is perhaps enough to call into question New York's resistance to adopting a statewide AMI mandate. Is it possible that New York's unwillingness to support comprehensive investment in the digital technology infrastructure will stall or delay its more comprehensive distribution system business model reform? For some states, AMI investment seems to be a necessary first step in ensuring a functional distribution system platform. Time will tell whether New York's focus on Reforming the Energy Vision will spur the development of a leading DER marketplace, or whether states like California, Massachusetts, and Hawaii will lead through more subtle business model reforms driven by comprehensive AMI technology, state policy leadership, and growing market forces. Indisputably, a true resource revolution is on the way in the electric industry. For now, across the nation, regulators accept that distribution utilities remain firmly at the center of the common infrastructure essential to DER industry success.

REFERENCES

Akibo, L., 2014. Charting a New Course. Keynote address delivered at the fourth Annual Hawaii Power Summit. Honolulu, December 3, 2014.

Bade, G., 2015. 5 Charts that explain U.S. Electricity Prices, Utility Drive. http://www.utilitydive.com/news/5-charts-that-explain-us-electricity-prices/378054/

Bronski, P., 2014. The Economics of Grid Defection. Rocky Mountain Institute.

California State Assembly, 2013. A.B. 327, Legislative Counsel's Digest, CA.

Competitive Priced Electricity Act, St. 2012, c. 209, and the Emergency Service Response of the Public Utilities Commissions Act, St. 2012, c. 216; Grid Modernization, Office of Energy and Environmental Affairs. http://www.mass.gov/eea/energy-utilities-clean-tech/electric-power/grid-mod/grid-modernization.html

Crosby, M., Cross-Call, D., 2015. New York and California are Building the Grid of the Future. http://blog.rmi.org/blog_2015_02_18_new_york_california_building_the_grid_of_the_future

De Martini, P., 2014. More Than Smart: A Framework to Make the Distribution Grid More Open, Efficient and Resilient. The Resnick Sustainability Institute at the California Institute of Technology.

Heck, S., Rogers, M., 2014. Resource Revolution: How to Capture the Biggest Business Opportunity in a Century. Houghton Mifflin Harcourt, Boston, MA.

HPUC, 2011. Framework for Integrated Res. Planning, Docket No. 2009-0108, Decision and Order, March 14, 2011.

HPUC, 2014a. Instituting A Proceeding To Review The Power Supply Improvement Plans For Hawaiian Electric Company, Inc., Hawaii Electric Light Company, Inc., And Maui Electric Company, Limited, July 8, 2014.

HPUC, 2014b. Policy Statement & Order Regarding Demand Response Program, Order No. 32054. April 28, 2014.

HPUC, 2014c. Integrated. Res. Planning, Docket No. 2012-0036, Order No. 32052, April 28, 2014.

Jones, K., Zoppo, D., 2014. A Smarter, Greener Grid: Forging Environmental Progress Through Smart Energy Policies and Technologies. Praeger, Santa Barbara, CA.

MA DPU, 2014. Investigation by the Department of Public Utilities on its own Motion Into Modernization of the Electric Grid. Order, June 12, 2014.

NYDPS, 2014. Staff Report and Proposal. Developing the REV Market in New York: DPS Staff Straw Proposal on Track One Issues. Case 14-M-0101, August 22, 2014.

NYDPS, 2015. Staff White Paper on Benefit-Cost Analysis in the Reforming Energy Vision Proceeding. Case 14-M-0101, July 1, 2015.

NYPSC, 2013. Proceeding on Motion of the Commission Regarding an Energy Efficiency Portfolio Standard (EEPS). Case 07-M-0548, December 26, 2013.

NYPSC, 2014a. Comments of Jon Wellinghoff, Stoel Rives, LLC and Katherine Hamilton and Jeffrey Cramer, 38 North Solutions, LLC. Case No. 14-M-0101, July 18, 2014.

NYPSC, 2014b. Proceeding on Motion of the Commission in Regards to Reforming the Energy Vision, Case No. 14-M-0101, Comments of David Gahl, Alliance for Clean Energy New York et al., July 18, 2014.

NYPSC, 2014c. Comments of the Joint Utilities (JUC). Case No. 14-M-0101, July 18, 2014.

NYPSC, 2014d. Comments of the Independent Power Producers of New York (IPPNY). Case No. 14-M-0101, July 18, 2014.

NYPSC, 2015. Proceeding on Motion of the Commission in Regard to Reforming the Energy Vision, Order Adopting Regulatory Policy Framework and Implementation Plan. Case 14-M-0101, February 26, 2015.

Pickering, M., 2015. Statement of CPUC Commissioner Michael Pickering at National Town Meeting on Smart Grid and Demand Response. Washington, DC, May 26, 2015.

Pyper, J., 2015. Forget Utility 2.0—The Power Sector Needs "Regulation of the Future". GreenTech Media. http://www.greentechmedia.com/articles/read/Forget-Utility-2.0-What-the-Power-Sector-Needs-is-Regulation-of-the-Future

Savenije, D., 2015. Hawaii Legislature Sets 100% Renewable Portfolio Standard by 2045. http://www.utilitydive.com/news/hawaii-legislature-sets-100-renewable-portfolio-standard-by-2045/394804/

Savenije, D., Cameron, C., 2014. Hawaii's Overhaul of the Utility Business Model. http://www.utilitydive.com/news/hawaiis-overhaul-of-the-utility-business-model/259923/

Trabish, H.K., 2014. Solar Installers Flee Hawaii as Interconnection Queue Backs Up. Utility Dive. http://www.utilitydive.com/news/solar-installers-flee-hawaii-as-interconnection-queue-backs-up/314160/

Trabish, H.K., 2015. HECO Clears PV Interconnection Queues in Maui, Hawaii Islands. http://www.utilitydive.com/news/heco-clears-pv-interconnection-queues-in-maui-hawaii-islands/380528/

Walton, R., 2015. Regulators: HECO Must Continue to Interconnect Rooftop Solar Systems. http://www.utilitydive.com/news/regulators-heco-must-continue-to-interconnect-rooftop-solar-systems/369795/

Wesoff, E., 2014. Grid Edge Live Keynote: "Rate Structures Are Making the System Less and Less Efficient". July 2, 2014.

Wesoff, E., 2015. How Much Solar can HECO and Oahu's Grid Really Handle? http://www.utilitydive.com/news/hawaiis-overhaul-of-the-utility-business-model/259923/

Chapter 22

The Fully Integrated Grid: Wholesale and Retail, Transmission and Distribution

Susan Covino, Andrew Levitt, Paul Sotkiewicz
PJM Interconnection, LLC, Audubon, PA, United States of America

1 INTRODUCTION

Distributed energy resources (DER) include demand response (dispatchable change in usage, either on the demand- or supply-side of energy markets), energy efficiency (passive reduction in usage), and onsite storage and generation, ranging from backup diesel to gas-fired microturbines, to rooftop solar, to combined heat and power.

DER installation is being done for a multitude of reasons: provide reliability during outages resulting from weather events, manage energy expenditures, meet customer desires to reduce their environmental footprint and/or support new evolving technologies. In the PJM or wider ISO or Regional Transmission Operator (RTO) context as shown in Fig. 22.1, recent extreme weather events, such as Superstorm Sandy have played an important role in DER adoption. The experience of an autumn outage of 2 weeks highlighted for many the value of onsite generation.[1] The coincidence of cheaper sources of natural gas from the Marcellus and Utica shale formations, in Pennsylvania and Ohio, has prompted a renewed interest by more customers in combined heat and power (CHP) technology.

From the perspective of an RTO, or generically a transmission system operator (TSO) operating the bulk power system, such resources are, by definition, connected to the distribution system, are "behind the meter," and are likely not

---

1. Customers in this context include state and local governments that are responsible for providing access to basic services during emergencies like Superstorm Sandy. See www.njeda.com/erb for information about the Energy Resilience Bank of New Jersey, which is funded by a $200 million grant from the Federal government. Maryland Energy Task Force (2014) outlines a plan for microgrid development in Maryland.

Future of Utilities - Utilities of the Future. http://dx.doi.org/10.1016/B978-0-12-804249-6.00022-1

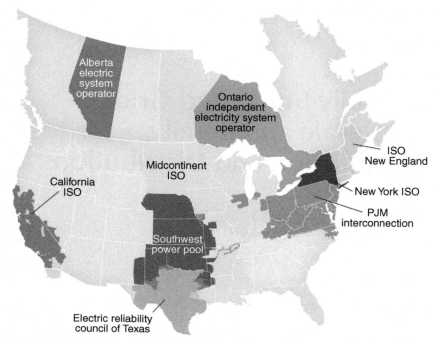

FIGURE 22.1   Map showing the independent transmission system operators of North America.
(Source: ISO/RTO Council (2015).)

under the direct control of the RTO/TSO.[2] Moreover, these resources do not generally participate directly in wholesale power markets operated by RTOs—they are not receiving or responding to prices in wholesale energy, ancillary service, or capacity markets, nor do they face any kind of market-related price signal at the retail level. However, DER adoption by customers large and small, regardless of the reason, does not take place in isolation. DER will impact power flows and infrastructure needs, across both transmission and distribution systems.

Within PJM, DER installations are becoming more prevalent. Behind the meter solar PV accounts for just over 1800 MW footprintwide. Much of this solar DER is driven by specific state solar energy carveouts, as a part of their larger renewable portfolio standards (RPS). There is more solar PV in the PJM interconnection queue, as shown in Fig. 22.2.

Demand resources providing demand response through the use of load modification or backup generation accounts for nearly 10,000 MW of capacity, and

---

2. In PJM the operator of the bulk power system and wholesale power markets for 61 million people in all or parts of 13 states and the District of Columbia, the line of demarcation between the bulk power transmission and distribution (wholesale and retail markets) generally occurs at the substation where voltage is stepped down from more than 100 kV to distribution level voltages.

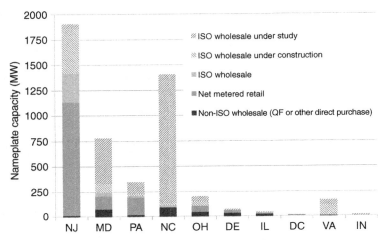

FIGURE 22.2   Solar deployment in PJM as of year-end 2014. Note: Graph includes data only for the portion of the state within the PJM footprint. *(Source: PJM Interconnection, LLC.)*

are primarily targeted at reducing capacity market expenditures. Energy efficiency resources account for another nearly 2000 MW of capacity. Together, these resources account for nearly 8.5% of PJM's all-time peak load of 165,000 MW.

DER, driven by reliability needs, differs from the energy efficiency and demand response described earlier. It is driven by customers' willingness to pay for energy up to very high prices, and therefore not market driven. Thus, it is a higher quality product, rather than one that offers cost savings. Using the example of Sendai microgrid in Japan, Marnay describes in chapter: Microgrids: Finally Finding Their Place, how a microgrid can deliver multiple power qualities to its customers. But reliability-driven DER also provides an opportunity to offset larger costs substantially, through market participation in wholesale energy and ancillary service markets.

And yet, PJM and its transmission owners are only beginning to understand there are many types of DER that exist, with which the transmission system has no interaction or visibility. The trend in DER is far from limited to PJM. According to a 2015 Utility Dive survey of 400 electric utility executives, utility leadership is anticipating a shift toward a more distributed grid: a plurality of respondents chose distributed energy resources as their utility's biggest growth opportunity in the next 5 years; a majority foresaw an increase in distributed energy resources as part of the generation mix; and 48% of utility participants are developing new business models for distributed generation.

From the perspective of an RTO as wholesale market and bulk power system operator, it is imperative to begin considering how to most effectively integrate DER into transmission system operations and wholesale power markets where price signals are used to maintain reliability. This chapter articulates a vision of a fully integrated grid, from a wholesale market perspective, where transmission and distribution are operated seamlessly in concert to enhance efficiency and reliability of both systems by explicitly accounting for increasing levels of DER.

This chapter is organized into five sections. Section 2 provides the high level vision for the fully integrated grid with the development of a Distribution System Operator (DSO) and pricing at the distribution level that corresponds closely to pricing methods used by RTOs/TSOs. Section 3 describes the role of the DSO, including functionalities and reliability benefits in operations, and benefits for short-term and long-term planning. Section 4 lays out the authors' vision of pricing, along with the problems with existing pricing schemes at the distribution level, as well as the innovation that can be driven by these envisioned prices. Section 5 concludes.

2 VISION FOR THE FULLY INTEGRATED GRID

The authors' vision of the fully integrated grid relies on the assumption of the full deployment of advanced metering infrastructure (AMI). AMI can, at a minimum, meter customer usage on an hourly interval basis, preferably at sub-hourly intervals, and possibly even have two-way communication capability. The fully integrated grid has two main components:

- The presence of a DSO that is independent of all market participants and carries out functions similar to an RTO or TSO
- The transmission of prices at the wholesale/transmission level, down to the distribution level, either directly or, at the very least as a dynamic translation of those wholesale prices, to retail prices that convey the most important information to individual customers

The fully integrated grid envisions close coordination between DSOs and the RTO/TSO, so that, from an operations and market perspective, the system is operated seamlessly as one grid, rather than the bifurcation between transmission and distribution that exists today.[3] Hanser and Van Horn, outline their view of how the distribution business will be reorganized, going forward in the context of the REV process in New York state. In our view, the DSO evolves to taking on the function, structure, and characteristics of ISOs and RTOs as they exist today, but at lower voltages.

In the context of the United States, there are jurisdictional regulatory issues to be considered that may not necessarily exist in other parts of the world. This vision of the fully integrated grid assumes such regulatory and jurisdictional issues are worked out. As a consequence, this chapter focuses on the operational, reliability, and market/pricing characteristics of the fully integrated grid.

---

3. This vision is shared by Gellings in his chapter in this volume. He notes that "[t]he Integrated Grid blends the most valuable generation, storage, power delivery and end-use technology tailored to meet local circumstances. …[and] enables true synergy between consumers and the grid."

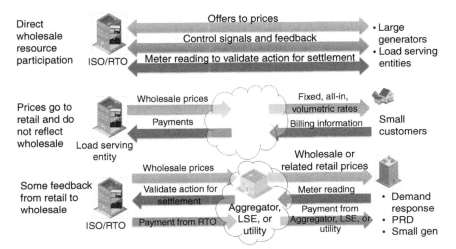

FIGURE 22.3   **Wholesale and retail are mostly separate today.** *(Source: PJM Interconnection, LLC.)*

2.1   How Close Is the Industry to This Vision?

When and how to show customers wholesale energy prices has long been discussed and debated. PJM's stakeholders, with the support of several state commissions, developed and filed at FERC in 2010 the Price Responsive Demand (PRD) option. PRD requires metering capable of providing hourly integrated values, dynamic retail rates triggered by nodal LMPs, and automated usage reductions at prices predetermined by the customer (PJM, 2011). While many market participants believe that automated price response is a critical option for residential customers, no Load Serving Entities (LSEs) have to date chosen the PRD option.[4] In this respect, the industry is still far from the vision articulated earlier and as shown in Fig. 22.3.

Given the broad deployment of AMI approved by state commissions, and funded in part by stimulus dollars, the grid can now verify each customer's response to price. According to the Federal Energy Regulatory Commission's most recent staff report on demand response and advanced metering, nearly 32% of the nation's customers have AMI, and industry projections indicate that more than 50% of US households will have AMI by 2015 (FERC, 2014, p. 4).[5]

---

4. The term Load Serving Entity (LSE) refers to any entity serving end-users within PJM. See the Reliability Assurance Agreement, Section 1.44.

5. Table 2-2 on page 4 indicates AMI penetrations greater than 50% in three North American Electric Reliability Council (NERC) regions: 52% in the Florida Reliability Coordinating Council (FRCC), 70% in the Texas Reliability Entity (TRE), and 51% in the Western Electricity Coordinating Council (WECC) (FERC, 2014, p. 4).

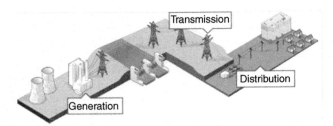

FIGURE 22.4 **Demarcation between transmission and distribution.** *(Source: PJM Interconnection, LLC.)*

Given this sizable investment, it is incumbent on utilities, regulators, and competitive LSEs to facilitate valuable efficiency gains by providing pricing options that incent customers to respond to prices, and enable them to realize savings from responding. However, in most cases, AMI is not even close to being fully utilized, let alone fully deployed, which is a necessary condition for the vision of the fully integrated grid.[6]

In the context of pricing, the California Pricing Pilots of 2003–04, as well as AMI policy developed in the province of Ontario, Canada, provided the data and analysis needed to support utility business cases for AMI deployment at the distribution level (Charles River Associates, 2005; Reid, 2005). Meters that record and communicate usage on at least an hourly interval basis enable customers to prove and benefit from responding to prices that reflect actual grid conditions at the transmission level and communicate them down to the distribution level as shown in Fig. 22.4. While such interval meters could also be used to bill on predetermined time-of-use (TOU) rates, rather than on dynamic rates linked to wholesale prices, the dependence of TOU rates on rough approximations of seasonal grid conditions limit their ability to achieve significant efficiency gains as discussed in the pricing section.

In short, the industry is quite far from the vision of the fully integrated grid, as articulated in this chapter. However, with the vision expressed herein, it is hoped that the market and reliability benefits to be gained from this chapter's vision of the fully integrated grid will push the ball forward.

3 ROLE OF THE DSO

The DSO is envisioned as an independent entity responsible for planning and coordinating resources on the distribution system in real-time operation, analogous to the transmission system operator as shown in Fig. 22.5 (Tong and Wellinghoff, 2014). The DSO is envisioned to manage the details of many

---

6. Jones et al. contrast and compare AMI deployment policies and implications between and among California, Massachusetts and New York.

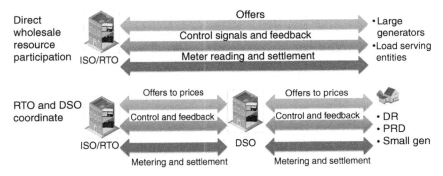

FIGURE 22.5   A DSO can help get the prices right, provide visibility.

thousands or millions of resources, aggregating these to the level of abstraction necessary for the RTO/TSO to have sufficient operational visibility about what is "behind the meter" with respect to DER. The DSO provides a means by which granular information regarding DER can then be utilized by the RTO/TSO to manage the larger system with better information than it currently has. Effectively, the DSO and RTO/TSO would coordinate and operate in concert to enhance reliability and market efficiency.

3.1   Details of Potential DSO Functionalities

The DSO potentially enables several functions:

- Maintaining a registry for distributed resources (such as demand response, combined-heat-and-power, or small batteries) that can serve multiple needs on the distribution or transmission system, and that can be aggregated up to the bus/node that connects the distribution and transmission system. In this way the DSO provides information to the RTO/TSO that appears as a single DER or load and reduces the information requirements on the RTO/TSO.
- Sending prices to DER and loads that are reflective of system conditions and are related to prices established at the wholesale level. These prices could be extensions of LMP, down to the distribution level, or even prices for distribution infrastructure that are akin to MW-mile methods used to price transmission infrastructure in many parts of the world.
- Carrying out a market operator function, similar to RTOs, to ensure fair, efficient, and nondiscriminatory compensation for grid services. These services might include energy transferred to or from customer premises, use of the fixed-cost portion of the distribution system and transmission system, sale of ancillary services to the bulk grid, and perhaps exchange of reactive power with the distribution system.

- Facilitating competitive third parties to participate in infrastructure investments and planning in a manner similar to that envisioned under FERC Order 1000. Allowing, for instance, a battery developer to provide a solution to address growing load on a feeder that will soon reach capacity during peaks.
- Managing interconnection of DER at the distribution level and coordinating this with interconnection at the transmission level.
- Integrating loads and power flows monitored at the distribution level into short- and long-term models used by transmission level forecast algorithms. This includes participation in modeling the sensitivity of customers with distributed energy resources to electricity prices, fuel prices, and weather parameters like temperature, cloud cover, humidity, and wind speed that can then be passed on to the RTO/TSO for bulk power system operations.
- Enabling some degree of aggregation in customer registration, bidding, control, and monitoring behind the substation. This allows not only a necessary abstraction of detail to enable feasible operations of the grid, but also the opportunity for aggregators to join complementary resources in order to meet grid requirements that none of them would meet alone.
- Sharing of critical grid data on retail customers with aggregators and with the transmission system operator, including nodal location, past, and future meter data, and load serving entity.
- Capturing resources' performance characteristics such as ramp rate, minimum and maximum output, and availability at the time of registration.
- Accommodating price responsive loads by providing the transmission system operator with daily elastic demand curves (rather than requiring the transmission system operator to forecast how load will respond to prices, as is sometimes done today).
- Increasing integration of distribution system dynamics into transmission system models. Current grid models typically begin with a static forecast of power flows to the distribution system, and then optimize wholesale generators to meet those power flows. In the future, the distribution system power flows will be more interactive, and therefore must form part of the dynamic optimization model.

Because many of these functions are part of fair, nondiscriminatory, and competitive outcomes, it is essential the DSO be independent from ownership of the distribution system and from competitive providers.

3.2 Benefits of an Independent DSO

The presence of a DSO that coordinates closely with the RTO/TSO benefits both the distribution and transmission system. On the distribution system, the DSO promises greater efficiency in infrastructure upgrades (or deferral of same), a decrease in losses, and better voltage control. On the transmission system, the

DSO aggregates information for the RTO/TSO to ensure better visibility of the entire grid in system operations.

From a market operations perspective, a DSO facilitates customer engagement through the expression of prices or rates that are directly related to wholesale market prices. These prices communicate the needs of the entire grid and facilitate participation of distributed energy resources, which can provide wholesale market services such as regulation and frequency response efficiently, and operating reserves, as well as participation in energy and capacity markets.

DSOs provide the RTO/TSO with increased visibility regarding conditions of the distribution system and availability of DER to enhance market efficiency and reliability in operations without the need for RTO/TSO to know the exact locations and features of every single DER "behind the meter." Such coordination reduces the computational burden on RTO/TSO systems, as well as allowing operators to see the bigger picture without getting bogged down in details.

3.3   An Example of Where DSOs Could Have Enhanced Operations and Reliability

Between Sep. 9–11, 2013, temperatures in the Mid-Atlantic and Midwest United States reached all-time highs for the month, up to 20°F higher than normal. As is common during the normally temperate Sep., planned outages for transmission and generation had been scheduled during this time (PJM, 2013).

As a result of this convergence of extreme weather and outages, PJM was forced to order load sheds in a few contained areas for no more than 9 h. While the load sheds were limited to a small fraction of the 61 million people in the PJM footprint, and, while they were a prudent response to avoid uncontrolled blackouts, such load sheds are extremely rare, and represent an action of last resort because of the disruption and inconvenience they cause consumers. PJM closely studied the event in order to learn how to improve training, tools, procedures and processes best, in order to ensure that PJM maintains its reliability goals.

One important lesson was the value of small distribution-level resources for management of emergencies on the transmission system. In this case, on Sep. 11, PJM identified a 6 MW behind-the-meter generator in the municipal utility of Sturgis, Michigan. The municipal utility started the generator and called for voluntary load reduction by its customers. Together, deployment of these resources eliminated the need for the critical load sheds which had occurred on Sep. 10.

Because the Sturgis generator was for use only during power outages, it was not previously registered with PJM. PJM has since worked with states and utilities to establish and document an approach for collecting information about behind-the-meter generation, as well as to manage dispatcher interactions with such resources better. This is exactly where a DSO, if in place, would have already had this information and thus could have prevented local load shedding and enhanced reliability. Moreover, with a full deployment of AMI and wholesale prices down to the distribution level, it is possible price responsive demand may have been available to alleviate the transmission emergency as well.

Other valuable lessons were learned from this heat wave. On Sep. 10, PJM also ordered load shed in Fort Wayne, Indiana, as a planned line outage coupled with unseasonably high loads put undue stress on in-service lines. Recommendations from the "Technical Analysis" relevant to this chapter included:

- Review PJM's overall approach to how and when to model and telemeter with a focus on the subtransmission system
- Review load management market rules to improve operational flexibility through the PJM stakeholder process to include shorter lead times; subzonal calls, calls outside emergencies, and shorter minimum run times
- Improve the flexibility of subzonal demand response by proactively defining DR subzones across PJM footprint and/or mapping DR resources to nearest substation
- Provide the dispatchers with better visibility of the location and amount (megawatts) of relief from DR (PJM, 2013, pp. 41, 42)

It is clear from this event and the subsequent technical analysis that better integration between transmission and distribution operations can avoid suboptimal outcomes like load sheds. As a result of this load shed, PJM improved the process for identifying local (ie, nodal or subzonal) demand response resources.

3.4 Role of DER and DSO in Providing Reliability Services

Electricity does not enjoy opportunities for storage in the same way as other commodities. Instead, every minute production needs to be almost perfectly matched to consumption. One way to measure this balance is through the system frequency which is 60 cycles/s (Hertz).[7] RTOs/TSOs set aside capacity to help maintain frequency at 60 Hz. This can be in the form of generation, storage, or demand response, and location is not relevant. This reserve provides an ancillary service called "regulation." In PJM today, there are effectively about 700 MW of capacity dedicated to providing frequency regulation. Today, demand response and storage can provide regulation service, but the majority is still provided by generation dispatched by the RTO/TSO. A DSO with information on DER "behind the meter" may be able to find sufficient resources so that large generators would not need to move to provide regulation, and allow them to operate more efficiently while maintaining grid reliability.

---

7. Frequency below 60 Hz implies load is greater than generation and requires some generation to increase output or load to decrease. Frequency above 60 Hz implies generation is greater than load and requires some generation to reduce output or load to increase. RTOs/TSOs combine these frequency changes with real-time measures of power flow to and from neighboring grids to determine the amount of internal balance that needs to be corrected.

Moreover, RTOs/TSOs also maintain reserves to ensure energy balance is maintained due to system contingencies, such as a large generator tripping offline, or the loss of a large transmission facility. Today in PJM, demand response that is "behind the meter" can provide reserves. But the presence of a DSO with visibility to DER of all types, and the ability to convey system information to DER at the distribution level, would have the ability to aggregate DER to provide reserves to the grid in the same manner an RTO/TSO does today. But, unlike regulation service, the location of reserves does matter as it is essential that reserves be in the right location to ensure they can be delivered where needed, due to transmission constraints. A DSO can ensure that DER assigned to provide reserves also satisfies this locational requirement. The use of "behind the meter" DER to provide reserves can also lead to more efficient outcomes as generation dispatched by RTOs/TSOs need not be backed down from what would otherwise be their optimal output to provide reserves. Moreover, using DER, rather than RTO/TSO, dispatched resources can also help avoid or postpone the need to go into reserve shortage conditions during extreme system peak conditions.

3.5 Role of the DSO in Increasing Short-Term Forecast Accuracy

With increasing levels of DER, the accuracy of short-term load forecasts becomes more difficult because these resources are not visible to the RTO/TSO. Improving the accuracy of such short-term forecasts enhances the efficiency of the day-ahead unit commitment process, and for scheduling and dispatching generation in real time. Currently, PJM utilizes an artificial neural network for forecasts that look forward 168 h and less. Artificial neural networks are a category of machine learning computer algorithms that are adaptive, nonlinear, distributed, and run in parallel. Neural networks are superior to traditional linear models at identifying patterns and relationships in complicated nonlinear systems, such as the dependence of electricity load on various weather and other parameters.[8]

A DSO can provide more granular data inputs associated with "behind-the-meter" generation such as solar, to improve the accuracy and granularity of forecasts. Moreover, for loads that are price responsive, and do reduce consumption when prices are high, but such activity is not visible to the RTO/TSO, the DSO can examine such information and possibly locate such price responsive demand to enhance real-time operations, especially during summer and winter peak events.

---

8. The neural network load forecast takes locational weather parameters (such as temperature, cloud cover, humidity, and wind speed), fuel prices, work and recreation patterns as inputs. Because the neural network adapts to new load patterns, it will ultimately reflect the underlying relationship between these parameters and load in the forecast.

4 COMMUNICATING PRICES FROM THE RTO/TSO LEVEL TO DER

In day-ahead and real-time energy markets in the US context, RTOs determine locational marginal prices (LMP) that determine the marginal cost of delivering one more megawatt-hour of energy to any location on the system. LMPs account for congestion, or the need to redispatch more expensive resources to maintain transmission security, and marginal losses. In PJM, these prices are determined every 5 min in real-time dispatch and on an hourly basis in the day-ahead unit commitment market.

Many, but not all, RTOs also operate a capacity market or resource adequacy construct for which prices are determined often on a locational basis, in order to satisfy resource adequacy targets defined by reliability organizations.[9]

With respect to transmission charges, there are a variety of cost allocation and pricing mechanisms in use around the world. However, one methodology that is widely used is based on the contributions of load and generation to flows on transmission assets during system peak, generically known as MW-mile methods.[10] In general, MW-mile methods allocate costs of transmission infrastructure to loads far from generation and generation that is far from the load center. But, in other cases, generation close to load can actually be paid as if they were providing capacity on the transmission system.

In this chapter's vision of the fully integrated grid, such locational and time-based pricing methods provide efficient signals regarding the location of generation and load, as well as appropriate timing of generation or consumption. In the ultimate end-state, pricing at the wholesale transmission level and the distribution level would be done accordingly.[11]

In applying LMPs to the distribution system, Sotkiewicz and Vignolo (2005) show that dispatchable distributed generation located at the end of a typical rural feeder can reduce losses by 37%, can reduce feeder voltage variability by 25%, and can earn 12% greater revenue under nodal pricing. Sotkiewicz and Vignolo (2007a,b) show how optimally located DER not only alleviate the need for new feeders, but can also be compensated for effectively freeing up capacity on the distribution system.

---

9. See the website (www.isorto.org) of the ISO RTO Council for information about similarities and differences among the member system operators in North America.

10. In PJM, this is known as the Dfax method.

11. The MIT Utility of the Future study brings the locational marginal price first formulated at MIT in the 1980s (Schweppe, Tabors and Kirtley, Homeostatic Control: The Utility/Customer Marketplace for Electric Power.) into the present and applies it to the distribution system based on two observations or insights. The first is the rise of distributed energy resources such as rooftop solar. The second is that customer engagement and nonutility actors are of increasing importance, both because distributed energy resources naturally accommodate competitive entrants and because the growth in variable renewable generation increases the value for the kind of flexibility that is potentially provided by greater customer engagement.

Still, it is important to understand this vision of the fully integrated grid really depends upon the existence of a DSO as described earlier. Moreover, the political and regulatory will to allow such a transition from the current pricing and rate design paradigm to the one this chapter envisions would also be required.

4.1 Flat Volumetric Rates Versus Pricing by Time and Location for Energy on the Distribution System

Unlike energy, capacity, and often transmission, where there exists a locational and time component to pricing, the current pricing at the distribution level often combines energy, capacity, and transmission under fixed volumetric prices that in no way reflect actual system conditions by time or location. Such a pricing mechanism makes sense when there is no AMI that can measure usage on an interval basis, or a DSO that can communicate wholesale prices. But technology has advanced to the point that such simplistic pricing can be replaced by more efficient pricing as envisioned earlier. The result of fixed, volumetric prices that do not consider time and location is that the decision to employ DER or buy energy directly from the grid is distorted in short-run decision making, in two crucial ways:

- By failing to reflect time-varying marginal costs
- By including long-run costs in the short-run price

For instance, at a time when the marginal cost for the grid to serve a customer is only $0.04 per kWh, that customer might face a $0.12 per kWh retail price for electricity, and thus decide to dispatch its behind-the-meter generator that costs $0.08 per kWh. At another time, the marginal cost to a second customer from the grid may be much higher, say $0.20 per kWh. Imagine the second customer has a less efficient behind-the-meter generator that costs $0.15 per kWh, and also faces a $0.12 per kWh retail price. Such a customer would decide to take power from the grid in lieu of running its generator. In both instances, the retail price is leading customers with distributed energy resources to make inefficient short-run decisions.

Knieps's chapter in this volume shares similar concerns about the current policies in European countries and their impact on grid integration of microgrids. The author concludes that "[a]s a consequence of an ignorance of locational and time-specific differences within the relevant electricity networks the wrong incentives for generation and consumption decisions are created."[12]

When there are few distributed energy resources, the impact of such inefficiencies is limited. However, with growing deployment of distributed energy resources, it becomes increasingly important that customers face correct price signals for consumption and production of energy. And getting these price signals can also unlock other potential means by which localized load shedding events,

---

12. According to Knieps "[i]nefficient network management instruments have to be replaced by a consistent market framework taking into account the pricing of the node-specific opportunity costs of network access."

as described earlier, can be avoided through sending prices that are reflective of local system conditions.

4.2   Customers Must See Nodal LMPs to Make Efficient Decisions for DER

The main goal of a system of locational marginal prices for energy on the distribution system is to move toward efficient make/reduce-versus-buy decisions by distributed energy resource owners.

Extending short-run marginal cost principles to the distribution system can simply be a matter of exposing retail customers to the bulk grid locational marginal price in the transmission zone the customer is served from. Indeed, many utilities offer such arrangements to large customers today and in some cases (eg, in Commonwealth Edison) even residential customers have access to marginal price–based rates.

However, such rates are suboptimal in several ways. First, they fail to include the granularity that can be present at different nodes (ie, substations) within a transmission zone. Such zones often cross major electrical boundaries and can contain significantly different congestion profiles. Second, they also fail to include features on the distribution system itself.

On a given feeder of the distribution system, marginal costs of energy exchange will vary by location. Very close to the substation, there are resistive and core losses in transforming the energy from the high-voltage transmission network to the medium-voltage distribution network. Further from the substation, those losses are added to resistive losses in the distribution wires. These losses are part of the marginal cost of energy, and would need to be included in prices for optimum efficiency.

A quick survey of distribution losses from select PJM utility member tariffs shows that average losses on the distribution system can be quite significant, in the 3–10% range. Marginal losses during peak times can be expected to be significantly higher, given the quadratic dependence of resistive losses on load. Distribution system losses will have a noticeable impact on nodal prices calculated on the distribution system, depending on the system in question. The effect will be slightly higher prices than on the transmission side of the substation, with prices increasing (on radial feeders) with distance from the substation. This disfavors customers sited at the end of the feeder, but rewards distributed generation there.

Fig. 22.6 shows an example of radial distribution feeders at a substation. Nodal prices at peak are shown in Table 22.1. These correspond to the numbered nodes on feeder A at increasing distances from the transmission system. The climbing price is due to increasing losses.

All distribution systems today are sized such that the peak loads can be met from the transmission system. Therefore the marginal costs of distributed generators located on the distribution system are not likely to set the locational price on the distribution system. By design, the next increment of load on the distribution

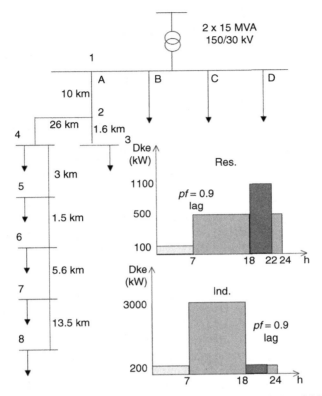

FIGURE 22.6   A rural distribution network with residential and industrial load profiles. *(Source: Sotkiewicz and Vignolo (2006).)*

TABLE 22.1 Nodal Prices at Peak for the Distribution Network in Fig. 22.6

Bus	Price ($/MWh)
1	30
3	31.503
4	35.118
5	35.571
6	35.742
7	36.183
8	36.732

Source: Sotkiewicz and Vignolo (2006).

system can always be met from the wholesale power markets associated with the transmission system, and this will therefore set the generation portion of the price on the distribution system. Distributed generators will, in all likelihood, be price takers, except in the unlikely event that they happen to set the price on the entire transmission system. Therefore, losses are the only parameter driving differences in nodal energy prices on the distribution system relative to the nodal price at the interface with the transmission system. However, in the future, it is possible that distributed energy resources will become so integrated into the distribution system infrastructure that this design maxim will no longer hold, in which case the congested distribution system could at times see a separate generation component to prices relative to the transmission system.

It is also conceivable to include other marginal costs, apart from losses and generation, such as the marginal cost of temporarily exceeding the nameplate limits on distribution system equipment. And, of course, this is not to mention pricing for nonenergy services, such as reactive power, ancillary services, or use of the distribution or transmission network.

New approaches to retail rate design for energy are being piloted and discussed, and in a few cases implemented, but this effort is in a nascent stage. The Pacific Northwest National Laboratory conducted a pilot on four distribution feeders of AEP in northeast Columbus, Ohio. The purpose of the pilot was to learn about the price responsiveness of customers on the feeder that received price signals that reflected not only the cost of wholesale power, but also the real time condition of the feeder.[13] The real-time retail price or "base price" is based on the PJM energy price. When the distribution feeder is constrained, then the cleared retail price can deviate from the base price (GridWise Architecture Council, 2013, p. 55). This early effort is among a number of research efforts designed, among other things, to cooptimize a customer's DER assets and power available from the grid.[14]

4.3 Considerations in Pricing to Reflect Fixed Costs of Infrastructure

One corollary of locational marginal pricing is that fixed network costs would no longer be billed on the basis of energy consumption, the result being an overall decrease in the financial attractiveness of distributed generation and energy efficiency. In the event that regulators desired to maintain the attractiveness of such technologies, explicit subsidies would have to be put into place. From an economic perspective, such an outcome is preferable; however, from a political perspective it may be an undesirable approach.

---

13. Programmable communicating thermostats monitor price signals and convert customers' desired temperatures and preferences for more comfort or more savings into bids for the next 5 min. The dispatch system adds together all of the bids on the feeder and clears the market where the supply and demand curves intersect. The clearing price is sent back to the participating homes where the smart thermostat adjusts the temperature. See GridWise Architecture Council (2013, p. 54).
14. More information about stimulus funded smart grid research can be found at www.sgiclearing-house.org

Experience with DER is resulting in diminishing distribution utility revenues, when customers charged per kWh consumed, install energy efficiency, offer load flexibility as demand response, and/or install onsite generation for reliability. The timing of this revenue loss unfortunately coincides with needed distribution grid modernization projects begun by many EDCs. If DER growth continues, the status quo will be unsustainable for EDCs and, in the longer run, for customers reliant on the value provided by the grid.[15]

4.4   Price Signals Drive Innovation and DER Technology Adoption

These price signals provide the price to beat for innovators—not only for new products, especially DER, and services, but also innovation that drives down the cost of providing DER, first perhaps to the largest customers, and later to smaller customers.

The transparent price record enables all comers to do risk analysis and to analyze price patterns associated with location or changes in the shape of load curves. Simply providing a level playing field for all resources invites and rewards successful innovators.

"The potential impact of technologies that automate customer energy choices is large. And, the barriers to faster realization of a future in which smart devices implement the preferences of ordinary consumers for comfort and lower bills are largely regulatory, not technological or economic" (Centolella, 2015). The grid of the future, as articulated in this chapter, shares this view of a world in which a customer's smart devices respond automatically to dynamic price signals on a continuous basis (Centolella, 2015, p. 28). And it will be this environment that provides customers with the greatest choice and control in terms of DER investments, as well as decisions to buy or reduce, or to make or buy, or to make, store, and buy. Over the longer term, the price signals envisioned in the grid of the future will drive DER adoption and investment decisions, and greater DER deployment will only accelerate the need for a DSO as discussed in this chapter. In this sense, the "right prices" provide a self-reinforcing mechanism of DER deployment, and the need for a DSO.

5   CONCLUSIONS

The vision of the fully integrated grid in the future requires prices being conveyed down to the distribution customer level, and even at the extreme, down to individual devices. The flexibility and market response available from DER can

---

15. The New York Public Service Commission's Reforming the Energy Vision (REV) has established the following goals for the New York distribution grid: "(1) Enhanced customer knowledge and tools that will support effective management of the total energy bill; (2) Market animation and leverage of customer contributions; (3) System wide efficiency; (4) Fuel and resource diversity; (5) System reliability and resiliency; and (6) Reduction of carbon emissions." It will be interesting to see how New York implements these goals in phase II of the proceeding.

enhance reliability and market efficiency, but also requires the development of DSOs to operate in concert, alongside RTOs/TSOs. In this sense, market and reliability outcomes will appear seamless across the distribution and transmission grids. This complex but efficient environment is achievable, and is really only an extension to the foundation of pricing and operations in existing organized wholesale markets, operated by RTOs.

However, there is a great deal of work to be done. The industry is quite far from the vision of the fully integrated grid, as articulated in this chapter, with AMI deployment, pricing, and the idea of DSOs only recently coming to the fore. But with the vision expressed herein, it is hoped that the market and reliability benefits to be gained from this fully integrated grid will push the ball forward to its achievement.

ACKNOWLEDGMENTS

The views expressed herein are those of the authors and not of PJM Interconnection LLC, its management, board, or members.

REFERENCES

Centolella, P., 2015. Next Generation Demand Response: Responsive Demand through Automation and Variable Pricing. March 17, 2015, p. 6. Prepared for the Sustainable FERC Project.

Charles River Associates, 2005. Impact Evaluation of the California Statewide Pricing Pilot – Final Report. Charles River Associates, March 16, 2005.

FERC, 2014. Assessment of Demand Response & Advanced Metering—Staff Report. Federal Energy Regulatory Commission, December, 2014, p. 4.

GridWise Architecture Council, 2013. AEP gridSMART Smart Grid Demo in Appendix A to the GridWise Transactive Energy Framework DRAFT Version, October 2013, p. 54.

ISO RTO Council website at www.isorto.org.

Maryland Energy Task Force, 2014. The Maryland Resiliency Through Microgrids Task Force Report, June 23, 2014.

PJM, 2001. Price Responsive Demand. PJM Staff Whitepaper, March 3, 2011. PJM Interconnection LLC, FERC Docket No. ER11-4628-000.

PJM, 2013. Technical Analysis of Operational Events and Market Impacts During the September 2013 Heat Wave. https://www.pjm.com/~/media/documents/reports/20131223-technical-analysis-of-operational-events-and-market-impacts-during-the-september-2013-heat-wave.ashx

Reid, L., 2005. Implementing Smart Meters in Ontario. Mid-Atlantic Distributed Resources Initiative's Metering Workshop. Ontario Energy Board, May 4, 2005. See the presentation on the MADRI website at http://sites.energetics.com/MADRI/

Schweppe, F.C., Tabors, R.D., Kirtley, J.L., 1981. Homeostatic Control: The Utility/Customer Marketplace for Electric Power, MIT Energy Laboratory Report MIT-EL 81-033, September 1981.

Sotkiewicz, P., Vignolo, J.M., 2005. Nodal Pricing for Distribution Networks: Efficient Pricing for Efficiency Enhancing Distributed Generation, August 2005, pp. 2 & 3.

Tong, J., Wellinghoff, J., 2014. Rooftop Parity. Public Utilities Fortnightly, August, 2014. http://www.fortnightly.com/fortnightly/2014/08/rooftop-parity

Subject Index

A

AC power systems, 57
AC's power quality, 56
Adaptability, 93
Advanced metering infrastructure (AMI), 59,
    287, 404, 420
  enabled dynamic pricing, 160
  infrastructure facilitates, 153
  technologies, 290
Advanced Metering Investments, 187
Advanced Microgrid Solutions (AMS), 379
AGL Energy, 113, 115
Amazon, 98, 103, 104
AMI. *See* Advanced metering infrastructure
    (AMI)
AMS. *See* Advanced Microgrid Solutions
    (AMS)
Annual household electricity cost, 112
Anza-Borrego Desert State Park, 65
Apple, 103, 104, 114
apps, 99
Arbor Day Foundation, 185
Arizona, power output, 141
Arizona Public Service, 199
Australia
  Future Grid Forum's analysis, 139
  load profile for holiday location, 132
  network pricing, for prosumer future,
    261–264
  photovoltaic (PV), 135
    capacity and generation, 13
  residential electricity tariffs, 122
  solar intermittency for residential
    building, 131
Australian Capital Territory (ACT), 212
Australian electricity markets, 113
Australian Energy Market Commission
    (AEMC), 122
Australian energy market developments, 110
  perfect storm, 110–113
  responses, 116
  utilities, commercial response of, 113
    AGL Energy, 114
    Energy Australia, 115
    Origin Energy, 115

Australian Energy Market Operator (AEMO), 7
Australian Energy Regulator (AER), 125
Australian National Electricity Market, 254
Australian PV capacity, electric power sector, 13
Australian residential tariffs, 111
Australian utilities, 110
  circumstances facing, 112
Automated teller machine (ATM) network, 40
Automatic Generator Control (AGC) system, 145
Average cost pricing, 112
Average cost tariffs, to demand-based
    tariffs, 122
Average demand, average annual growth, 111
Average weekday household consumption, 117

B

Backup generation resources, 57
Back-up power, 290
Bankymoon, 319
  new platforms, 319–320
Barstow System, 195
Basic Generation Service (BGS) charges, 207
Batteries
  charging and discharging rules, 222
  charging rules, 222–224
  electrochemical, 365
  with hydro pump storages, 212
  and thermal storages, 212
  storage systems, 328
Bid-based security-constrained, 261
Big Bang Disruption, 94
Big Data, 99
Bitcoin, 319
Blackouts, 179
Borrego Springs microgrid, 66
Borrego Springs project, 65
Bring-your-own-thermostat demand-side
    management programs, 375
Brittle power, 78
Business models
  innovation, 93
  for power system flexibility. *See* Power
    system flexibility, business models
  transitions, 107

C

California
  electricity demand through 2040, 9
  marginal energy cost, 44
  projected overgeneration in electric power
    sector, 15
  renewable portfolio standard by 2024, 15
  retail competition, 178–179
California Energy Commission (CEC), 9
California Independent System Operator
    (CAISO), 378
California IOUs, 392
California Public Utilities Commission
    (CPUC), 44, 178
Capacity remuneration mechanisms (CRMs), 231
  designs, evaluation criteria for assessing, 240
  risk distorting, 235
Capacity schemes, 235
Capacity tariffs
  load shifting, anticipated customer response
    of, 221
  load smoothing, anticipated customer
    response of, 221
Capacity tickets, 241
  ROs, comparison, 242, 244
Capital intensive industries, 195
Case study, German utilities, 323
Centralized grids, 345
Cheap airline seats, 22
Cheap kilowatt-hour unit costs, 77
China's electricity demand growth rates, 14
Clean Energy Regulator (CER), 217
Clean Line Energy Partners (CLEP), 296
Clean Power Plan, 288
Cloud, 98
CO_2 emissions, 39
Combined cycle gas turbine (CCGT), 286
Combined heat and power (CHP) systems, 64
Command economy, 324
Commercial and industrial (C&I)
    consumers, 374
  distributed storage targeted, 377
  Motor's batteries, 377
Commission's Bureau of Investigation, 181
Commodity distribution, 92
Commodity Infrastructure Service, 100
Company structures, 336
Competitive ercot retail market
  product and service differentiation, 160
Conseil International des Grandes Réseaux
    Électriques C6.22 Brochure
    (CIGRÉ, 2015), 52
Consumer Maturity Model (CMM), 105

Consumers bills, changes, 206
Consumer-sited storage, 31
Consumers mature, on continuum, 105
Consumption band, proportion of customers,
    119
Contracts, 236
Conventional generation sources, 145
Conventional utility distribution system, 388
Cost of energy, 41
Cost reflectivity, 347
  tariffs. *See* Tariffs
CPP. *See* Critical peak pricing (CPP)
Critical peak pricing (CPP), 218, 252
CSIRO study, 117
Curtailment, 368
Customer-centric view, of electricity service
  automation, role of, 85
  grid as balancing resource, 82
  grid services, financial ways, 85–86
  low cost electricity, 80–81
  monopoly, 84
  overview of, 75–76
  pricing, 86
    bill, rearrangement, 87
    real-time, 87
    scheduled load, 88–89
  promise of, 80
  reliability/resilience, 83
  traditional model, 76–79
Customer–grid interaction, 85
Customer loads, 22
Customer segments, buying decisions of, 329
Customer-sided energy efficiency, 188

D

Data analytics, 99
Day-ahead bids, 270
DC transmission, 283
Decentralized, emerging paradigm, 55
Decentralized interaction, 275
  of renewable energy, 274
Decentralized reliability options (ROs), 231,
    232, 235, 241
  advantages, 238–243
  background, 232–235
  convention design, 243–245
  electricity supply, 231
  mechanics, 236
  Strawman design of, 237
Declining grid-based electricity demand, 109
Deindustrializing, 7
Demand-based distribution network tariffs, 248

Demand-based tariff, 86, 262
Demand falling, California projects, 9
Demand growth rates
    in different states, 5
    GDP for United States, 8
Demand management services, 286, 367
Demand resources, 418
Demand response (DR), 26, 186, 371
    operations, 374
    resources, divisibility of, 32
Demand-side regulation, 326
Department of Public Service (DPS), 37
Department of Public Utilities (DPU), 406
DER. *See* Distributed energy resource (DER)
Desert Knowledge Australia Solar Centre
    (DKASC), 131
Diesel generators, 76
Diesel gensets, 66
Digital-based smart energy, 289
Disaggregated nodal pricing, 272, 273
DisCo's burden, 70
Disruption dynamics
    impact of megatrends, 285
Disruptive innovations, 327, 328, 338
Disruptive retail products, 170
Distributed energy resource (DER), 6, 14, 25, 75,
        92, 169, 193, 211, 213, 249, 383, 417
    anticompetitive effects, 402
    benefit–cost analysis (BCA), 401
    characteristics of, 202
    customer demand, 411
    distribution utility structure, 385
    DSO, role of, 426–427
        nodal LMPs, 430–432
        RTO/TSO level, communicating prices, 428
        short-term forecast accuracy, 427
    empowered consumers, 87
    enabling higher penetration, 34
        environmental benefits, higher
            penetration, 37
        higher penetration enhances resiliency, 36
    environmental groups, 403
    generators fear, 404
    installation, 417
    integration challenges, 385
    ownership, 402
    package, 83
    period-to-period, 390
    physical infrastructure limitations, 390
    rise of, 14–16
    scenario-based analysis, 389
    self-reinforcing mechanism, 433
    supporting policies, 193

    technology adoption, 433
    utilities, 39–40
Distributed Generation Interconnection Plan
    (DGIP), 411
Distributed resources, 194
Distributed System Implementation Plan
    (DSIP), 405
Distributed System Platform (DSP), 38, 401
Distributed utility
    California's plans, for reforming, 405–406
    demonstration projects, 404
    DERs/DSP, ownership of, 402–403
    efficiency programs, 404
    Hawaii, DER boom, 408
        demand response, 411–412
        history of reform, 409–410
        integration of renewables, 410
    Massachusetts, current DER reform, 406
        electricity vision, 408
    microgrids, 404
    New York's REV, overview of, 400
    overview of, 399
    platform and provider, 401–402
    renewables, 404
    REV policy
        benefit-cost considerations in
            implementation, 403–404
        implementation, 405
Distribution grid, at microgrid node, 273
Distribution hosting capacity, 36
Distribution management system (DMS), 31
Distribution networks, 272
Distribution services platform provider
    (DSPP), 385
Distribution system
    flat volumetric rates *versus* pricing by time
        and location, 429
Distribution system operator (DSO), 368, 420
    operations/reliability, 425–426
    role of, 422
Distribution tariff, 87
Distribution utilities (DisCos), 51, 53, 54
    impact, 145, 146
DSO. *See* Distribution system operator (DSO)

E

eCommerce, 103
Efficient decision-making, 249
Efficient investment decisions, 250
Efficient price signals, 247
Electrical energy storage technologies, 365
Electric bill, components of, 201

Electric Discount and Energy Competition Act
(EDECA), 202
Electricity blocks, 88
Electricity consumption, disruptive
technologies/pricing
batteries
charging rules, 222–224
hydro pump storages and thermal
storages, 212
modeling approach, 213–215
overview of, 211
results, 225–228
tariff description/anticipated customer
response, 215–219
Electricity demand, 4
grinding down of actual and projected, 112
growth rates, 5
Electricity distribution networks, 247
Electricity from everywhere, 81
Electricity market, trading demand-side
flexibility services, 370
Electricity, nancial analogies, 86
Electricity network impacts, 143
Electricity price, by selected countries,
2007–13, 110
Electricity systems architecture
changing market architecture, 272–274
disaggregated nodal pricing
basic principle of, 271–272
impact on, 271
transactive energy, 272–274
Electricity tariffs modeling
distribution networks, optimal pricing
of, 215
history of, 195–196
subsystems, 215
Electric kettle preferences, 352
Electric Power Research Institute (EPRI), 44
Electric power sector
Australian PV capacity, 13
average sales growth, 4
California
demand falling, 9
projected overgeneration in, 15
China's electricity demand growth rates, 14
demand growth rates
in different states, 5
GDP for United States, 8
distributed energy resources, rise of, 14–16
double whammy, 6
falling demand, 6–13
rising tariffs, 6–13
energy storage, next growth business, 23

future of, 3, 17–23
future utility, 17
Japan's electricity consumption
2006–2013, 11
NRG vs. E.ON, 21–22
solar business, growth, 23
total demand in OECD, 13
UCSD's Microgrid, 18
United Kingdom
per capita electricity consumption,
2008-13, 9
projections of demand, 11
US demand growth
NERC assessment, 12
US investment in transmission and
distribution, 5
zero net energy (ZNE) home, 16
Electric power system, connectivity of, 33
Electric Reliability Council of Texas
(ERCOT), 183
demand response surveys, 163
implementation, 154
market's marginal fuel source, 154
Electric Transmission Texas (ETT), 296
Electric vehicle (EV), 219
adoption, 355
effectual demand, 357
innovation gallery, 357
integration, 408
Electrification, 93
transformative power of, 344
e-mail, 93
Empower, 14
End-customers, 247
End-to-end heating, 115
Energy Australia, 113
Energy consumers
responses, 346
roles, 344
Energy-consuming devices, 34
Energy demand, 329
Energy efficiency assistance, 184
Energy Information Administration
(EIA), 10
Energy management systems (EMS), 87
Energy-only market, 232
Energy Resilience Bank, 202
Energy service companies (ESCOs), 157,
159, 402
Energy storage, growth in, 23
Energy transformation, 283, 286
blueprinting the future, 292
business model innovation, 291

emerging business models, across value
chain, 294
political and regulatory concerns, 290
Energy transition, challenge
areas of disruption, 284
capital crunch, avoiding, 291
competition, 286–287
customer behavior, 284–286
distribution channels, 289–290
government/regulation, 290
need for innovation, 290–291
production service model, 287–288
future utility business models, 292
gentailer model, 294–295
grid developer model, 296
network manager models, 296–297
partner of partners model, 298
product innovator model, 297
Pure play merchant model, 295
traditional core business, 293
value-added enabler model, 298–299
virtual utility model, 299
overview of, 283
Energy usage, for all-electric
household, 350
EnerNOC, 376
Engie (GDF-Suez), 306
Entrepreneurial thinking, 323
Environmental development, 333
stability and predictability of, 331
E.ON's management, 316
Europe
independent system operators (ISOs), 296
transmission system operators (TSOs), 296
European electricity interconnection, total
installed capacity in 2014, 367
European electricity market designs, 267
European Emissions regulation, 325
European energy. *See also* European utilities
acquisitions of, 305
market, 231
utilities, categorization of strategies, 316
European Target Model, 243
European utilities, 303
categorization of, 315–317
cultural prerequisites, 317
Bankymoon, new platforms, 319–320
Getec Group, decentralization, 317
Google/Nest, customer focus/
lifestyle, 318
lessons from new entrants, 317–320
LichtBlick, flexibility, 319
International Energy Agency (IEA), 303

internationalization strategies, 304
differences, 304–307
observations, 307–309
strategies toward utility 2.0, 309
assets management, 310–311
management of information, 311–313
observations on, 313–315
alliances as prerequisite, 313
crossing the chasm/fail-fast, 313
leveraging regulatory incentives, 315
wide developments, 326
EV. *See* Electric vehicle (EV)

F
Facebook, 103, 104
Federal targets, 325
Feed-in-tariff (FiT), 6, 95, 109, 346
Financial Transmission Rights (FTRs), 255
FiT. *See* Feed-in-tariff (FiT)
Fixed-for-floating, 240
Flame shell, 350, 351
Flat tariff
anticipated customer behavior in BAU, 223
battery charging and discharging rules, 224
Flexibility, 93, 130, 369
service, 368
strategies, 336
TSOs/DSOs procure, 368
Fossil–fuel-based power production, 94
Fossil-fuel dominated generation assets, 379
Framework Guidelines, 239
Free power plans, 185
Fully integrated grid
DER/DSO, role of, 426–427
nodal LMPs, 430–432
RTO/TSO level, communicating prices,
428
short-term forecast accuracy, 427
DER technology adoption, 433
distributed energy resources (DER), 417
distribution system
flat volumetric rates *versus* pricing by
time and location, 429
DSO, operations/reliability, 425–426
DSO, role of, 422
independent DSO, benefits of, 424–425
North America, independent transmission
system operators of, 418
PJM, solar deployment, 419
potential DSO functionalities, details of,
423–424
price signals drive innovation, 433

Fully integrated grid (*cont.*)
  reflect fixed costs of infrastructure, 433
  vision for, 420
    customers wholesale energy prices, 421
Future Grid Forum, 228

G

Generalized merit order rule, 272
Generation Attribute Tracking System
    (GATS), 205
Gentailers, 113
German energy sector, 338
German regulatory framework, 326
Germany's zealous energy policy, 324
Getec Group, 317
  business model, 318
  decentralization, 317
GetUp!, 116
Gigawatts, 288
Going off-grid, 118
Good, Lynn, 76
Google, 103, 104, 114
Google/Nest, customer focus/lifestyle, 318
Google's motivation, 318
Great East Japan Earthquake (GEJE), 51
Greenhouse-gas emissions, 283
Greenhouse Gas Policy, 114
Grid. *See also* Integrated grid
Grid bypass, 176
Grid-connected customers, 79
Grid connectivity, 25
  primary benefits of, 26
Grid-customer interface, 85
Grid developers, 296
Grid DG
  storage options, 354
Grid electricity, 352
  vision in electricity, 91
Grid-hostile renewables, 61
Grid-light, 17
Grid loses market share, 80
Grid modernization, 406
Grid's reliability, 59
Grid Utility Support System (GUSS), 212
GridWise Transactive Energy Framework, 268
Grinding down, 111

H

Hawaiian Electric Company (HECO), 408
Hawaii, DER boom, 408
  demand response, 411–412

history of reform, 409–410
integration of renewables, 410
Hawaii Public Utility Commission, 408
Hedging network pricing risk, 255–257
Hedging service, 259
HePQR opportunity, 57, 70
Heterogeneous power quality and reliability
    (HePQR), 52
High penetration intermittent generation
    (HPIG), 134
  CSIRO's definition, 135
High ramp rates, 141
Hosting capacity, 36
Houston Lighting and Power Company, 155
HPIG. *See* High penetration intermittent
    generation (HPIG)
Hurricane Sandy
  experience, 64
  heat and power plant, 65
Hybrid, differentiated reliability, 83

I

Iberdrola, 307
Illinois, retail competition, 181–182
Impax Asset Management, 305
Inclining block tariffs (IBT), 252
Information and communication technologies
    (ICT), 268, 367
Information systems, 364
Infrastructure to Services Transition, 92
Institute for Electric Innovation (IEI), 12
Insurance policy, 86
Integrated grid, 26, 28, 29, 31
  added value of, 26
    grid support, 27
  approach, 42
  average residential consumer, 47
  costs and benefits, 43
  deferring capacity upgrades/reducing system
      losses, 33
  DER
    enabling higher penetration, 34
    environmental benefits, higher
        penetration, 37
    higher penetration enhances
        resiliency, 36
    utilities, 39–40
  distribution utilities, future of, 37
    REV market framework, 38
  enabling connectivity, 33–34
  enabling demand response, 31–33
  EPRI launched, 47

grid connectivity, benefits of
  consumers, 25–26
grid's boundaries, 40
optimizing distribution operations
  distributed storage, 29–31
  microgrids, 28–29
overall attributes/value of, 40
  average residential consumers, impact,
    44–47
  avoided costs, 44
  EPRI's integrated grid study, 47
overview of, 25
providing ancillary services, 31
telecom-like transition, 39
Integrated Interconnection Queues (IIQs)
  system, 412
Integrated Resource Planning (IRP), 409
Intelligent universal transformer (IUT), 40
Interconnection, 25
Intermittency
  countermeasures, 147–148
  framing intermittency, 134–135
  negative solar intermittency impacts,
    evidence of, 139–143
  operations impacts, 144–147
    load following, 145
    load-frequency control, 145
    operating reserve, 146
    ramping rate, 145
    unloadable generation, 146
  potential power system impacts, 134–135
  renewable energy, 136–139
  for stakeholders, 143–144
  utilities, 131–133
Intermittency timescale
  potential power system, 135
International Energy Agency (IEA), 134, 303
Internet, 104
Internet of Things (IoT), 91, 372
  accelerating change, 93
  disintermediation, 93
  disruption, 93
  ecosystems, 114
  electric utilities, 92
  innovation platform, 95–99
  telecom/electric utility transformation,
    100–107
    electricity evolution, 103–104
    PEAAS model
      applying, 104
      getting, 104–107
    telecom evolution, 101–103
Internet-ready fridges notwithstanding, 350

Intraday markets, 363
Inverters
  distribution grid, 41
Investments, in decentralization, 193
Investment Tax Credit under Section 25D, 206
Investor owned utilities (IOUs), 405
IoT. *See* Internet of Things (IoT)
Irradiance ramps
  timescales for loss, 144
  without high-resolution data, 143
Isolated power systems, 53

J

Japan's electricity consumption 2006–2013,
    10, 11

K

Knieps, 121

L

Lawrence Berkeley National Laboratory
    (LBL), 12
LED lighting, 94
LichtBlick, 319
  flexibility, 319
Li-ion battery, 65
Load demand, 141
Load-frequency control, 145
Load intermittency, residential house load
    profile, 131
Load profile, customers in PSE&G
    service, 195
Load serving entities (LSEs), 421
Load shed, PJM geographic display of, 33
Local generation, primary benefits of, 27
Locational marginal prices (LMP), 32, 428
Location Net Benefits Methodology
    (LNBM), 392
Low-income consumer, 207
Low-Income Home Energy Assistance
    Program t (LIHEAP), 203

M

Machine-to-machine (M2M)
    communication, 311
Massachusetts, current DER reform, 406
  electricity vision, 408
Megagrid, 54
Megagrid treadmill, 61
Mercury and Air Toxics Standards (MATS), 288

Metered-based pricing, 96
Mgrids. *See* Milligrids
Micro combined heat plants (micro CHP), 328
Microgrids, 6, 17, 28, 51, 53, 60, 63, 404
  benefits of, 29, 30
  Borrego Springs, 66
  defined, 63
  diesel gensets, 66
  energy
    distribution network, 279
    real-time allocation of, 278
  evolution of, 53–56
  examples, 64–68
  overview of, 51–53
  power quality and reliability (PQR), 56
    pyramid, 59–60
    resilience, 56–58
    treadmill, 60–62
  prosumer, 270
  regulation/policy, emergence, 68–71
  welfare maximizing incentives for, 277
Microgrids necessitate real-time nodal pricing
    within distribution networks, 277
  objectives for, 277
    self-sufficient commons, 277
    welfare maximizing incentives, 277
  real-time generalized merit order,
    relevancy of
    wholesale markets on distribution
    networks, 278–279
Microphasor measurement units (PMUs), 393
Milligrids, 55
  definition, 54
Missing contracts, 233
Mobile devices, 99, 353
Mobility, 103
Modeled scenarios, 214
Modernization, of grid, 27
MP2 Energy, 170

N

Nanogrids, 55
Natural gas service, 68
Nest Lab, 358, 375
Net energy metering (NEM), 6
  generation, 13
Net load variability, 141, 364
  in Californian grid, 142
  by hour of day, 143
  reducing, 147
Net-metering, 193, 194
Network hedging instruments, 255

Network intelligence, 271
Network-price hedging instrument, 257
Network pricing, for prosumer future, 247
  in Australia, 261–264
  efficiency, in retail contracts, 249–251
  efficient network pricing, 260–261
  example, 257–259
  hedging network pricing risk, 255–257
  hedging temporal, 253–255
  locational pricing risk, 253–255
  retail competition, 252–253
  retail tariffs, 252–253
Network pricing structures
  shifts, 123
Network tariffs, 213
Network Transformation Roadmap (NTR), 139
New Energy and Industrial Technology
    Development Organization (NEDO)
    microgrid demonstration projects, 67
New energy technology adoption, barriers
    to, 359
New Jersey
  households eligible, 204
  utilities, network and supply costs, 199
New York
  average electricity sales growth, 4
  REV, overview of, 400
New York Independent System Operator
    (NYISO), 38
New York Public Service Commission
    (NYPSC), 3, 400
  staff, 404
NextEra Energy, 294
Next generation utility concepts, 372
Nodal prices, at peak for distribution
    network, 431
North America, independent transmission
    system operators, 418
North American Electricity Corporation
    (NERC), 12
  assessment, 12
Northeast region, fixed monthly charges for
    utilities, 199

O

OECD
  economies, 6
  total demand, 13
Off-grid
  net financial position, 121
  reliability, 355
Office of Clean Energy (OCE), 201

Ohio
  distribution utilities, 182
  retail competition, 182–183
One Big Switch, 116
Operating and maintenance (O&M) costs, 41
Optimal spatial spot pricing, 271
Optimal wholesale pricing, 247
Optimize energy expenditures, 318
Origin Energy, 113
Overgeneration, 15

P

Pacific Northwest National Laboratory, 432
Partner of partners model, 298
PEAAS model, 100, 103
Peak demand, 117, 233
  average annual growth, 111
Pennsylvania-Jersey-Maryland (PJM), 239
  footprint, 426
  system, 32
  utilizes, 427
Pennsylvania, retail competition, 180–181
Perfect storm, 110
Performance-based regulation, 189
Personal Energy as a Service (PEAAS), 101
Personal Energy marketplace, 107
Personalization, 102
Personal service revolution, 95
Personal Telecommunication as a Service
      (PTAAS), 101
Photovoltaic (PV), 6, 27, 328. *See also* Solar
      photovoltaic (PV)
  energy storage, 30
  grid support value of, 28
  output profiles
    zip codes in Western US, 134
  penetration, expected factors limiting, 140
  rooftop, 6
  self-generation, 16
  smart inverters with communications and
      control, 35
  solar rooftop, 7
  US rooftop solar, 19
  volt-VAR control with high-penetration, 35
PJM. *See* Pennsylvania-Jersey-Maryland (PJM)
Plain Old Electric Service (POES), 101
Plain Old Telecom Service (POTS), 101
Platform business models, 97, 98
Policymakers, 260
  retail tariff, 252
Positive energy buildings, 96
Post-liberalization power system, 363

Post POES Stage, 103
Post POTS Stage, 103
Power Analytics EnergyNet platform, 99
Power Purchase Agreement (PPA), 114
Power quality and reliability (PQR), 59
  burden on megagrid, 61
  microgrids, role of, 62
  pyramid, 59, 60
Power sector, impact of megatrends, 285
Power Supply Improvement Plans (PSIPs), 409
Power supply, traditional paradigm, 54
Power system behavior, 31
Power system flexibility, business models, 364
  flexibility-enabling resources, 365–368
  new business models, 372, 380
    demand-side management, aggregation
        for, 374
    future of utilities, 379
    market design innovation, 378–379
    partial taxonomy, 373
    software developers, 376
    storage providers, 376–377
    thermostats, as demand-side management
        tool, 375
  overview of, 363
  trading flexibility services, 368, 371–372
    designing contracts for, 370–371
    next generation utilities, 371–372
Power utility company revenues, erosion
      of, 284
PQR. *See* Power quality and reliability (PQR)
Price responsive demand (PRD), 421
Price risk, hedged, 232
Price-to-beat (PTB) price, 154
Pricing in transition
  alternative pricing models, development of,
      121–126
  Australian energy market developments, 110
    perfect storm, 110–113
    responses, 116
    utilities, commercial response of, 113
      AGL Energy, 114
      Energy Australia, 115
      Origin Energy, 115
  coexistence, 116
  electricity price
    countries, 2007–13, 110
  overview of, 109
Prosumer, 14, 92, 105, 267, 363
Public Service Comission regulatory
      authority, 39
Public Utilities Commission (PUCT), 183
  PowerToChoose website, 160, 184

Public Utilities Regulatory Policies Act
  (PURPA), 196
PV. *See* Photovoltaic (PV)
PwC Global Power & Utilities, 17

Q
Queensland households, 116

R
Ramping rate, 145
Ramp rate control, 146
Rate adjustments, winners and losers, 208
Rate design, fundamental principles of, 197
Reactive compensation, advanced inverter
  for, 29
Real time pricing (RTP), 180, 252
Reforming the Energy Vision (REV), 3, 31,
  37, 399
  Commission's staff, 37
  goals of, 37
  implementation, 405
  policy, benefit-cost considerations in
    implementation, 403–404
  regulatory changes, 400
Regional distribution grid operators, 311
Regional transmission operator (RTO), 417
Regulatory risk, diversification of, 308
Reliability, 57, 58
  of electricity supply, 231
Remotely controlling devices, 376
Renewable and Distributed Systems
    Integration (RDSI) program, 65
Renewable energy intermittency matter, 136–139
Renewable energy sources (RES)
    generation, 270
Renewable Energy Target (RET), 216
Renewable generation, 188
Renewable portfolio standard (RPS), 202, 410
Renewables, options to manage variability, 365
Reposit Power in Australia, 289
REPs. *See* Retail electric providers (REPs)
Repurposed distribution utility (RDU), 384
  advanced distribution operations,
    developments, 394–395
  advanced distribution planning
    recent developments, 392–393
  comparison of conventional, 391
  current situation, 384–386
  expansion forecast, 389
  operations, comparison of conventional, 394
  overview of, 383

planning infrastructure, to maximize value,
    390–391
  transition
    framing, 386–388
    planning, 388
  transition operational practice reforms,
    393–394
  transition system characteristics, 387
Residential consumers, 374
  by CPS Energy and Austin Energy, 167
  tariff, quantifying change, 203
Residential consumption projection
    scenario, 114
Residential demand charges, 262
Residential Electricity Tariffs, in
    Queensland, 217
Residential rate design/death spiral, for
    electric utilities
  cost recoveries via fixed/variable
    components, 195
  electricity tariffs, brief history of, 195–196
  fixed *vs.* variable, 196–201
  overview of, 193–195
  quantifying tax effects
    for New Jersey State Government,
      206–207
  winners/losers, 202
    DER, supporting policies/incentives, 202
    tariff for residential consumers
      quantifying change, 203–205
Residential rate offerings, 161, 162
Residential usage, current consumer bills, 204
Residual balancing services, 82
Residual services, 80
Resilience, 56
Retail competition, 155, 176, 187–188
Retail contracts, 251
Retail customer choice
  background, 154
  market share, retailers, 155
  market structure, establishment of, 154
  retail competition, areas, 155
Retail electricity markets, rehabilitating, 175
  future energy economy, 175
  restructuring, 176
  retail competition, 176, 187–188
    California, 178–179
    Illinois, 181–182
    Ohio, 182–183
    Pennsylvania, 180–181
    Texas, 183–187
    United States, 176–178
  utilities, as clean energy partners, 188–189

Retail electricity prices, Queensland
    Competition Authority, 216
Retail electricity sales, year-on-year change
    in, 194
Retail electric providers (REPs), 153
    approximate residential and commercial
        market share, 156
    load-serving entity, 156
    market-based generation costs, 159
    programs, plans, and technologies,
        160–162
    State's renewable energy policies, 156
    stylistic observations, 169
    value-added services, 162–163, 167
Retail market
    competitive ercot
        product and service differentiation, 160
    competitive strategy, product differentiation,
        158–159
    integration, 91
    pricing, 157–158
    regulatory and market factors impacting,
        156–157
Retail sector, competitive pressures and the
    response, 168–170
REV. *See* Reforming the energy
        vision (REV)
Robotics, imagined futures in, 358
Rocky Mountain Institute (RMI), 348
Rokas Group, 305
Rural distribution network
    with residential and industrial load
        profiles, 431

S

Samsung, 114
Sandia National Laboratories, 139
San Diego Gas and Electric's (SDG&E)
        Borrego Springs project, 55
Self-sufficient commons, 277
Senate Bill 7 (SB 7), 154
Sendai Microgrid
    energy center, 67
    2011 Japanese earthquake and tsunami, 68
Sensors, 34
Set and forget experience, 298
Shift demand, 125
Signaling information, 216
Sioshansi, 363
Smart appliances, 358
Smart-charging electric vehicles, 80
Smart, end-user equipment, 343

Smart grids, 86, 267, 269
    costs, possible consumer implications, 45
    for real-time allocations, 269
    and transactive energy (TE) frameworks,
        269–271
Smart Grids Begs Disaggregated Nodal
        Pricing, 429
    architecture. *See* Electricity systems
        architecture
    evolution of, 267
    generalized merit order reconsidered, 275
        high voltage, 275
        medium voltage, 276
    microgrids, outside options for, 276
    network management instruments, 268
    power flows, market driven incentives
        for, 275
Smart Grid technologies, 104
Smartification, of grid, 326
Smart market, 260
Smart meters, 358
Smart phones, 99
Smart Power Infrastructure Demonstration
        for Energy Reliability and Security
        (SPIDERS), 56
Smart technologies, 358
Smeg kettles, 358
SnapChat, 93
Social Networks, 103
Social networks, 99
Societal Benefits Charge (SBC), 201
Software-as-a-service suppliers, 376
Solar alternative compliance payment (SACP)
        schedule, 202
Solar business, growth in, 23
SolarCity, 19, 170, 189
Solar customers, 115
Solar deployment, in PJM, 419
Solar FiT, 222
Solar generation power profile, residential
        rooftop, 132
Solar intermittency, 129, 147
    negative impacts, 139–143
    for residential building, 131
Solar photovoltaic (PV), 109, 211
    anticipated customer response, in BAU
        case, 220
    Arizona, power output, 141
    battery system, 120
    conflicting outcomes, 140
    electricity demand and implications, 114
    energy systems, 129
    financing models, 110

Solar photovoltaic (PV) (*cont.*)
  generation, 223, 383
  growth rate of, 111
  installation rates of, 113
  onsite generation, 94
  output profile, 133
  as partial grid-substitute, 111
  peak consumption, 117
  solar FiT
    annual cost of electricity, 225
    for apartment, 227
    for detached house, 226, 227
    system size and city location, 119
    uptake of battery storage, 214
Solar PPA prices, 84
Solar Renewable Energy Certificates
    (SRECs), 202
Solar system output, self-satisfy, 120
Staff Working Document, 235
State energy efficiency, 179
State ownership *vs.* internationalization, 308
Storage technologies, classification, 365
Strategy development, 332, 333
  blueprint strategy, 335
  conflicts in resource allocations, 340
  development of, 339
  parameters, 334
SunPower solar cells, 176, 378
Supplier, total load supplied, 177
Sustaining innovations, 327, 328, 338
Swedish communications technology, 311
Swiss utility Axpo, 305
System Average Interruption Duration Index
    (SAIDI), 57
System Average Interruption Frequency Index
    (SAIFI), 57
System capacity utilization, 112

T

Target Model, 235, 239
Tariffs
  anytime maximum demand, 125
  based on the peak demand, 263
  building blocks of, 198
  cost reflectivity, 248
  declining block, 125
  demand-based, 262, 263
    distribution network, 248
  energy charge for, 218
  falling appliance prices, 351
  modeled residential electricity, 218

for residential consumers
  quantifying change, 203
  stability and predictability, 126
  volume-based electricity, 346
Technology-based forecasts, 355
Technology drives change, 102, 103
Telecom, 102
Telephone monopolies, 102
Tesla batteries, 176
Tesla's Powerwall batteries, 19, 348
Texas, retail competition, 183–187
Thriving despite disruptive technologies, 323
  best practice strategies, 335
    cooperation, 337–338
    diversification, 337
    efficiency/effectiveness, 335
    flexibility/options, 336
    sustaining/disruptive innovation, 338
  blueprint strategy, 335
  building blocks, for strategic choices, 331
    differentiated strategy, developing, 333
      energy and energy efficiency, 335
      energy grid, 334
      energy sales, 334
      Eurelectric, 334
      production, 334
      trading, 334
    positioning, 332
  conflicts, in resource allocations, 340
  environment, 324
    challenges, 331
    customers' needs, 329–330
    dynamics via partners/competitors, 330–331
    energy policy, 324–325
    regulatory measures, 325–327
    societal development, 324
    technology, 327–328
  German utilities, 323
  strategy development/realization, success
      factors for, 339–340
Time-management, 336
Time-of-day pricing, 197
Time-of-use (ToU), 213, 252
  rates, 422
    prices right, provide visibility, 423
  tariffs, 219
Tohoku Power Co. (ToPo), 67
Trading demand-side flexibility services, 370
Traditional grid, pros and cons of, 77
Traditional retailer, 115
Transactive energy (TE), 88
  frameworks, 269

Transactive interaction, 91
Transmission and distribution (T&D)
    demarcation, 422
    network, 195
Transmission and distribution utility (TDU)
    Senate Bill 7 (SB 7), 157
Transmission grid substations, 272
Transmission network, 275
    nodes, 274
Transmission system operator (TSO), 296,
        368, 417
    policies, 245
Transmission tariff, 87
True microgrids, 55
TruSmart, 187

U
United Kingdom
    fridge-freezer, 10
    per capita electricity consumption, 2008-13,
        9, 10
    projections of demand, 11
United States
    challenge for power companies, 291
    demand growth, NERC assessment, 12
    electricity demand growth and GDP, 5
    independent system operators (ISOs), 296
    investment
        transmission and distribution,
            1997–2012, 5
        in transmission and distribution, 5
    retail competition, 176–178
    utilities, cost components of generation,
        transmission and distribution, 46
Universal Service Fund (USF), 201
Unloadable generation, 146
US Department of Energy, 46
Utilities, intermittency, 131–133
Utility 2.0, 310
    challenges on, 315
    RWE uses, 313
Utility business models, 3, 4, 6, 53, 189

Utility customer, of future
    business-as-usual, 344–346
    distributed grid, 346–347
    electricity utilities, 343
    electric vehicle (EV) adoption, 355–357,
        360
        grid saving, 355
    grid defection, 347–348
    grid—fate and its drivers, 344
    grid parity in prices/product, 353–355
    preferences/Maslow's basement/coopetition/
        energy ecosystems, 349–353
    S-bends, 355
    S-curves, 355
Utility of the Future, 399

V
Variable-peak pricing (VPP), 252
Virtual audit, 186
Virtual Power Plant, 276
Virtual trading models, 121
Virtual utility model, 299
Vivino, 93
Voltage fluctuations, 136
Voltage, PV array, 137, 138
Volt–VAR control, with high-penetration
        PV, 35
Volume risk, 233

W
Wholesale, 363
    electricity marketplace, 178
    electricity prices, 179
    markets, on distribution networks
        real-time generalized merit order,
            relevancy of, 278–279
Wright System, 195

Z
Zero marginal cost—electrons, 15
Zero net energy (ZNE) home, 16

Printed in the United States
By Bookmasters